TENACIOUS OF LIFE

TENACIOUS OF LIFE

The Quadruped Essays of John James Audubon and John Bachman

JOHN JAMES AUDUBON AND JOHN BACHMAN

EDITED AND WITH ORIGINAL COMMENTARY

BY Daniel Patterson AND Eric Russell

University of Nebraska Press

LINCOLN

∞

Publication of this volume was assisted by
the Virginia Faulkner Fund, established in
memory of Virginia Faulkner, editor in chief
of the University of Nebraska Press.

Library of Congress Cataloging-in-Publication Data
Names: Audubon, John James, 1785–1851, author.
| Audubon, John James, 1785–1851. Quadrupeds
of North America. | Bachman, John, 1790–1874,
author. | Patterson, Daniel, 1953– editor.
Title: Tenacious of life: the quadruped essays of John
James Audubon and John Bachman / John James
Audubon and John Bachman; edited and with original
commentary by Daniel Patterson and Eric Russell.
Description: Lincoln: University of Nebraska Press,
[2021] | Includes bibliographical references and index.
Identifiers: LCCN 2020037873
ISBN 9781496213341 (hardback)
ISBN 9781496226723 (epub)
ISBN 9781496226730 (mobi)
ISBN 9781496226747 (pdf)
Subjects: LCSH: Audubon, John James, 1785–1851. |
Bachman, John, 1790–1874. | Mammals—North America.
Classification: LCC QL715 .A94
2021 | DDC 599.097—dc23
LC record available at https://lccn.loc.gov/2020037873

Set in Garamond Premier by Mikala R. Kolander.
Designed by N. Putens.

The editors dedicate this volume to the rising generation:

Noah Jonathan Russell

Thomas Daniel Patterson

Ava Carol Torosian

Harrison Charles Torosian

Harper Lindsay Huntoon

Michael Hayes Patterson

and those who will come after.

May there always be wild quadrupeds in your lives.

CONTENTS

ILLUSTRATIONS

PREFACE

The first half of the nineteenth century in the United States was a period of unprecedented and rapid growth, change, uncertainty, and hope. The world's newest nation was yet to come into focus, and its inchoate character was much debated and contested. The Louisiana Purchase doubled the size of the territory of the new country. Wars were fought on the Atlantic, in the Floridas, and in the West against Mexico. The presidents of the "land of the free" held people in bondage. Steam power made upriver travel and trade much easier and more profitable. The Anglo-American population along the Ohio River valley doubled, tripled, and quadrupled. The cotton gin and the Napoleonic Wars made slave labor more profitable than ever before while also setting northern states to oppose slavery all the more resolutely. The Native American population in the Southeast was dramatically reduced through forced marches westward following the Indian Removal Act. Canals and railroads increased exponentially the speed and reach of the new commerce. Still, there were intermittent panics, depressions, and bank failures. And in the international world of science, Charles Darwin was considering whether to allow himself to publish what he saw.

Amid all this, the French American artist-naturalist John James Audubon and the German American minister-naturalist John Bachman (pronounced "Backman") struggled with (and sometimes against) one another to create and publish the images and words, the graphic art and its literary voice, that comprise *The Viviparous Quadrupeds of North America*. Under the same title, Audubon and his son Victor Gifford published in two volumes the 150 lithographic colored plates, known as the Imperial Folio edition (New York, 1845–48), and in three volumes the essays (or "letterpress") to accompany the plates (New York, 1846–54). Subsequent to his father's death in 1851, Victor published in 1854 the first edition to pair the plates with their essays; this is the smaller "octavo" edition, and it was much more affordable for many. Dropping the unpoetic and unhelpful word "viviparous" (all North American mammals bear their young live), Victor adopted a leaner title: *The Quadrupeds of North America*.

Citizens of the United States were eager for cultural distinctions and accomplishments peculiar to their particular place and independent of Europe. And European naturalists and their patrons were eager to discover America's yet-unknown natural wonders. Audubon had already supplied, against all odds, the 435 plates of *The Birds of America* (1827–38) and the five-volume letterpress *Ornithological Biography* (1831–39) to universal praise. The quadruped plates and essays appeared to much praise but less enthusiasm than had poured forth for those in *The Birds of America*. And they appeared amid the intensifying pressures that would lead into the Civil War. Audubon's death in 1851 took some of the wind out of the sails of the publicity campaign that Audubon had been the driving force behind. The Audubon mammals would always lie slightly out of view in the shadow of the birds.

Despite the rise of attention to American nature writing beginning in the late 1980s, however, the relative neglect has continued. No study of the quadruped essays has been published, and they have rarely been mentioned, and then only in passing, in ecocritical studies. The main purpose of the present volume is to bring this neglected one-thousand-page body of prose essays about North America's mammals into the light of critical attention and into the hands of general readers. Our hope is that a greater familiarity with these essays will more fully inform future studies of the history of American nature writing.

Similarly, John Bachman has been neglected as an American nature writer. Because of Audubon's fame and the much greater value of his name on various permutations of republications of *The Quadrupeds of North America* throughout the twentieth century, Bachman's name on the title page seems never to have been taken seriously, as if he was nothing more than an assistant to the great man. The recent work of three scholars, however, has provided a basis on which this volume and others yet to come can restore Bachman to his rightful place among American nature writers (Jay Shuler, *Had I the Wings*; *John Bachman: Selected Writings on Science, Race, and Religion*, edited by Gene Waddell; Lester D. Stephens, *Science, Race, and Religion in the American South*).

As explained in the introduction, each of the original quadruped essays consists of eight parts. Because we designed this volume as a literary anthology, we consistently present the "Habits" section and occasionally also the "General Remarks" section. The other sections, being highly technical, have value for historians of nineteenth-century science; the sections we present herein generally contain the best writing, imagery, and stories of a given essay. In selecting the thirty-six essays for inclusion, we have attempted to be representative of the whole while also offering readers the essays we find the most interesting and engaging. Enjoy.

ACKNOWLEDGMENTS

Tenacious of Life is something like a companion volume to the fine, full-color reproduction of the quadruped plates published by the Jule Collins Smith Museum of Fine Art, Auburn University, as *Audubon's Last Wilderness Journey: The Viviparous Quadrupeds of North America* (London, 2018). Because the set of plates held in the Louise Hauss and David Brent Miller Audubon Collection is one of the best-preserved sets in existence, the then director of the museum, Dr. Marilyn Laufer, recruited experts and the publisher D. Giles Limited, London, to make this important contribution to Audubon scholarship. In our hope that *Tenacious of Life* will do for the essays what *Audubon's Last Wilderness Journey* has done for the plates, we thank Cynthia Malinick, the new director of the Smith Museum, for permission to publish the plates that accompany the selected essays herein, as well as Danielle Funderburk, registrar, for supplying the images.

Matt Bokovoy, senior acquisitions editor at the University of Nebraska Press, supported and guided this project from our first conversation about it and shepherded it when called upon. The state of Audubon studies is healthier for his involvement in it. Thank you, Matt.

For informed commentary on the manuscript and for insightful suggestions, we thank Laura Dassow Walls and Ron Tyler.

For permission to publish the portrait of John Bachman, we thank Jennifer McCormick, archivist and collections manager at the Charleston Museum, Charleston, South Carolina.

For permission to publish John Woodhouse Audubon's 1843 portrait of his father, we thank Gregory Raml, special collections librarian, American Museum of Natural History, New York, New York.

We also thank the Department of English Language and Literature at Central Michigan University for funding for the Bachman portrait.

We owe our greatest debt of gratitude, of course, to our spouses, families, and friends, who encourage, enable, and—yes—tolerate.

INTRODUCTION

When John James Audubon first conceived of a mammal project to complement his *Birds of America*, his motives were mainly fame, ambition, and opportunity. He knew his art would bring America's quadrupeds to life much more fully than any previous illustrations had managed to do. After he teamed up with John Bachman, however, and work on the essays got under way in earnest, the purposes for such a work multiplied and gave it much deeper significance.

The most general purpose, of course, was to advance natural history studies, but, more specifically, Bachman and Audubon knew they would be contributing to a natural history of their still-youthful country. The premier American naturalist of the preceding generation, Charles Willson Peale, had articulated the felt need already in 1799: "Natural History is not only interesting to the individual, it ought to become a NATIONAL CONCERN, since it is a NATIONAL GOOD."[1] Especially in the first decades following the American Revolution, identifying as a native of America depended upon knowing American flora and fauna and how the species differed from European species. A common knowledge of plant and animal species could help bind together Americans originating from diverse European countries.

Our authors also wanted to encourage people to become more effective observers of nature and to find the study of nature interesting. To that end, many of the essays in effect coach readers in the art of scientific experimentation and inference. Time after time, the collective plural voice of the essays encounters an unanswered question and proceeds to demonstrate the logical steps they followed to discover the answer. Certainly, the "Virginian Opossum" is a prime example of this in its narrative of the series of examinations made to determine the period of gestation in North America's only marsupial. Especially for younger readers, the essays were modeling how to become a naturalist.

Bachman particularly felt the need and the opportunity to clarify what had become the generally muddied state of mammalogy in his country. As the reigning expert on hares, as one example, Bachman complained, "Our different species of Hares [. . .] have been so much mixed up in the accounts of authors, that great confusion exists in regard to their habits, and their specific identity."[2] In so massive a project, both Bachman and Audubon came to see themselves as achieving recognition as reliable, authoritative American naturalists not unworthy of the attention of Alexander von Humboldt and the leading European naturalists and as contributing accurate knowledge at the top tier of American natural history.

The Quadrupeds of North America is also intended to educate readers so they can make more informed and therefore helpful decisions about which mammals to consider as pests and to correct misconceptions and prejudices often acquired in childhood and from reading inaccurate reports. Furthermore, in many essays the authors describe and thereby preserve a knowledge of America's fast-fading former wildness and bounty, providing in so doing a baseline against which later attempts at ecological restoration might be measured.

Finally, the quadruped essays of John Bachman and John James Audubon comprise a rich body of nineteenth-century American science, travel, and nature writing worthy of your close attention.

Before introducing the essays more fully, however, we offer a brief characterization of the plates—the visual art—in *The Quadrupeds of North America* and of nineteenth-century natural history.

The Plates

For his transition as an artist from drawing feathers to fur, Audubon employed the same methods he had used for his paintings of birds. According to his granddaughter Maria R. Audubon's edition of his journals, he explains this in an entry dated Edinburgh, January 7, 1827:

> No one, I think, paints in my method; I, who have never studied but by piecemeal, form my pictures according to my ways of study. For instance, I am now working on a Fox; I take one neatly killed, put him up with wires, and when satisfied of the truth of the position, I take my palette and work as rapidly as possible; the same with my birds. If practicable, I finish the bird at one sitting,—often, it is true, of fourteen hours,—so that I think they are correct, both in detail and composition.[3]

To achieve the most realistic and convincing effects in fur as he did in feathers, he "used a mixed medium of pencil, watercolor, ink, and scratching," texturing his quadrupeds practically hair by hair.[4] In his most successful mammal plates, the hair or fur is so finely or richly textured that it is almost distracting. From among the plates reproduced in this volume, this is especially true of *American Wild Cat* (plate 1), *Grey Fox* (plate 21), and *Virginian Opossum* (plate 66), but the quality of the rendering of hair and fur is high throughout the plates in *The Quadrupeds of North America*.

Also as he did with the birds, Audubon enlivened his subjects by posing them in animated scenes representing each species' natural habitat or behavior. *Virginian Opossum*, for example, presents a mature female and a seven-month-old male sinuously posed on branches of a persimmon tree reaching for the fleshy fruit while grasping, in the case of the male, with hind hands and the prehensile tail characteristic of the sole marsupial species in North America. Audubon's and his sons' commitment to creating lively and compelling figures and not allowing them to grow stiff like museum specimens is apparent in most of the plates through compositions that draw the viewer in and through the subtle interactions of individuals depicted. *Leopard Spermophile* (plate 39) is a compelling composition: in the broad triangular space occupied by the two ground squirrels on the mound of their burrow, the lower of the two seems to be waiting for some communication from the dominant one, whose head and shoulders rise above the vast prairie landscape. That landscape ingeniously contrasts with the foregrounded and enlarged male and female, which are among the smallest of the prairie mammals. The irony is satisfying. These prairie creatures are fully alive and engaged in their natural habitat.

While the plates of *The Quadrupeds of North America* have not been as popular or as well known as those of *The Birds of America*, many of them, mostly those completed by Audubon himself rather than by his sons, reward close scrutiny and analysis. One of the moles in the foreground of *Common American Shrew Mole* (plate 10) is, as its essay explains, making a crunching sound as it chews an earthworm; the farmer depicted plowing his field in the distance below, not yet having read the essay, still believes moles damage the roots of his crops, although they dine exclusively on worms. The subtle feature of the feather wafting downward above the upturned face of the *Grey Fox* (plate 21) creates a narrative in which the lithe fox has just attempted to pounce on a wild bird, which barely escaped, leaving just the one trace. Many plates not reproduced in this volume, of course, are equally engaging. In *Black Rat* (plate 23), for example, the group of old and young eating eggs in a nest in a coop all appear on the verge of alarm, suggesting the species lives in constant fear of being detected and killed. *Say's Squirrel* (plate 89) narrates the transformation of forest to farmland, as well as the unintentional transformation of the squirrels from wild forest creatures to pests. The two males in the foreground are associated with the nuts on the ground, their traditional food, while looming—and advancing—behind them are the stumps of felled forest trees and the log cabin of the farm family that will remove the stumps and plant the corn that will cause enmity between human and squirrel. The wall of forest in the distance is becoming a memory.

Despite the excellent qualities of many of the quadruped plates, they have not achieved the popularity or reputation of Audubon's birds. Typically, the main reason for this is held to be that the work of John Woodhouse Audubon, who is generally credited with having drawn about half of the quadruped plates, is somewhat inferior to that of his father.[5] With regard to *White American Wolf*, for example, Sarah Boehme writes: "John lacked the sure touch of his father. His modeling of the figure renders the anatomy slightly less convincing in general, and in the instance of this wolf, the head seems wooden and not well connected to the body." Boehme continues: "The plates that are based on Audubon's drawings have graceful compositions and lively representations. Those plates that are not by the elder Audubon have representations that are stiffer, less lifelike, and less original in composition."[6] One sees the "stiffer, less lifelike" quality especially in several of the largest mammals. The heads and snouts of *Rocky Mountain Sheep*, *Prong-horned Antelope*, and *Rocky Mountain Goat*, for example, are unnaturally narrow and pointed. That several of these later plates were painted from mounted museum specimens no doubt contributed to the difficulty of representing the living animal.

Nevertheless, what his sons accomplished is worthy of the legacy of the great artist. Although 90 of the 150 plates are inscribed as the work of John James Audubon, only 76 or so were exclusively his work.[7] Audubon's typical practice was to leave the backgrounds for his elder son, Victor, to paint and the larger mammals for John Woodhouse, who worked in oils. As Lois Elmer Bannon explains, "Although some of John Woodhouse Audubon's works are attributed to him in the legends, many are not."[8] Thus, it seems likely that some fifteen or more plates attributed to Audubon *père* are actually the work of the younger son and frequently with help from his brother. And much of their work is superb. If we allow that one measure of the success of a given plate is how much money people are willing to pay for it, we must see the success of John Woodhouse's artistry in the two bison plates (56 and 57) that Sarah Boehme attributes to him in the fact that those two plates demand the highest purchase prices of all the quadruped plates.[9] All in all, the vast project of the quadruped plates was a collaboration of "Audubon, his sons, and his printmakers," and as Boehme concludes, "In the end, the beauty of the work is greater than the sum of its parts."[10]

Nineteenth-Century Natural History

Nineteenth-century natural history was a set of practices directed toward understanding animals, plants, the earth, and people, particularly non-Europeans. Although many of today's specialized scientific fields can trace their roots through these practices, in Audubon and Bachman's antebellum era, natural history was not yet professionalized to the point of rigid disciplinary adherence: although most naturalists' expertise was directed toward one or two particular fields, polymathic naturalists like Alexander von Humboldt (1769–1859) were still widely respected. Nonetheless, the grand systematizing of Enlightenment figures like Linnaeus (1707–78) and Buffon (1707–88) was giving way to more focused projects of study.

Narrowing the scope of naturalists' chosen subject matter yielded a much larger emphasis on study outdoors in nature itself, to the benefit of American practitioners. "Closet naturalists" was how Audubon routinely referred to those who produced natural history solely from museum collections and reading in libraries, and while his criticism stemmed in some measure from his lack of formal education, the increasing currency of field observation meant that more people could participate in natural history than ever before. Certainly those with the status and money for a college degree still dominated the discourse, but natural history was becoming a more dynamic network of field naturalists, specimen collectors, and informants from across the socioeconomic spectrum—"a democracy of facts," to use Andrew J. Lewis's phrase. This increasing diversity led to more robust and informative natural history, of course, but it also ensured competition among scientific institutions and naturalists themselves. Such a cultural-scientific climate made it possible, for example, for Audubon to challenge Alexander Wilson as America's foremost ornithologist, and it nurtured the information network undergirding Audubon and Bachman's mammalogy. And while American naturalists were often motivated to outperform their European forerunners and counterparts, natural history remained a transatlantic enterprise of necessity, as Europe was still home to the scientific canon for much of the nineteenth century.

The Making of the Quadruped Essays

On Monday, October 17, 1831, the Reverend John Bachman had, of course, no inkling of what was about to jostle into his life. It was his day off. He generally took Mondays and Tuesdays off from his duties as pastor of St. John's Lutheran Church in Charleston, South Carolina. Earlier this day, John James Audubon, in Charleston for the first time, had delivered his letter of introduction to the Reverend Samuel Gilman, minister of the Unitarian church that stood next to Bachman's Lutheran church. Audubon's letter would have informed Gilman that the man before him was the creator of the plates of *The Birds of America* and the author of *Ornithological Biography*,

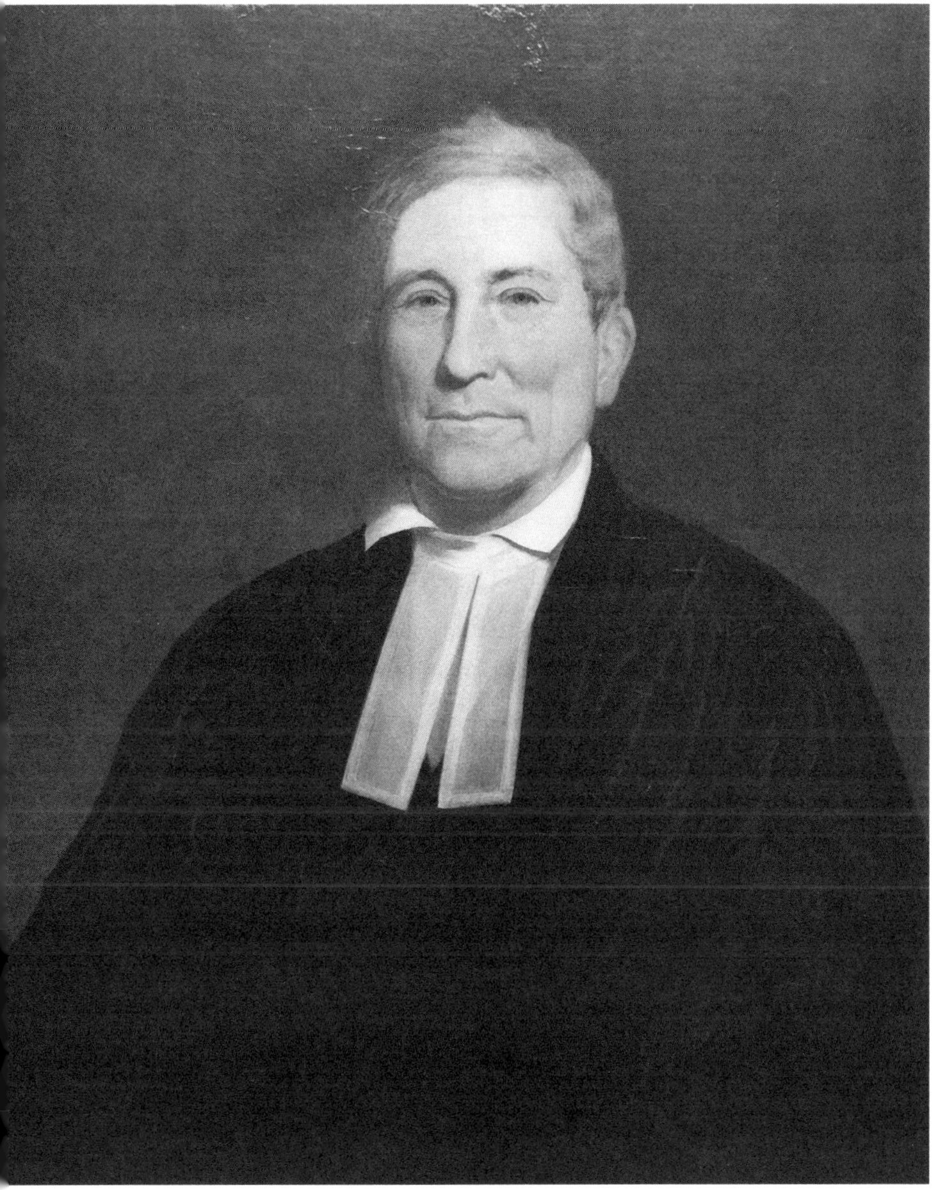

1. John Woodhouse Audubon, *John Bachman*, ca. 1837. Courtesy of The Charleston Museum, Charleston, South Carolina.

the first volume of which was newly published, and that he would appreciate any assistance as he continued to advance his work on the *Birds*. Because of Bachman's well-known and long-standing passion for natural history, Gilman would have understood quickly that of all the people in Charleston whom Audubon, the "American Woodsman," should meet, his colleague Bachman was the first. In the street before Bachman's home, the romantic artist-naturalist and the scientific pastor-naturalist were introduced, and *The Viviparous Quadrupeds of North America* had found its coauthors.[11]

vJohn Bachman (1790–1874) had moved south from his childhood home in the forested hills north of Albany, New York, near the town of Rhinebeck, at least partly because the climate relieved the symptoms of tuberculosis from which he suffered. He was a young, new minister when he made the move at age twenty-four. After briefly considering law as a profession, following some formal training in languages and natural history, he trained for the ministry. His choice of profession, however, did not completely cloak his identity as a naturalist. His passion for the methodical study of nature had been a constant in his life. As he himself wrote in 1858, "From my earliest boyhood [. . .] I had an irrepressible desire for the study of Natural History. At the age of fourteen, I had made an extensive collection of plants, birds and quadrupeds of my native State." Trained and assisted by a man named George, who was held in slavery by Bachman's father, Bachman became a capable hunter and trapper of wild animals. He knew, however, that few saw value in what he was most dedicated to: "Such were the prejudices existing in the community in those early, unenlightened days, against the supposed trifling pursuits of Natural History, that I pursued my investigations by stealth, and labored without those guides which numerous scientific works now present." To acquire copies of the published works of natural history he needed in order to develop as a more scientific naturalist, he devised a plan: "I will earn the money and buy the books. George will help me to catch the beaver and other fur-bearing animals, and I can sell the skins." His mother as well encouraged his scientific pursuits by making "room for my treasures, and on very cold, rainy evenings I had a fire in the 'spare room,' where I could study in quiet."[12]

Beyond his passion for nature study, it was Bachman's precocious methodicalness that made his parents confident that he would flourish as a student in Philadelphia no matter what profession he might pursue. As the seat of science in early America, Philadelphia nurtured the passion for natural history shared by many of the young nation's first naturalists: John and William Bartram, John Edwards Holbrook, John Kirk Townsend, Thomas Nuttall, Alexander Wilson, Charles Willson Peale, to name some of the best known. Upon his move there, the twelve-year-old Bachman turned to his studies even amid the noise and bustle of the narrow streets. The formal botanical garden established by John Bartram and then occupied by his more famous son William helped Bachman build a firm foundation for his knowledge of botany and provided thorough hands-on training in the methods and concepts of Linnaean taxonomy. At the garden he also met and befriended Alexander Wilson, who, not many years later, would become Audubon's chief rival in the development of American ornithology and avian illustrations. Wilson at the time was teaching at Gray's Ferry, a mile from the garden, and was in the beginning stages of his work that would result in the hard-won publication of his bird drawings and natural history essays entitled *The American Ornithology* (1808–14).

Because the novice Bachman had learned to identify many bird species by both sight and song, and because Wilson could do neither, this brief friendship provided both a significant boost to the young naturalist's confidence in his abilities and needed assistance to Wilson. Wilson would return the favor with a job recommendation when Bachman was looking to earn an income to cover his living expenses while he prepared for the ministry. With the appearance of Wilson's first volume in 1808, his status as an important naturalist was established. When Bachman was in Philadelphia several months later, he called on his friend. Wilson suggested he apply for the teaching post he himself had formerly held at Elwood School, a few miles northwest of the city. Bachman was offered and accepted the post, resumed his assistance to Wilson in the field to find and identify more new birds for his second volume, and thereby cemented the bond between the devoted young naturalist and his now-well-known mentor.

Philadelphia also provided the setting and the opportunity for one of its youngest aspiring naturalists to make the acquaintance of the man who was en route to becoming the most accomplished Western naturalist of the nineteenth century before Darwin: Alexander von Humboldt (1769–1859). At the end of nearly five years of travel devoted to exploration, the advancement of the natural sciences, and the gathering of information about Latin American countries, Humboldt sailed to Philadelphia before returning home. He wanted specifically to meet President Jefferson, whose political philosophy he had admired for years, as well as to have some experience of the world's youngest nation, in which he perceived much promise for the future of liberty and science. Despite being only fourteen years old, Bachman was invited to join a small group of the city's leading naturalists to greet Humboldt at Peale's museum of natural history. From this first encounter and several subsequent days of conversation with Humboldt, Bachman was inspired by the man's brilliance, of course, but also by the crucial role he placed upon the scientific method and the art of inference from empirical evidence. Another aspect of Humboldt's scientific mind that left an impression on Bachman's still-developing mind was his constant reference to the whole of nature, to how any given fact of nature operated in the larger system. "His mind dwelt upon the great laws of nature, comprehending the whole circle of the sciences," Bachman wrote in 1869.[13] Thus, Bachman's conception of nature from his youngest years was essentially dynamic rather than mechanical and static, and his conception of science was as a means to comprehend the whole.

Before leaving Philadelphia, Humboldt recorded the name of the promising young naturalist and thereafter for the rest of his life shipped his publications to him "as they successively appeared, mostly in the French language, with the exception of his 'Aspects of Nature', which was in German."[14] Thus, Humboldt's influence was a constant throughout Bachman's adult life. When Bachman voyaged to Europe in 1838, Humboldt greeted him in Berlin and introduced him to many of Europe's prominent naturalists. When, at Humboldt's invitation, Bachman addressed an international assembly on the state of natural history studies in America, his place as one of America's leading naturalists was secured.

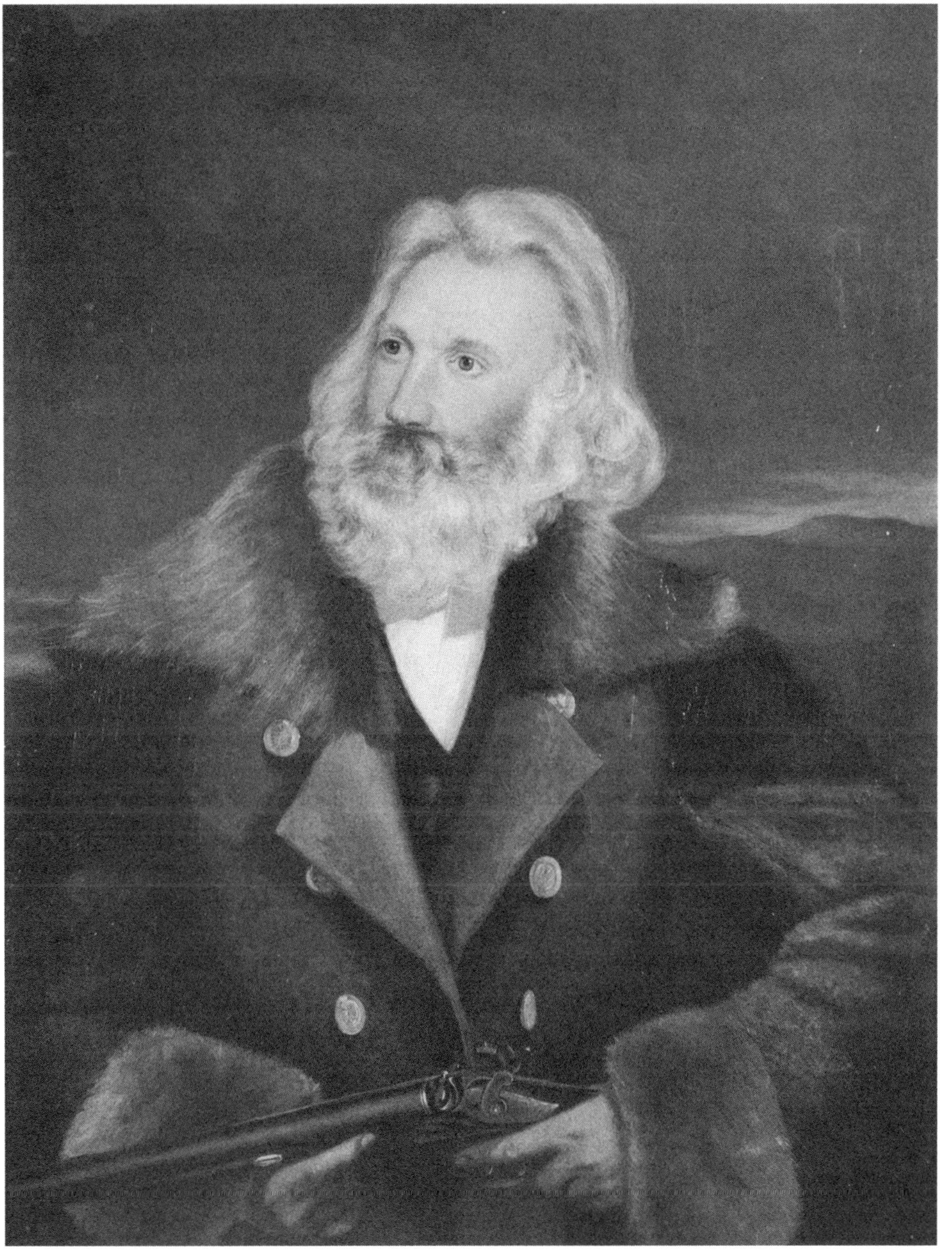

2. John Woodhouse Audubon, *John James Audubon*, 1843. Image #1498, American Museum of Natural History Library.

When Bachman first met Audubon, he had devoted the preceding fifteen years to building his ministry and a family. Whatever nature studies he pursued lacked a context or motivation to lead to publication. Audubon became the catalyst Bachman may not have been looking for but to which he responded quickly. The reignition of Bachman's passion for nature study was so rapid and complete that his first scientific publication appeared in 1833 and inaugurated a stream of contributions to natural history that flowed without interruption until the late 1850s, when concerns about the approaching Civil War distracted him.[15]

John James Audubon (1785–1851) had over the preceding three years or so become famous and much praised as an American naturalist on a level with European naturalists. Accounts of his travels, patronage, and ongoing publications appeared in newspapers and journals. Word was spreading of the unprecedented excellence and scale of the plates of his *Birds of America*. And he already entertained dreams—after completing his *Birds of America* and *Ornithological Biography*—of beginning his work on the mammals of North America and undertaking an expedition west to the Rocky Mountains and beyond. Thus, when he first met Bachman, Audubon stood between his recent success and fame and his irrepressible ambition for more.

Audubon laid the foundation for his success with his voyage to England in 1826.[16] Prior to this he had failed in several attempts at business while always continuing to develop his skills as a painter of birds. Within days of his arrival at Liverpool with 250 of his paintings and several dozen letters of introduction, Audubon was dining with Liverpool's leading families, that of Richard Rathbone primary among them, and receiving visitors in his rented rooms to view his paintings. The first European exhibit of his paintings occurred in the halls of Liverpool's Royal Institution on July 31, 1826. The praise for his work was widespread and universal. Simultaneously, he began to cultivate a public image as the American woodsman, donning a deerskin hunting shirt and leggings and wearing his hair long.

By the end of 1826, the Edinburgh engraver William Home Lizars had engraved and printed the first of what would grow to be the 435 plates comprising *The Birds of America*. Lizars produced the first ten plates, but when his workers struck for better wages, Audubon arranged with the London engraver and publisher Robert Havell Jr. to take on his monumental project, and the momentum of Audubon's success accelerated. He traveled extensively to win subscribers for both the plates and the prose essays of *Ornithological Biography*, which he soon realized he must produce. All in all, he was away from his family and country for three years before returning. During that time, his wife, Lucy Bakewell, was successfully conducting a school on a plantation in Louisiana and had come to doubt that their marriage would survive. Their two sons, still educating, began to realize that their lives likely would be subsumed by their father's work.

After approximately a year together in England, Scotland, and France, Audubon left Lucy in Philadelphia to be escorted by their son Victor Gifford across Pennsylvania and down the Ohio River to Louisville, Kentucky, where she would reside with her younger brother William Bakewell.[17] To pursue the work of obtaining the new bird species he needed to observe and draw, Audubon headed south with two assistants: Henry Ward, a young taxidermist from England, and George Lehman, a landscape painter from Switzerland. Audubon had left his engraver and publisher in London enough original bird paintings to keep him and his studio of colorists occupied for some months.

The first meeting of Bachman and Audubon was fateful both for them and for others. Over the next decade, the relationship between these coauthors and their respective families would go through many developmental phases.[18] Audubon first, of course, had to complete the paintings for *The Birds of America* and the essays for *Ornithological Biography*, for which much work remained to be done. But in his correspondence and later visits with Bachman, the mammal project was a steady presence.

Bachman immediately offered Audubon and his assistants rooms in his home for lodging and work space. Audubon's regard and respect for his new friend were at least partially encouraged by the fact that he had already imagined undertaking a work about North America's mammals, for which he knew he would need expert scientific assistance. It is remarkable and a sign of Audubon's passion and ambition that already in January 1831 he had written to Charles Bonaparte of his intention to complete a work on the mammals.[19] Over time, the two became much better acquainted,

Audubon continued his extensive travels in North America and Europe, Bachman established his scientific reputation, and the two families formed numerous bonds.

Audubon returned to Charleston to spend the winter of 1836–37 making natural history excursions, but now, beyond birds, his attention turned increasingly to quadrupeds, as his wife wrote later: "This winter he began the studies in Natural History, which led to the publication of the Quadrupeds of North America, in connection with Dr. Bachman."[20] The next bond between the two families occurred shortly thereafter when the younger Audubon son, John Woodhouse, married Bachman's daughter Maria on June 24, 1837. Throughout this time, Bachman maintained his ministry while also conducting experiments that produced new knowledge about mammals, which he published. The next year, when he conducted his tour of the important seats of natural history in Europe, Bachman managed to spend three weeks in Edinburgh helping Audubon with the fifth and final volume of *Ornithological Biography*, the final obstacle to Audubon's turning fully to the work on quadrupeds, the final plate of the *Birds* having been printed on June 20. Despite his having become more qualified than ever before, now holding a doctor of philosophy degree from Berlin University, Bachman had not committed to join Audubon as coauthor. At this time, Audubon was motivated by financial need to solicit subscriptions to *Quadrupeds*, but Bachman's family responsibilities and pastoral obligations made such a commitment difficult for him.[21] Nevertheless, his interest in the project ran deep, as is apparent in all the careful advice he gave Audubon as he laid the foundation for the work while not yet having a clear plan for it. Bachman's interest would run even deeper when the elder Audubon son, Victor Gifford, married Bachman's second daughter, Mary Eliza, on December 4, 1839.

Before he finally agreed to join Audubon on the title page of the work, Bachman's role became that of cautionary advisor. He saw that Audubon did not see some of the difficulties that lay ahead: "The books cannot aid you much. Long journeys will have to be undertaken. [. . .] The western Deer are no joke, and the ever varying squirrels were sent by the old boy himself, to puzzle the naturalists."[22] Even though the specific terms of their cooperation and possible coauthorship had not been settled, Bachman insisted to Audubon, "The Book must be original & credible—no compilation & no humbug."[23] The final straw for Bachman seems to have been his exasperated realization, a few days after the marriage of his second daughter, that "you cannot do without me when you come to the animals I know full well—so you will have to come and pay me a visit soon."[24]

Audubon's "visit" followed on January 2, 1840, in the form of a grand promise: "I believe that such a publication will be fraught with difficulties innumerable, but I *trust* not insurmountable, provided We Join our Names together, and you push your able and broad shoulders to the Wheel, I promise to you that I will give the very best figures of all our quadrupeds that ever have been thought of or expected."[25] Ultimately, Bachman saw what he had not been able to see before because of Audubon's great fame: that he himself had built such a reputation as a scientist that his name on the title page with Audubon's might even enhance sales and thereby help provide for his daughters' futures. He wrote on January 13, 1840: "I am willing to have my name stand with yours, if it will help the sale of the book. The expenses and the profits shall be yours or the boys. I am anxious to do something for the benefit of Victor and John, in addition to the *treasures* I have given them—and this is all I can do."[26] Late the next year, Bachman would believe of himself, "I know more of our quadrupeds than anyone else."[27]

Audubon the showman had procured his scientist.

The conditions under which Bachman would compose the essays, however, were far from ideal. The opening months of their new professional relationship were not propitious for the project. In April, Audubon traveled to Charleston with Bachman's consumptive daughter Maria, John Woodhouse, and their toddler, Lucy. Bachman was not prepared to see his daughter Maria so severely ill; his distress was great. The household was also under stress because Bachman's wife, Harriet, was ailing and required much care. Audubon exacerbated the stress and worries by drinking every day and offending everyone with foul language and causing disturbances that he did not remember the next day. Despite Audubon's resolve to stay as long as his daughter-in-law was so sick, a clash with Bachman at dinner on June 6 settled things. The elder naturalist stopped drinking, retreated into only perfunctory responses, and started for home that afternoon. When he reached Minnie's Land (his family estate on the Hudson River north of Manhattan) six days later, he refused to speak to Lucy about Bachman or his time in Charleston. While it pained Bachman severely to do so, he banned his revered friend from his home unless he drank only moderately. Over time, the coauthors and fellow fathers-in-law resumed their friendship, but Audubon did not again see Bachman in his home.

This home in which Bachman wrote the needed essays became one of nearly constant mourning. Maria died there in September. Daughter Mary Eliza died the next year. Finally, the tuberculosis that killed her adult daughters ended the suffering of Harriet, but not until July 15, 1846. Over their thirty years of marriage, Harriet Bachman bore fourteen children, losing five in infancy. After yet another daughter died the following summer, Bachman was moved to write to Victor, "One by one I have seen my children swept to the grave; we have not been out of mourning for years; and now I know not but that further and equally heavy trials await me."[28] That Bachman one day finally completed the essays for *Quadrupeds* almost defies belief.

While Bachman worked assiduously at home, Audubon traveled far and wide soliciting subscribers for three family publishing projects: *The Birds of America*, the Royal Octavo edition of the *Birds*, and the *Quadrupeds*, for without a stream of money coming in, the Audubons could not pay the publication costs. As he traveled, he regularly asked people with knowledge of mammals to write anecdotes or observations for potential use in the essays in a notebook he carried.[29] With him he also carried the first four plates of *Quadrupeds*, for he worked assiduously at home as well, painting the mammals to be printed. Often working fourteen hours a day, he reported on December 30, 1841, "Since the 8th of May, I have drawn 61 Species comprising 115 figures. I have now on hand Specimens enough to enable me God willing to have drawings of 100 Species by the 1st of May next."[30]

Apart from his paintings, Audubon's greatest effort for the *Quadrupeds* project was his expedition up the Missouri River in 1843.[31] He was away from home for eight months, spent a great deal of family money, produced relatively little new scientific knowledge, but did indeed, as Bachman would later concede, write up "the best account of the habits of the buffalo, beaver, antelopes, bighorns, &c., that were ever written."[32] Bachman's final and lasting assessment of the expedition's value, however, left Audubon little space in which to deny how inefficient the trip was: "But after all your expenses were greater than the knowledge was worth."[33]

Throughout the years of composition, Bachman had to navigate his way through another problem. Audubon's assurance that he and Victor would send Bachman all the material he would need (e.g., Audubon's journals and notes, published works of natural history, preserved specimens to describe, editorial assistance) was only indifferently realized. Shortly after Audubon returned from the upper Missouri River, Bachman became so frustrated that he stopped working on the essays for nearly a year. In a letter to Victor late in 1844, Bachman was at times stern, almost hysterical. To write the essays "I must have books for reference," books that were not available in Charleston but were in New York and Philadelphia: "I would not have you buy them; but could you not copy for me such articles as we need?" He pleaded that he be told at least a month before a painting was given over to the engraver so that he could produce the written description, but he also had to have the specimen: "I cannot describe without it; I will guess at nothing." Besides, "writing descriptions is slow and fatiguing work. I cannot, in the careful manner that I am doing them, write more than three in a week." He found "the labor greater than I expected, and fear that I may break down and, therefore, cry in time, 'Help me Cassius or I sink!'"[34] More evidence of the Audubons' greater interest in the images than the essays appears in Bachman's response to their request that he shorten some of them: "I cannot shorten the articles, many of them I ought rather to lengthen." His "daily practice" was to begin writing each morning at 4:00 a.m. and work until breakfast, but when his pastoral duties permitted, he continued until 3:00 p.m. While he complimented the "brush of my old friend, Audubon" as "a truth-teller" and praised "his drawings as the best in the world," he cautioned against "any errors of description that have crept in on the plates." When he originally accepted the responsibilities of scientific editor for the project, he made clear that his standards would be high: "no compilation & no humbug." Despite his frustrations during the process of composing the essays, he would not lower those standards:

"There is but one principle on which a just man can act; that is always to seek the truth and to abide by it."[35]

Audubon did, however, contribute much of the writing that went into the essays, mostly, though, in the form of his raw, unedited journals and field notes. Early in the process he worked apparently tirelessly to aid Bachman's writing. In a letter dated January 8, 1845, he announced to his friend that he had "written Ten articles already" in as many days.[36] Evidence suggests that later, however, he was generally relieved to be able to leave the bulk of that work to so highly qualified and capable a partner as Bachman. Through the 1840s, Bachman's story is one of frustrated, resourceful perseverance through neglect from the Audubons, deaths in his family, illness, and injury. By 1847 no more letters bear Audubon's signature; Victor alone communicated with Bachman. When he traveled to Minnie's Land in April 1848, Bachman learned that what he had suspected was mostly negligence was rather the result of a form of dementia: "Alas, my poor friend Audubon! the outlines of his countenance and his form are there, but his noble mind is all in ruins." By contrast to a home in ruins, he thought, "the ruins of a mind once bright and full of imagination, how much more inexpressibly melancholy and gloomy."[37]

Bachman soldiered on to fulfill his commitment to his two deceased daughters to complete the project. His ailing coauthor died on January 27, 1851. Victor was managing the ongoing publication and sending his father-in-law occasional materials while always encouraging him to keep up the writing. Finally, on March 13, 1852, Bachman could write to friend Edward Harris, "My Dear Sir: Rejoice with me, the book is finished. I did not expect to have lived to complete it. But Victor Audubon came on, and I made him hold the pen, while I dictated with specimens and books before me, and we went on rapidly; we worked hard, and now we are at the end of our labors."[38] When the third volume of the essays appeared in 1854, the entire *Quadrupeds* project, which one reviewer had already dubbed a "Great National Work," was complete.[39]

The Essays

Throughout the quadruped essays, two very different authors speak in one collective plural voice: "Reader, let us carry you with us to the boundless plains over which the prong-horn speeds." This merging of the more impetuous, expressive, eager voice of Audubon and the more decorous, dignified, reserved voice of Bachman creates a kind of rhythm or dance between the two as they guide their readers through the lives of North America's mammals and their interactions with humans. Audubon noted in his 1826 journal that friends had admonished him for a writing style that sometimes became "Bombastick."[40] Since Bachman was the final editor or composer of the essays in *Quadrupeds*, he did not allow such style to make its way into print. Throughout *Ornithological Biography*, Audubon addressed readers directly with phrases such as "Kind Reader." When Bachman saw similar phrases in galley proofs that Victor had not shown him first before publication, he complained that either Audubon or Victor had added to Bachman's prose "worn out old sayings—such as 'kind reader'" which he deemed "rather below the dignity of such a work."[41] Audubon's affection and talent for vivid imagery, however, regularly enlivens the essays. In "Brown or Norway Rat," Audubon provided Bachman with scenes of daily life amid the spring floods, or "freshets," on the Ohio River near Shippingport, Kentucky, in his earlier days there: "We once [. . .] went out in a skiff, during a freshet [. . .] and after rowing about over the tops of fences that were secured from rising with the waters [. . .] we came to the orchard attached to the garden, and found the peach trees full of rats, which seemed almost as active in running among the branches as squirrels" (94–95).

While it is difficult to say with certainty which author actually wrote a given passage or essay, internal evidence often indicates with which author a particular observation or narrative originated. Settings in Kentucky, Louisiana, the northern prairies, and Labrador or on the Ohio, Mississippi, or Missouri Rivers would come from Audubon. Bachman's settings are northern New York State, regions near Philadelphia, and the southeastern states. Both authors can be seen to contribute to the overall vividness and interest of the essays. Since the bobcat hunt narrated in "Common American Wild Cat" is set near Charleston, it is clear that Bachman wrote this scene. His use of the present-time narrative brings the reader figuratively almost into physical contact with the bobcat: "Here he comes, almost brushing our legs as he dashes by and disappears in the bushes" (10). Many of the essays take readers to multiple places and suggest that both authors contributed different observations or anecdotes, sometimes with comic effect. In "Missouri Mouse," for example, it becomes apparent that Bachman was admonishing Audubon for not collecting enough specimens of this mouse when he was on the northern prairies: "At that time being in quest of antelopes and large animals, we did not give it that close attention, which we should have done" (277). In "Raccoon," the collective voice seems to have ganged up to shoot a raccoon: "Drawing a pistol from our holsters, we shot it dead" (139).

Audubon spent roughly the years 1827 through 1839 producing, with some kind of assistance from several people, the 3,100 published pages of essays about bird species in *Ornithological Biography*. When writing about birds, he found himself developing a subtle argument for a change in his fellow humans' view and treatment of birds; his argument was that birds lead complex lives, they display evidence of consciousness, they deserve less prejudicial treatment from humans, and nature is perfect in its design and inner workings, even though humans see certain animals as inconvenient or harmful.[42] When he turned with Bachman to writing about mammals, however, different ideas arose. The ethic implied and expressed in *Quadrupeds* assumes the complete justification of the human domination, use, and killing of mammals. As just one representative example, in a scene in "The American Cross Fox," the fox has been pursued into its burrow by a dog and "our hunter," who is calmly building a fire in the mouth of the burrow: "There is no escape for the Fox—an enemy worse than the dog or the gun is destroying him; he dies a protracted, painful death by suffocation! [. . .] If not quite dead, the Fox is at least exhausted and insensible; this is sometimes the case, and the animal is then knocked on the head" (22). Whereas in *Ornithological Biography* there is often an awareness that readers might disapprove of some of the killing narrated, in *Quadrupeds* that is missing.[43]

A contemporary reviewer sheds light on the different ethics of *Ornithological Biography* and *Quadrupeds* by describing a popular conception held by many at the time of an essential difference between birds and mammals: "Birds are the type of aspiration. They address the imagination. They appeal to what is exulting and exalting in us. [. . .] Animals or quadrupeds, on the other hand, are types of the sensuous life. They appeal to our material, lower and purely animal instincts." When predatory birds kill prey, death is instantaneous. Mammals, by contrast, prolong the deaths, dally with their prey, and "gloat upon its agonies," thereby increasing their appetite before devouring. Humans relate to mammals primarily as the hunter to the hunted.[44] In this context, it is not surprising that one of the dominant motifs throughout *Quadrupeds* is that of the hunting story.

Quadrupeds acknowledges this popular bias while also arguing that it ought to be qualified or ameliorated, because nature is perfect. The opening two paragraphs of "Leopard Spermophile" make this argument: "We believe it is generally supposed that 'birds,' with their varied and pleasing forms, their gay and beautiful plumage, their tuneful throats, and their graceful movements through the air, present greater attractions to the student of nature than 'quadrupeds.'" Nevertheless, "it should be remembered that all the productions of nature are the work of so infinite a wisdom, that they must, in every department of the physical world, excite our greatest interest and our admiration" (70). The authors of a thousand-page body of essays about mammals found it necessary, as have all nature writers, to train readers to find nature interesting. While good hunting stories are hard to beat for excitement, suspense, and adventure, the life history of a ground squirrel can provide a window onto the marvelous workings of creation.

That *Quadrupeds* sold very well, especially in the Royal Octavo edition, which combined the plates with the essays, was no doubt largely due to the hunting stories and tales of the chase and related dangers. While some of the more memorable hunts originate with Audubon's upper Missouri River journals, Bachman supplied many more subtle tales of men and boys devising methods and technologies to take many species of mammal; embedded in many of these narratives are glimpses into daily life in the first half of the nineteenth century. Bachman shared his memory of hunting chipmunks as a boy in northeastern New York State: "The lads, full of delightful anticipations, usually meet half an hour before the time appointed. They come with their 'shining morning faces,' full of glee and talking of their anticipated success. In lieu of fire-arms they each carry a stick, about eight feet long. They go along the old fashioned worm-fences that skirt the woods, a crop of wheat or of buckwheat has just been gathered, and the little Hackee is busily engaged in collecting its winter store" (42–43). Bachman's gentlemanly voice wraps up the scene with a charming admonition: "Reader, don't smile at this group of juvenile sportsmen; older and bigger 'boys' are often engaged in amusements not more rational, and not half so innocent" (43).

The quadruped essays also reflect Bachman's rather complicated and somewhat conflicted theory of nature and humanity, which is shaped not only by his study and observations of nature but also by his religious beliefs. No other nineteenth-century American nature writer left behind so thorough a scientific treatise as Bachman's *Doctrine of the Unity of the Human Race Examined on the Principles of Science* (1850). While the nominal purpose of the treatise is to prove that the five major races of humans all belong to the same species (contra the polygenists), in doing so, Bachman explicated his theory of biological species and their varieties.

Two beliefs form the foundation for his philosophy of nature: first, since one Author created both the Bible (the revealed word of God) and nature, there can be no discrepancy between the two. Second, in one special act of creation, God made all the species of animal life that occupy our planet, and no new species can emerge from a union of two distinct species.[45] Any offspring born from a cross between species would be infertile hybrids and never could, he insisted, become a new species. Bachman, however, was a tireless observer of animal species and meticulous recorder of data relevant to identifying and distinguishing species. Over time, as he gathered information about species over ever-broadening geographic regions, he had to account for the variations within a given species—which he saw as adaptations needed to survive in a particular environment—in a way consistent with his theological principle that new species do not arise in nature. Regional differences in the Virginian deer, for example, are the result of "food or climate" and do not comprise a distinct species.[46] His fail-safe strategy for denying that any variety of a species could ever be named a new species is to refer to ambiguous cases as "permanent varieties." He admits the black squirrel as a "true species" somewhat reluctantly, he explains, "from the consideration that if it be no more than a variety, it has by time and succession been rendered a permanent race."[47] Thus, "permanent race" or "permanent variety" becomes his hedge against new species.

Bachman came to associate the concept of species with geographic or ecological boundaries, which implied variation from a neighboring species or ancestor, otherwise known as "common descent": "This species [the Texan skunk] apparently takes the place of the common American skunk [. . .] in the vicinity of the ranchos and plantations of the Mexicans."[48] Neighboring species show close resemblances in form, habits, diet, and reproduction because of common descent. As he wrote Audubon before his 1843 Missouri River expedition, Bachman expected "every species" to be "new in the far west."[49] Similarly, in discussing the range of the bridled weasel, Bachman writes, "As we have not heard of the existence of our Ermine in Mexico, we are inclined to the belief that this species takes the place of the Ermine in the South, and that with similar roving and predacious habits it has a more extended geographical range than is at present known."[50] Many of his statements imply a causal relationship between a given ecological region and the species suitably adapted to live there; his observation that a physical environment seems to call up its own species occasionally comes close to challenging his basic theological belief that God created and distributed all the species.

Because he studied many specimens of many species from many regions in North America and Europe, Bachman saw larger patterns of relationships existing among genera, species, and varieties. Darwin did the same thing at roughly the same time: the 1830s and 1840s. The similarities between Bachman's thought and Darwin's are fascinating, showing as they do scientific minds struggling to draw sound conclusions from the empirical evidence even when those conclusions threaten the cultural or scientific status quo. Bachman, for example, elegantly inferred the rudiments of what Robert H. MacArthur and E. O. Wilson a century later would denominate "island biogeography," a set of principles that lay at the basis of Darwin's work.[51] Bachman's succinct statement is uncannily Darwinian: "If an island or a continent were suddenly to rise up by a convulsion of nature, from the bosom of the Atlantic or Pacific, would God give either of them races of animals or plants by a new creation? We think not: they would be dependent for their supply on species now in existence, which would gradually adapt themselves to the climate."[52] When Darwin contemplated the first species that would arrive on a new landmass and the changes to their anatomy and behavior that would necessarily occur, he saw that the study of "the zoology of Archipelagoes" could "undermine the stability of Species."[53] The crucial difference between the two scientists is that for Bachman, the animal species newly adapted to the newborn landmass would be "permanent varieties," not new species.

Bachman displayed the same reasoning in a consideration of whether mammal species in Europe and America, those that do not enter the arctic regions, could be identical. His conclusion was no, because his God gave organisms the ability to adapt to their particular environments: "We admit that were an animal restricted to the temperate climates on either continent, and not found in the polar regions, there would be a strong presumptive argument against the identity of closely allied species existing in Europe and America."[54] He thus conceived a kind of biological island on which species "closely allied" to species in Europe would be distinct from their European cousins, as it were, because Bachman clung to his theological belief that there was only one act of creation.

Finally, a tension between Bachman's science and his religion remains unresolved among all his statements about species and varieties. Precisely what he understood "closely allied" to indicate about similar species is uncertain. That his God created all species as eternally distinct from one another raises a question about the nature of that "alliance." Where his language, his phrasing, suggests an acceptance of common descent among species, it implies change over time and thereby challenges the fixity of species. His religion, however, allowed him to go no further than "permanent variety." And herein lies a crucial difference between Bachman's conception of nature and Darwin's: whereas Bachman saw only continual variation within species, all returning ultimately to original forms, always anchored to original forms, and thereby denying that anything new will arise, Darwin saw ongoing creation, which inspires humans with awe and wonder.

All of Bachman's thought and writing about animal species in his treatise, however, are subordinated to his larger purpose of arguing for the unity of the five races of humanity in order—lamentably—to justify the practice of racial slavery in the United States. He argued against the polygenists, who attempted to justify slavery by claiming that Black Africans were not of the same species as white Europeans and were thereby undeserving of the moral or ethical consideration due humans as the pinnacle of God's creation. The same method and principles he applied to his discussion of animal species he applied also to his discussion of the human species.

Bachman's *Unity*, as Darwin said of his *On the Origin of Species*, is "one long argument."[55] But Bachman opens *Unity* with an unsound premise, the "fact" of the inferiority of Black Africans: "The fact that nature has stamped on the African race the permanent marks of inferiority—that we are taught by their whole past history the lesson of their incapacity for self-government, and that the Scriptures point out the duties both of masters and servants, should be sufficient to dispel every improper motive in an unbiassed search after truth alone."[56] Bachman argued that varieties or races of humans emerged after the biblical flood, which left the only surviving humans in "the high mountains of Western Asia," whence in groups they began to migrate. Those who migrated southward turned gradually darker due to climate changes, and those who migrated northward grew lighter in color. Bachman believed that after the biblical Fall, humans degenerated, physically, intellectually, and morally, from the original humans and that over time, because it was God's will and under the influence of "civilization and Christianity," the European white race developed, a race of a lighter color than Adam and Eve, thereby rising from their "state of degradation and [. . .] restored to the image of his Maker." The Black race developed in the opposite direction under the influence of "a tropical climate, parched by a burning sun, in some parts laden with moisture" and "for ages sunk in superstition and barbarism." Thus, "the European is, in form and colour, as much of an improved race, as the African is a degenerate one."[57] This is the gist of his argument for the superiority of whites over Blacks.

The Christian minister's justification for the enslavement of Black Africans follows suit. Merciful God, he claimed, made races suitable to every climate: whites aid inferior Africans by giving them conditions to live in that keep them healthier than they were in Africa and by teaching them Christianity and offering a chance at eternal life. God has made it the mission of southern slaveholders "to call forth the exercise of our minds and the display of our benevolence." Slaves are "a portion of our household; we are therefore responsible for the manner in which our duties are performed in promoting their usefulness and comfort." "In this particular, the situation of the African is unquestionably more favorable in our Southern States than it was in his native land, where to slavery was added despotism, heathenism, and the most degrading vices." While God does convert "evil into good," his ways "are inscrutable": if "Africa is ever [. . .] civilized and christianized," it will be because the southern states of America shone the "first dawn of light" on her. Ultimately, he concluded, all of humankind will have "reason to bless the wise provision of heaven in so constructing the human frame in the various races of men, that they in their several gradations can mutually benefit each other, by cultivating every soil, the products of which are so necessary to the support of man, and by this means binding together the whole human family in one bond of universal dependence and brotherhood."[58] By Bachman's argument, then, the enslavement of Black Africans by whites in the American South was the will of God.

While we wish that Bachman and his coauthor, Audubon, had been morally clear enough to have opposed slavery in their day, they were not and they did not. Bachman's case can be seen as tragic when one considers his facility with the modern scientific process and methods and the contributions he might have made had he not allowed his thinking to be constrained by his religion and by his long residence in slaveholding states. And so always, as in so much of American literature and history, many threads in the fabric or aesthetic of the quadruped essays are colored by the taint of the white enslavement of Black Africans and their descendants. The numerous "servants" who are present in many of the narratives of hunts and who assisted Audubon and Bachman in various ways were not in a position to give their version of the same events.

The Format of the Essays

John Bachman told the man he called "Old Jostle" from the beginning that producing a good work of mammalogy would not be easy, and Bachman himself couldn't have known how right he would be. Notwithstanding the agonies of maintaining a professional partnership with the Audubons from over seven hundred long miles away, Bachman did manage to ensure that *The Quadrupeds of North America* was a work of sound empirical science, a work that satisfied his expert vision, which was reflective of the astonishing amount of field study and research it required.

Bachman's vision for how *Quadrupeds* should be structured was not purely his own conception, of course, but like all his scientific ideas, it was built on a foundation of tenacious research. The origins of the organization of the essays lay in major zoological works published in the 1820s, particularly Richard Harlan's *Fauna Americana*, John Godman's *American Natural History*, and Sir John Richardson's *Fauna Boreali-Americana*. Bachman knew these works well: he cited them frequently and sometimes challenged them in *Quadrupeds*, and from them he gleaned the necessity of providing species dimensions, dental formulas, physical descriptions, and accounts of previous naturalists' investigations. Bachman's (and Audubon's) most significant improvement on these works' methodologies would be providing more extensive narratives of each animal's behaviors and interactions with humans. Audubon had already offered such narratives in his *Ornithological Biography*, but *Quadrupeds* would present a more formalized, multisection structure for not only distinguishing but also synergizing empirical and narrative approaches to studying species.

The natural history text that most closely resembles *Quadrupeds* in form is John Edwards Holbrook's *North American Herpetology*, the first volume of which was published in 1836; an entire five-volume edition was published in 1842. Like Bachman, Holbrook was an important figure in Charleston's circle of naturalists, and also like Bachman, he placed a premium on form that complemented content. Each essay in *North American Herpetology* includes eight subheadings: "Characters," "Synonymes," "Description," "Colours," "Dimensions," "Geographical Distribution," "Habits," and "General Remarks." Almost always, the "Description" section—the detailed account of the reptile's physical constitution—is the most extensive. Occasionally, however, the "General Remarks" are longest when species identification requires more explication. Holbrook's practices point toward his empirical emphases, and while Bachman and Audubon would retain these emphases in *Quadrupeds*, particularly in Bachman's sometimes lengthy "General

Remarks," "Habits" would routinely take precedence in their essays. The eight-section format utilized by Holbrook would be copied by Bachman, with the exception of "Geographical Distribution" appearing after rather than before "Habits," but the significant expansion of "Habits" suggests a desire to reinforce the "history" component of natural history. One could justifiably surmise that the habits of reptiles are harder to track than those of mammals or even that they aren't as interesting or as extensive as the habits of mammals, but Bachman and Audubon clearly believed that when a story was available, it should be told.

Bachman's preference for his chosen format was not based solely on either a professional nod to Holbrook or a vision of comprehensive science, however; it turned out to be a pragmatic decision as well. As Lester D. Stephens has observed, by the mid-1840s, Bachman had realized that he could rarely depend on publication-ready letterpress material from the Audubons, and he could not rely on receiving requested articles and specimens that they had access to. Up until this point, Bachman had hoped to simply insert the Audubons' written contributions into his own, ostensibly made easy by the chosen format, but he was having to rewrite everything they sent because it was underdeveloped, inaccurate, or both. Bachman was exhausted with working nearly ten hours a day on the essays, and in 1845 he complained to Edward Harris, a member of Audubon's 1843 Missouri River expedition, that all the Audubons seemed to care about were subscriptions and getting plates published. The height of Bachman's frustration seems to have come when Victor had the audacity to alter some of Bachman's writing, which nearly caused Bachman to quit the project, as he was advised to do by some other naturalists.[59] And Bachman was trying to produce the most current and scientifically accurate mammalogy while keeping up with his duties to his Lutheran congregation and enduring the deaths and illnesses of family members.

But Bachman persisted in his desire to discover and share scientific truth. To illustrate how the form of the quadruped essays framed their content and to show that Bachman's hard work paid off, we will analyze the first essay of the first volume, "Common American Wild Cat."[60] A description of the genus *Lynx* precedes the essay: such descriptions are common when different genera are introduced in the three volumes. As usual, Bachman describes the Greek origin of the generic name and the global distribution of the genus (eight species across four continents, in this case). He also points to previous naturalists' errors concerning this genus: for example, he explains that Constantine Samuel Rafinesque (a figure garnering little of Bachman's professional respect) "proposed the genus LYNX, but gave no detailed characters, although he states that he had increased the species of this genus from four to fifteen! In which supposition, alas, he was sadly mistaken."[61]

Also of note in the genus description is the dental formula, which had become a key factor in determining mammalian genera. "With the exception of there being one less [tooth] on each side, in the upper jaw," genus *Lynx* has the same formula as genus *Felis*; Bachman points out that the great number of species in genus *Felis* had recently required reconsideration and separation into different genera.[62]

These observations on other naturalists and genus formation influence whom Bachman credits with establishing the animal as a distinct species. The title and headnote, "Characters," and "Synonymes" of the essay appear as follows:

LYNX RUFUS.—GULDENSTAED.

COMMON AMERICAN WILD CAT.—BAY LYNX.

PLATE I.—MALE.

L. Cauda capite paullo breviore, ad extremum supra nigra, apice subalbida; auribus pagina posteriore maculo sub albido nigro marginato distinctis; hyeme et auctumno rufo-fuscus; vere et æstate cinereo-fuscus.

CHARACTERS.

Tail nearly as long as the head, extremity on the upper surface black, tipped with more or less white; a whitish spot on the hinder part of the ear bordered with black; general colour reddish-brown in autumn and winter, ashy brown in spring and summer; soles naked.

SYNONYMES.

BAY LYNX, Pennant, Hist. Quadr., No. 171. Arctic Zool., vol. 1., p. 51.

FELIS RUFA, Guld. In Nov. Comm. Petross. xx., p. 499.
FELIS RUFA, Temm., Monog., &c., vol. 1., p. 141.
LYNX FASCIATUS, Rafin. In Amer. Month. Mag., 1817, p. 46.
LYNX MONTANUS, Idem, Ibid., pp. 46, 2.
LYNX FLORIDANUS, Idem, Ibid., pp. 4, 64.
LYNX AUREUS, Idem, Ibid., pp. 46, 6.
FELIS CAROLINENSIS, Desm., Mamm., p. 231.
FELIS RUFA, Godm., Amer. Nat. Hist., vol. iii, p. 239; Fig. in vol. I.

Johann Anton Guldenstaed (or Güldenstädt), with his name following the Latin binomial, receives credit for establishing the species, yet his name follows Thomas Pennant's in "Synonymes," and there, Guldenstaed is recorded as identifying the species as *Felis rufa*. This apparent discrepancy can be explained by a passage in the volume's introduction. Bachman states that it has been necessary that he and Audubon "examine numberless papers, published in different cities of Europe and America," in order to correctly identify species and that "we have, in all cases, sought to discover and give due credit to every one who has in this manner made known a new species."[63] Pennant may have been the first to describe the species in print, but Guldenstaed was the first to publish a description of it as a *distinct* species. Additionally, Guldenstaed's description from the 1770s was clearly published before genus *Felis* became too large and had to be reconsidered. Thus, Guldenstaed established *Felis rufa* and gets credit for *Lynx rufus*.

Despite Bachman's evident concern for scientific authority, professionalism, and the sensibilities of fellow naturalists, these initial sections indicate his intention to appeal to a more general readership as well. As is the case with all of the quadruped essays, in addition to the Latin binomial, Bachman includes mammals' common names; furthermore, he first describes the animal's key identifying characteristics in Latin but then roughly translates that Latin into English in "Characters"—the Latin appears as title matter, while the English gets its own subheading. In *North American Herpetology*, Holbrook does not include this Latin with his English "Characters," so Bachman's use of Latin should be understood as a deliberate acknowledgment of, as Pamela Regis describes it, "Linnaean Latin" as the "langua [*sic*] franca among naturalists," a "deliberately static" type of Latin "shorthand" devoid of verbs wherein the "unexpressed, unwritten verb implied in all of the descriptions is 'to be.' In the lists [of species], things merely are."[64] The Latin characteristics and their English corollaries are thus intended to briefly convey physical attributes for quick, efficient identification but for different audiences: those who are knowledgeable of natural history's lingua franca and those who are not. This bilingualism positions Bachman as a naturalist who can scientifically communicate in different registers. Bachman wants *Quadrupeds* to be accessible to any transatlantic naturalist, as well as any literate English speaker: the Latin characteristics, as part of the title matter, would have been expected and readily understood in function and meaning by educated naturalists, whereas this function and meaning is spelled out for the uninitiated, so to speak, in a section with the subheading "Characters."

"Synonymes" is a snapshot of a species' presence in natural history publication, but it is also a reference list, and as is the case with scholarship today, the reference list functions as a credential, a demonstration of familiarity with what others have said on the topic. But "Synonymes" has a foreshadowing role as well, particularly as readers become more familiar with Bachman's professional priorities throughout *Quadrupeds*: "Synonymes" very often points toward what Bachman will address in "General Remarks": the naturalists he will take to task; the level of astuteness required in his scientific detective work; the problematic identification of some species and varieties—but more on that below.

The "Description" section of "Common American Wild Cat" and of all the quadruped essays is essentially the preliminary remarks of an autopsy report. Whereas "Characters" offers a way of identifying the species at a glance in the wild, "Description," along with "Colour" and "Dimensions," provides many details that can only be gleaned from a specimen that is completely immobilized (which in almost all cases in antebellum natural history means "dead"). In "Description," we read observations like "hind-feet webbed to within five-eighths of an inch of the claws. [. . .] Mammæ eight; four pectoral and four abdominal." Here, the animal is merely an object, but "Description" does regularly assure us that the authors—more often Bachman—have inspected this species with their own eyes and hands; the same cannot be said for some zoologists writing before Bachman and Audubon. The "Dimensions" section provides similar empirical exactitude: "From point of nose to root of tail—30 inches. [. . .] Anterior length of ear—1¾," and so on.[65]

The "Colour" section should be seen as an extensive continuation of Bachman's scientific findings from the late 1830s. Bachman's "Observations on the Changes of Colour in Birds and Quadrupeds," published in *Transactions of the American Philosophical Society* in 1838 and read before the society the year prior, marks a significant moment in zoology because it describes molting patterns in birds and pelage changes in mammals derived from several years of seasonal observation. Up until this point, a lack of understanding of these patterns and changes had caused many naturalists to erroneously identify new or separate species based solely on coloration.[66] Accordingly, Bachman observes of the bobcat that its black markings "are brighter and far more distinct" in summer and fall compared to "those killed in the winter and spring months." He adds, "There are, however, at all seasons of the year, even in the same neighbourhood, strongly-marked varieties, and it is difficult to find two individuals precisely alike." This caveat occurs after four paragraphs describing the species' general coloring and markings; Bachman then warns naturalists that although there are consistencies, too much should not be made of variation: "If the various supposed new species of Wild Cat described by RAFINESQUE, HARLAN, DESMAREST, &c., are entitled to a place in our Fauna, on account of some peculiarity of colour, we have it in our power, from specimens before us, to increase the number to a considerable extent; but in doing so we think we should only swell the list of synonymes, and add to the confusion which already prevails in regard to some of the species belonging to this genus."[67]

"Habits," on the other hand, gives readers information they *can* rely on and presents the bobcat in three different modes, modes common for carnivorous and omnivorous animals throughout *Quadrupeds*: as predator; as prey; and, in a type of liminal state between predator and prey, as a wild animal averse to domestication. Up until "Habits," the bobcat is presented as an immobilized and inspected object, a dead specimen; now, it figuratively comes to life through narrative accounts of its behavior—and very often dies again. This literary resurrection of the animal through field observation narration (and the redeath that often accompanies it) is a central function of "Habits" in *Quadrupeds*. Additionally, "Habits" regularly contains anecdotes that characterize the human relationship to the animal, a relationship that almost always consists of violence.

Bachman's written treatment of the bobcat is conditioned in large measure by its status as a clever adversary to humans, primarily because it wreaks havoc in the poultry yard. Thus, human "vengeance" on the bobcat is presented in picturesque hunting scenes and justified because it is conducted "fairly" by gentlemen planters and their obedient hounds (8). As Daniel Justin Herman notes, driving with hounds harks back to aristocratic practices in England; the order for fairness is a form of antebellum etiquette suggestive of the ways hunters, according to Herman, "balanced the passion and aggression of their sport with the republican virtue of self-restraint" so that they could thus consider themselves in league "with reformers, men who would curb the baser traits within themselves and their society." On the other hand, Marti Kheel observes that such ethical codes of sport and restraint amount to little more than discursively renouncing personal responsibility for violent behavior.[68] However readers might conceive of this battle of attrition between bobcat and human, the planters defended their property in a way they felt benefited them and felt becoming of them as southern American gentlemen.

Crucial to this hunting picturesque is a motif that occurs throughout *Quadrupeds*: Bachman's rather saccharine representation of slavery. Slaves are never called "slaves" in *Quadrupeds*; they are most often "servants," as is the case with the drivers of the hounds in the bobcat hunt. "These men," Bachman relates, "who are so important to the success of the chase, are possessed of a good deal of intelligence and shrewdness, are usually much petted, and regarding themselves as belonging to the aristocracy of the plantation, are apt to look down upon their fellow-servants as inferiors, and consider themselves privileged to crack a joke with their masters" (8). Bachman's description of the drivers is condescending in its flattery, presenting the slaves as dramatically ironic, somewhat comic figures; they are described as content with their position and complicit in the reinforcement of the plantation slavery hierarchy. Bachman could present such an idealized portrait of slavery because, as he troublingly asserts in his *Doctrine of the Unity of the Human Race Examined on the Principles of Science*, slavery in the southern states actually improves the constitution, intellect, and morality of Blacks, who comprise an otherwise degenerate variety of the human species in their native Africa or as freemen in North America.[69] Thus, in Bachman's view, the drivers owe any perceived superiority they experience to the institution of slavery itself, and their role in the hunting picturesque is evidence of that improvement.

Near the end of "Habits" comes Bachman's account of his failed, darkly comedic attempt to domesticate a young bobcat, a creature despised, of course, precisely because of its predatory intrusions into the pseudodomestic realm of the poultry yard. As it turns out, this bobcat becomes just such a vicious scourge on Bachman's chickens, exhibits no desire whatsoever to be with or near people, and is killed because of its behavior (14). Any reader can see the injustice of keeping a wild animal against its will and then killing it because of its feral habits, but this anecdote demonstrates that Bachman saw experiments in domestication as his right; it was also his presumed right to dispose of a nonhuman animal in any way he saw fit. For him, the experiment and the lessons learned from it are the matters of import, not the life of the animal itself.

After the brief section "Geographical Distribution," in which readers learn that the lynx can be found from Texas to Nova Scotia, from the upper Missouri River to Florida, Bachman moves on to "General Remarks," the final section of the essay. Throughout *Quadrupeds*, the "General Remarks" section provides the closing comments on a species and elucidates research methodologies—the legwork and reasoning that go into definitively establishing mammalian species—and it serves as confirmation of the authors' (primarily Bachman's) scientific authority.[70] A key concern in this section of "Common American Wild Cat" is whether a species of lynx discovered on the west coast by Lewis and Clark (which came to be called "*Felis fasciata*" by other naturalists, including Rafinesque, as listed in "Synonymes") and a species, also out west, identified as "*Felis rufa*" by Richardson are actually bobcats or truly different species. Because Bachman has been unable to examine these western specimens and has only read descriptions of them, what he *can* say with complete certainty is that the bobcat and the Canada lynx are the only two species living "east and north of the Mississippi": this certainty is established because "we have carefully examined several hundred specimens in the museums and private collections of Europe and America, and have, at this moment, upwards of twenty lying before us, that were obtained in various parts of the country, from Texas to Canada."[71] Readers are offered a glimpse into Bachman's and Audubon's study—"at this moment [. . .] lying before us"—so they can be assured that these naturalists make decisive claims only about species they have personally inspected or observed and that they are transparent about the limits of their research.

Reflecting on the quadruped essays' organization sheds some light on how Bachman was able to present a rather diffuse array of information in an intelligible way. Furthermore, it shows that each formal component of an essay serves a specific rhetorical function. Even when doing something seemingly mundane, like listing sources in "Synonymes," Bachman is seeking to persuade. He wanted and accepted nothing less than for his and Audubon's work to illuminate the truth about North American quadrupeds, and he employed the form he felt most efficiently conveyed that truth and his and Audubon's qualifications for telling it.

The Tenacity of Life

Without question, violence is the most pervasive theme throughout *Quadrupeds*. Bachman and Audubon offer accounts of nonhuman predation, but more often the blood is on their hands or those of other people. Taken together, these scenes of death and injury represent a central condition of mammalian life, human and nonhuman: various forms of hunting and scientific experimentation position violence as a primary mode of human interaction with animals, and the dynamic interplay of predator and prey is a principal lens through which Bachman and Audubon understand mammalian species. Closing the introduction to *Quadrupeds* with a description of their intentions, Bachman and Audubon state that their aim is not only to "clear away the obscurity" surrounding some species and to familiarize readers with the habits and habitats of these animals but also to describe "the mode of hunting or destroying such as are pursued either to gratify the appetite, to furnish a rich fur or skin, or in order to get rid of dangerous or annoying neighbours."[72] This promise marks the end of the introduction, pointing toward just how prevalent violence will be in the pages that follow and figuring quadrupeds as food, commodity, or nuisance.

Concepts of agency manifest in *Quadrupeds'* scenes of violence are reflected in this volume's title, *Tenacious of Life*. Although Bachman and Audubon's view of mammals is decidedly speciesist—that is, they assign a different set of rights to nonhumans and conceive of humans as superior—more often than not, "tenacious of life" is an utterance indicating their awe or frustration in the face of an animal's ability to survive various forms of violence.

The phrase "tenacious of life" appears seven times in *Quadrupeds*, which may seem meager for such a large work, but it is a phrase that carries considerable conceptual weight. In two instances, both in "Virginian Opossum," Bachman conducts experiments to determine how long young opossums can survive without pouch protection and sustenance. In the first experiment, Bachman removes an opossum from the mother via Caesarean section "near the time of parturition" and then reports that it is "remarkably tenacious of life," as it "moved several inches on the table by crawling and rolling, and survived two hours; the thermometer in the room was at the time standing at 66° Fahrenheit" (160–61). In the second experiment, week-old opossums are "very tenacious of life" because "they remained alive on the floor without any covering through a cool night, in a room containing no fire, and still exhibited a slight motion at twelve o'clock on the following day" (164). The cool detachment with which he describes essentially torturing these animals is striking, but to be sure, Bachman's experiments are part of a larger contemporary natural historical conversation concerning the hitherto mysterious gestation of the opossum, and he is the first naturalist to describe the animal in utero.[73] Nonetheless, he is impressed by the animals' short-lived struggle to survive, which conflicts with the mere helplessness he seemingly expected to witness.

In the case of "Common American Shrew Mole," the "tenacious of life" observation appears rather abruptly but in a way that demonstrates the close relationship between mammalogical observation and violence for Bachman. He describes the mole's snout, a curious appendage capable of versatile movement, and explains how the mole can put its snout in its mouth—apparently to clean it—and extricate it "with a kind of smack of the lips; this habit we observed three times in the course of a few minutes." The very next sentence marks a seemingly sudden shift: "The Shrew Mole is exceedingly tenacious of life; it cannot easily be put to death, either by heavy pressure or strangling, and a severe blow to the head seems to be the quickest mode of despatching it" (54). One moment he is curiously fixated on the mole's snout; the next moment he is crushing, strangling, and beating the animal. Since most people at the time considered moles "annoying neighbours," as *Quadrupeds'* introduction puts it, Bachman's intentions are pragmatic. But the mole's fascinating physicality extends far beyond its flexible snout to its agency in the face of overwhelming violence: Bachman's assault must be persistent in order to surmount the tiny mole's robust constitution.

In the essay on the mink and contrary to "Common American Shrew Mole," Bachman strictly plays the role of observer in the "tenacious of life" spectacle of violence. In his characteristic desire to helpfully inform, Bachman explains through experience that although minks will dive upon seeing a gun's flash, percussion caps provide too quick of a discharge for them to escape. When it comes to trapping minks, however, Bachman is less experiential: "We have usually seen the traps baited with the head[s]" of different types of birds, he tells us; he does no such baiting himself. We are then told, "The Mink is exceedingly tenacious of life, and we have found it still alive under a dead-fall, with a pole lying across its body pressed down by a weight of 150 lbs., beneath which it had been struggling for nearly twenty-four hours."[74] We are not privy to how long the mink suffered thereafter, but Bachman knows how long the mink has suffered thus far. As with the opossums, length of survival is a key point of observation, but this time, Bachman presents himself more casually—he does not describe setting up the experiment, only reaping the informative fruits of it. Yet, Bachman's mere gaze is what makes this spectacle so grotesque: he passively watches the torture, like a bored spectator at the Coliseum desensitized by too much martyring.

The threats of guns and dogs confirm that the fox squirrel is also "tenacious of life," though Bachman is more precise in qualifying these threats. Against a dog, Bachman tells us, the fox squirrel "defends himself with great spirit, and is often an overmatch for the small terriers which are used for the purpose of treeing him." Just as larger dogs are better for subduing the squirrel, so is the rifle better than the shotgun: "He is very tenacious of life, and an ordinary shot gun, although it may wound him repeatedly, will seldom bring him down from the tops of the high pines to which he retreats when pursued, and in such situations the rifle is the only certain enemy he has to dread."[75] What is interesting and troubling is that in Bachman's view, the squirrel does not have to "dread" being maimed by a shotgun; it only has to "dread" the rifle that will kill it. In his *Unity*, Bachman claims that humans are the only animals governed by reason and that other animals were created to serve human purposes.[76] His disregard for the squirrel's nonfatal injury and suffering functions as a demonstration of his Cartesian philosophy: since the squirrel is not a rational creature, it is not owed moral consideration. One wonders if Bachman was aware of Montaigne's assertion in 1588 that animals deserve moral consideration because they can suffer and enjoy.[77] If he were, he certainly did not have room for such a stance in his Christian, anthropocentric worldview.

As with the fox squirrel, guns and dogs help designate the fisher (or Pennant's marten) as "tenacious of life." And as in "Common American Wild Cat," discussed earlier, slavery has a role in the hunting episode concerning the fisher, and Bachman once again describes a spirit of cooperation between slave and master. During the night, a "servant" has treed what he believes is a raccoon and asks Bachman to shoot the animal down, as the tree is too big for the slave to climb or fell. When Bachman arrives at daybreak, he finds that the "racoon" is a fisher. He does not shoot the animal right away, because he wants to study it, so he teases it by shaking the vines in the tree, frightening and infuriating it. Bachman reports, "He was brought down after several discharges from the gun. He seemed extremely tenacious of life, and was game to the last, holding on to the nose of a dog with a dying gasp" (77). In this vicious spectacle, the naturalist contributes to the violence, then lets the violence run its course. Though Bachman compliments the fisher's fortitude, its death is a scene of defeat in the face of nearly insurmountable odds. Recalling the bobcat hunt described earlier, this spectacle is a violent picturesque in which master, slave, and domesticated animals work in harmony to subdue a wild creature.

The final use of "tenacious of life," in "American Black Bear," is rather redundant and curiously placed. After two extensive stories demonstrating the fortitude of the black bear when attacked by dogs, humans, or both, Bachman begins a paragraph, "The Black Bear is very tenacious of life, and like its relative, the Grizzly Bear, is dangerous when irritated or wounded" (338). The rest of the paragraph describes the nests bears will make when no suitable den is available. One wonders if Bachman was simply exhausted writing the third volume of *Quadrupeds*; indeed, "American Black Bear" is one of the final extensive essays in the work. Yet, Bachman goes on to criticize bear baiting and relates a fascinating story about a young farmer forced to partially strangle an attacking bear by shoving his arm deep into its throat (339). If Bachman was tired of writing, he still managed to draw his readers in.

When an animal is "tenacious of life" in *Quadrupeds*, it has exhibited vitality in a manner bordering on the sublime; that is, it has in some way impressed Bachman and Audubon with its fortitude, and it has prompted consideration of the limits of human strength and power. As illustrated above, the narratives of this tenacity are sometimes comprised of picturesque elements, showing that *Quadrupeds* is a study of the diverse ways that animals functioned as conduits for aesthetics and meaning in the antebellum United States.

Conflicted Views of American Nature

Readers of the essays contained in this volume will find the dualities of *Quadrupeds* hard to ignore: beauty and sublimity; violence and caretaking; nature as teacher and nature as dominated; predators and prey. These dualities constitute the conception of nature that informs *Quadrupeds*.[78]

Speciesism and ecology, side by side, condition how most Westerners perceive the natural world. In addition to examining this perception in *Quadrupeds*, we briefly consider two other important contemporary examples of nature writing—Susan Fenimore Cooper's *Rural Hours* (1850) and Henry David Thoreau's *Walden* (1854)—in order to situate *Quadrupeds* in an antebellum conversation about how nature in the United States should be viewed. Predictably, this consideration occurs via Cooper's and Thoreau's thoughts regarding mammals.

Rural Hours, in the guise of a single year, presents two years' worth of Cooper's observations on the natural world surrounding her village, Cooperstown, New York, and her and the villagers' complex relationship with that natural world. It would not be inaccurate to describe Cooper as a "nature lover," as one of her primary goals in *Rural Hours* is to demonstrate how an intimate understanding of and deep appreciation for one's natural environment can improve one's existence in that environment by making it more sustainable. Yet she harbors some prejudices when it comes to the nonhuman world. The latter portion of *Rural Hours* is marked by Cooper's uneasy preoccupation with the presence of a "panther," or cougar, near Cooperstown.[79] This preoccupation is so intense that, other than deer, no other mammal in the book receives as much attention.

Cooper does not want to believe there is a cougar in the neighborhood because in her view, it isn't supposed to be there. Given that it has been half a century since one was spotted in the county, she concludes that the "several respectable persons" who claim to have seen it must "have been deceived." Six days later, two more "respectable persons" report sightings, but Cooper's willful disbelief persists: "It would be passing strange, indeed, if a panther were actually roving about our woods!"[80]

After nearly a month, Cooper begins to believe the stories as they persist, especially since one of the sightings occurs in a place she had recently visited. Now she begins to wish that she could "have caught a glimpse of it—just near enough to decide the point" and to be able to brag about it. Cooper's perspective comes into closer focus at this point: "Bad as their reputation is, they seldom, I believe, attack human beings unless exasperated; and of course we should have been satisfied with a distant and brief interview; for no doubt we should have been very heartily frightened." Then, two weeks later come the newspaper reports of a cougar killed just north of Cooperstown, but instead of expressing relief, Cooper is perturbed: "It is rather mortifying that he should not have been killed in this county, where he chose to show himself repeatedly; but in fact, our sportsmen were too much afraid of being hoaxed to go out after him; they only began to believe the truth of the story when too late."[81] Cooper chastises her county's hunters for the very same incredulity she exhibited, but just as the cougar had become a point of twisted local pride for her, so should her people have had the honor of killing it.

Two weeks after the cougar is killed, the tables seem to turn. Cooper describes a rather conjectural newspaper story wherein a New Hampshire trapper unintentionally comes across two cougars, tries to backtrack his way out of the discovery, and then, when he trips over a dead tree, is mauled and killed by the cougars. Cooper twice emphasizes the terror the trapper must have experienced when he was killed.[82]

Although the vast majority of *Rural Hours* promotes a generative appreciation of nature, Cooper's feelings concerning cougars—proceeding from disbelief to fearful thrill, then to annoyance, and finally to pity and horror—reveal that deep within her otherwise fairly protoecological conception of nature lay flickers of speciesism. A cougar has nothing beautiful to offer, no uplifting lessons to teach, unlike virtually all other natural phenomena in *Rural Hours*.

Perhaps it is due in part to the machismo of the authors, but in *Quadrupeds*, the cougar is not represented as an animal to be feared. The authors relate many fascinating tales—from other informants and sources—of cougars attacking people, but Bachman and Audubon want their readers to know that such cases occur rarely and under extreme circumstances and that such stories should be taken with a grain of salt, "making a due allowance for a natural disposition in man to indulge in the marvelous" (266). Bachman and Audubon would have viewed the story of the New Hampshire trapper in this light: in fact, cougars are rather cowardly when it comes to people, and all one needs in order to safely kill one is "moderate courage," a good rifle, good aim, and a few good dogs (267). Additionally, readers are told that cougars are devoted mothers, and the seldom-seen spots on the cubs are "beautiful," which further distances Bachman and Audubon's conception of the cougar from Cooper's. If the cougar is the exception to the rule for Cooper, for Bachman and Audubon, it simply reinforces the rules: although there is beauty to admire in God's creatures, humans are top predators, and their relationship with those creatures is most often violent.

Better known than Cooper's *Rural Hours* is *Walden* and Thoreau's effort to peacefully coexist with the mammalian species he encounters in the woods outside Concord. The one exception, of course, is the woodchuck Thoreau kills and eats because it tears up his bean field, which gives him "momentary enjoyment, notwithstanding a musky flavor," though he decides such a practice is not desirable.[83] *Quadrupeds*, similarly, informs us that woodchuck meat is "tolerably good, although a little strong."[84] Additionally, we should also remember that Thoreau bought and ate a little pork while he was at Walden Pond.[85] We can reasonably speculate that Thoreau would have objected to being called a speciesist, but like many people, his speciesism emerges through self-interest and convenience.

Even if all parties agree that woodchuck is okay, not great, eating, Thoreau's rare moment of deliberate, violent speciesism stands in stark contrast with his encounters with other animals, such as *Mus leucopus*, the American white-footed mouse of *Quadrupeds*. Although Thoreau sent a specimen of this mouse to Louis Agassiz, who was very intrigued by it, the species was already old news to Bachman and Audubon: other than the common mouse, no species was more numerous in North America. Despite this discrepancy, Thoreau's account of the mouse's willingness to eat the crumbs near his feet and nibble on cheese in his hand is of a piece with the mouse's gentle image in *Quadrupeds*.[86]

Bachman and Audubon would have approved of Thoreau's willingness to observe animals so closely, but they would have shaken their heads at his lack of familiarity with common species, and they might have wondered why a man living in a cabin alone wouldn't continue killing and eating woodchucks, especially if they were "tolerably good" and threatening his crops.

The Quadrupeds of North America frequently features the hunting picturesque in which humans and dogs combine their efforts against feral quarry: relatedly, "Winter Animals" in *Walden* contains Thoreau's remarks on the dynamics among humans, hounds, and foxes. Whereas Bachman and Audubon present fox hunts in terms of theriocidal cooperation and excitement, Thoreau employs fox hunting as the basis for exploring the concepts of pursuit and discovery writ large. More specifically, Thoreau suggests different motivations and fulfillments in people and hounds when it comes to hunting foxes and leaves the implications thereof for the reader to decide. Generally, Thoreau observes, humans use hounds to acquire the trophy of the fox's tail, the symbol of a completed feat, but he is much more interested in cases in which humans and hounds are separated and following differing agendas.[87]

In the first case of a hound hunting alone, the owner from Lexington comes to Thoreau's cabin inquiring after the dog, which "had been hunting for a week by himself." Even though Thoreau repeatedly tries to answer the man, he keeps getting interrupted because the man becomes more curious about what Thoreau is up to out on Walden Pond by himself. "He had lost a dog, but found a man," Thoreau concludes.[88] Here, Thoreau seems to suggest that the object of pursuit isn't necessarily what someone is actually pursuing. Read alongside *Quadrupeds*, *Walden* might suggest to readers that both man and hound are poorly trained. In essays such as "American Cross Fox" and "Gray Fox," we encounter nothing but hunters and dogs well versed in performing their duties in sync: even if a clever fox occasionally evades hunters and dogs, dogs are not allowed to run amok through the woods without their masters' permission (17–25).[89]

Next, Thoreau relates a secondhand story from an "old hunter" about a bitch and her pups "hunting on their own account," but this story describes a much more complex relationship among hunter, hounds, and fox. The old hunter hears hounds approaching, and when the fox they are tracking crosses the road in front of him, he fires at the fox but misses. Later in the day, in a different location, he hears the hounds again and awaits the arrival of the fox. When the fox finally comes into view, it rests on a rock, knowing the hounds are far behind but unaware of the old hunter's presence. "For a moment compassion restrained [the hunter's] arm; but that was a short-lived mood, and as quick as thought can follow thought his piece was levelled, and *whang*!—the fox rolling over the rock lay dead on the ground." When the hounds arrive on the scene they are confused, but as the old hunter approaches they understand, and after he skins the fox, they return to the woods. Later that night, the old hunter is visited by the owner of the hounds—"a Weston squire"—and learns they have been out hunting alone for a week; the hunter relates the story and offers the owner the fox skin, but the owner declines and departs.[90]

Though Thoreau maintains speciesist silence regarding the old hunter's decision not to exercise compassionate restraint, Bachman and Audubon may have wondered how the old hunter saw it as his right to kill and skin a fox he put no effort into pursuing, though they would have approved of the old hunter's offer of the skin to the Weston squire. And as with the first story above, they might have wondered why the Weston squire had such a hard time controlling and finding his hounds, even if they understood that, for Thoreau, the animals themselves weren't the point but rather what animals could suggest about being human.

Bachman and Audubon would have seen reflections of their own work when Thoreau related what he had learned about fox behavior from hunters, particularly foxes' crafty evasion strategies. Additionally, they would have concurred with Thoreau's observation that the gray fox is no longer found around Concord.[91] However, they would have smirked when hearing, "They tell me that if the fox would remain in the bosom of the frozen earth he would be safe."[92] In "American Cross Fox," Bachman and Audubon make it clear that just because a fox is concealed in its den does not mean it is safe from hunters who use smoke to suffocate it (22).

For Cooper, Thoreau, Bachman, and Audubon, knowing nature was unequivocally a good thing. But whether the motive was community improvement, self-improvement, or science, speciesism always emerges in the midst of that knowing, sometimes in small glimpses and sometimes in full, blatant fury. And whereas *Rural Hours* primarily offers us one woman's effort to make nature a larger part of American communities, and whereas *Walden* mostly shows us one man's demonstration of how nature might more philosophically inform American minds, *Quadrupeds* suggests that nature is already inextricable from American life, that we have to understand it in and of itself before we can make it a philosophical springboard, and that such inextricability and understanding can be harmonious but are more often contentious.

Coda: The Partial Eclipse

Following the Civil War and then Bachman's death in 1874, historians lost sight of America's preeminent mammalogist. Just as Humboldt's renown and work were eclipsed by Darwin's work on evolution, so was Bachman's work on species. Even though Bachman was observing the rise of new species on Earth in the existing evidence, his theology did not allow him to draw that logical, empirical conclusion. The 1850 book in which he laid out all his evidence and reasoning for the development of "permanent varieties" of animals over time in changing conditions was never widely known, and after Darwin's 1859 *On the Origin of Species* and the Civil War, *The Doctrine of the Unity of the Human Race* slipped irretrievably into scientific oblivion. It may just as well have been burned by Sherman's troops along with Bachman's papers and library. Bachman's obscurity was further guaranteed by his consistent loyalty to the Southern cause leading up to and throughout the Civil War and by his thoroughgoing and unbending justification of the enslavement of Africans and their descendants.

Audubon's fame and reputation, by contrast, were not extinguished by the Civil War. George Bird Grinnell's naming of the Audubon Society after him (and his wife, Lucy) in 1886, subsequent biographies, his granddaughter's edition of his journals (*Audubon and His Journals*) in 1897, and more affordable editions of *The Birds of America* assured his rather vivacious place in the history of American culture. His learned friend, however, could barely be seen in Old Jostle's long shadow.

NOTES

1. Peale, *Introduction*, 12.
2. Audubon and Bachman, *The Quadrupeds of North America*, 1:96.
3. Audubon, *Audubon and His Journals*, 1:202–3.
4. Dwight, *Audubon*, 51.
5. On the work of John Woodhouse Audubon, see, for example, Herrick, *Audubon the Naturalist*, 2:405.
6. Boehme, "Omega," 62, 65.
7. Butler, "Checklist," 269–75.
8. Bannon, *Handbook*, 67.
9. Boehme, "Omega," 63–65; Clark and Clark, *Handbook of Audubon Prints*, 109–13.
10. Boehme, "Omega," 65.
11. For accounts of the coauthors' first meeting, see Audubon, *Letters*, 1:143; Shuler, *Had I the Wings*, 1–4; Bachman, *John Bachman*, 94–97; and Audubon, "Letter." All accounts agree that both men were elated by their first meeting.

Unless otherwise specified, general biographical information about Bachman is from Shuler.

12. Bachman, *John Bachman*, 94, 17–18.
13. Bachman, *John Bachman*, 396.
14. Bachman, *John Bachman*, 391.
15. For a complete bibliography of Bachman's publications, see Bachman, *Selected Writings*, 359–69.
16. The full story of his rise from obscurity to success is told in Audubon, *Journal of 1826*.
17. For the facts of Audubon's life, we rely on Herrick unless otherwise stated. Subsequent biographies interpret the facts variously and for a variety of purposes, but Herrick is the scholar who laid the factual foundation for all that would follow. (For a discussion of many of those biographies, see Audubon, *The Missouri River Journals*, 211–33.)
18. Ron Tyler's "Publication" (a model of scholarship based upon exhaustive research in the primary evidence) provides a full account of the working relationship between Audubon and Bachman during the writing of the essays in Audubon and Bachman, *The Viviparous Quadrupeds of North America*, the Royal Octavo edition. See also Patterson, "*Quadrupeds* Timeline."
19. Ford, *A Biography*, 449n26.
20. Audubon, *The Life of John James Audubon*, 400.
21. Tyler, "Publication," 121.
22. Tyler, "Publication," 123.
23. Shuler, *Had I the Wings*, 169.
24. Bachman, *John Bachman*, 179.
25. Audubon, *Letters*, 2:229–30.
26. Bachman, *John Bachman*, 182.
27. Shuler, *Had I the Wings*, 186.
28. Bachman, *John Bachman*, 247.
29. One of Audubon's subscription notebooks is among the Audubon Archives in the Stark Museum of Art Library, Orange, Texas (cat. no. 11.1.115). It includes several observations about mammals written in as many different hands. The journal Audubon maintained during his travels of 1840–42 contains similar observations. See Audubon, *Journal of John James Audubon*.
30. Tyler, "Publication," 139.
31. See Audubon, *The Missouri River Journals*.
32. Audubon, *Audubon Reader*, 598.
33. Bachman, *Selected Writings*, 348.
34. Shakespeare, *Julius Caesar*, 1.2.

35. Bachman, *John Bachman*, 202–3.
36. Audubon to Bachman, January 8, 1845, Vault Graff 109, Newberry Library, Chicago.
37. Bachman, *John Bachman*, 255.
38. Bachman, *John Bachman*, 276–77.
39. Webber, "The Quadrupeds," 625.
40. Audubon, *Journal of 1826*, 56.
41. Stephens, *Science, Race, and Religion*, 48.
42. See Patterson's discussion of Audubon's call for the ethical considerability of birds in Audubon, *Ornithological Biography*; and Audubon, *The Missouri River Journals*, 267–97.
43. In "The Roseate Tern," for example, after narrating having "carried destruction among them" (i.e., the terns), he thinks to obviate possible criticism from his readers: "And now I had my cap filled to the brim with specimens. You may rest assured that I took precious care of those which I had procured, but not another individual was robbed of life on that excursion" (Audubon, *Ornithological Biography*, 3:296–97).
44. Webber, "Viviparous," 273. Webber reworked this material at greater length in the first chapter of his *Hunter-Naturalist*.
45. Bachman, *The Doctrine of the Unity*, 7, 15.
46. Audubon and Bachman, *The Quadrupeds of North America*, 2:239.
47. Audubon and Bachman, *The Quadrupeds of North America*, 1:263.
48. Audubon and Bachman, *The Quadrupeds of North America*, 2:20.
49. Bachman, *Selected Writings*, 338.
50. Audubon and Bachman, *The Quadrupeds of North America*, 2:73.
51. MacArthur and Wilson, *The Theory of Island Biogeography*.
52. Bachman, *The Doctrine of the Unity*, 38.
53. Darwin, *Correspondence*, 1:xix.
54. Audubon and Bachman, *The Quadrupeds of North America*, 2:66.
55. Darwin, *Origin of Species*, 459.
56. Bachman, *The Doctrine of the Unity*, 8. Of the several biblical references to servants, one suffices here: "Servants, obey in all things your masters . . . in singleness of heart, fearing God. . . . Knowing that of the Lord ye shall receive the reward of the inheritance: for ye serve the Lord Christ" (Colossians 3:22–24).
57. Bachman, *The Doctrine of the Unity*, 160, 157, 158, 156.
58. Bachman, *The Doctrine of the Unity*, 209, 210, 211.
59. Stephens, *Science, Race, and Religion*, 44–50.
60. The animal that Bachman referred to as the "Common American Wild Cat" or "Bay Lynx" is now popularly known as the bobcat.

61. Audubon and Bachman, *The Quadrupeds of North America*, 1:2.
62. Audubon and Bachman, *The Quadrupeds of North America*, 1:1.
63. Audubon and Bachman, *The Quadrupeds of North America*, n.p.
64. Regis, *Describing Early America*, 21.
65. Audubon and Bachman, *The Quadrupeds of North America*, 1:3, 1:4–5.
66. See Bachman, *Selected Writings*, 134–76.
67. Audubon and Bachman, *The Quadrupeds of North America*, 1:4.
68. Herman, *Hunting and the American Imagination*, 156–57; Kheel, "License to Kill," 110.
69. Bachman, *The Doctrine of the Unity*, 209–12.
70. Exceptions to the rule of "General Remarks" closing the essays include "American Bison" (101–23) and "Canada Porcupine." Bachman does not include "General Remarks" when there is little or no controversy in the identification of a species.
71. Audubon and Bachman, *The Quadrupeds of North America*, 1:14–15.
72. Audubon and Bachman, *The Quadrupeds of North America*, n.p.
73. For an account of the opossum's provocative presence in natural history before *The Viviparous Quadrupeds of North America*, see Parrish, "The Female Opossum."
74. Audubon and Bachman, *The Quadrupeds of North America*, 1:257–58.
75. Audubon and Bachman, *The Quadrupeds of North America*, 2:137.
76. Bachman, *The Doctrine of the Unity*, 27, 289.
77. Montaigne, *Essais*, 176–83.
78. Our thinking in this section is influenced by Bruno Latour's conception of modernity. Latour has defined modernity as, on the one hand, a belief that human culture and nonhuman nature are separate entities; simultaneously, on the other hand, modernity is also about recognizing networks in which humans and nature are inextricably intertwined (see *We Have Never Been Modern*, 10–11).
79. For all of Cooper's attention to language, particularly with American place-names and her desire for them to accurately reflect their physical locales (see especially the extensive entry for February 7), Bachman would have taken issue with her calling *Felis concolor* "panther" rather than "cougar." Bachman shared Cooper's views on place-names; they both felt that naming American towns Rome, Carthage, or Athens, for example, demonstrated a lack of creativity and insight (262; Cooper, *Rural Hours*, 300). Regarding the cougar, they both even reported that early colonists believed cougars were lions (262; Cooper, *Rural Hours*, 252–53). However, for Bachman, calling the cougar a "panther" was an example of the "ignorance or want of imagination" in which American animals and towns were named after Old World species and places.

80. Cooper, *Rural Hours*, 236, 241.
81. Cooper, *Rural Hours*, 263–64, 282.
82. Cooper, *Rural Hours*, 309–10.
83. Thoreau, *"Walden,"* 44.
84. Audubon and Bachman, *The Quadrupeds of North America*, 1:18–19.
85. Thoreau, *"Walden,"* 43, 45.
86. Thoreau, *"Walden,"* 152, 152n6; Audubon and Bachman, *The Quadrupeds of North America*, 1:301, 304.
87. Thoreau, *"Walden,"* 186.
88. Thoreau, *"Walden,"* 186.
89. Audubon and Bachman, *The Quadrupeds of North America*, 1:164–71.
90. Thoreau, *"Walden,"* 186–87.
91. Thoreau, *"Walden,"* 188; Audubon and Bachman, *The Quadrupeds of North America*, 1:172.
92. Thoreau, *"Walden,"* 186.

BIBLIOGRAPHY

Audubon, John James. *Audubon and His Journals*. Edited by Maria R. Audubon. 2 vols. New York: Charles Scribner's Sons, 1897.

———. *The Audubon Reader*. Edited by Richard Rhodes. Everyman's Library 284. New York: Alfred A. Knopf, 2006.

———. *John James Audubon's Journal of 1826: The Voyage to "The Birds of America."* Edited by Daniel Patterson. Lincoln: University of Nebraska Press, 2011.

———. *Journal of John James Audubon Made While Obtaining Subscriptions to His "Birds of America" 1840–43*. Edited by Howard Corning. Boston: Club of Odd Volumes, 1929.

———. "Letter from Audubon to the Editor." No. 1. *Monthly American Journal of Geology and Natural Science* 1 (1832): 358–63.

———. *Letters of John James Audubon*. Edited by Howard Corning. 2 vols. Boston: Club of Odd Volumes, 1930.

———. *The Missouri River Journals of John James Audubon*. Edited by Daniel Patterson. Lincoln: University of Nebraska Press, 2016.

———. *Ornithological Biography*. 5 vols. Edinburgh: Adam Black, 1831–39.

Audubon, John James, and John Bachman. "Descriptions of New Species of Quadrupeds Inhabiting North America." *Proceedings of the Academy of Natural Sciences of Philadelphia* 1 (October 1841): 92–103, and *Journal of the Academy of Natural Sciences of Philadelphia* 8, no. 2 (1842): 280–323.

———. *The Quadrupeds of North America*. The Royal Octavo edition. 3 vols. New York: Published by V. G. Audubon, 1849–54.

———. *The Viviparous Quadrupeds of North America*. The Imperial Folio edition. 2 vols. New York: Published by J. J. Audubon, 1845–48 [the 150 lithographic plates only].

———. *The Viviparous Quadrupeds of North America*. 3 vols. New York: Published by J. J. Audubon and V. G. Audubon, 1846–54 [text only].

Audubon, John Woodhouse. *Audubon's Western Journal: 1849–1850*. Edited by Frank Heywood Hodder. Cleveland, 1906.

Audubon, Lucy. *The Life of John James Audubon, the Naturalist*. New York: G. P. Putnam and Son, 1869.

Bachman, C. L. *John Bachman, D.D., LL.D., Ph.D.: The Pastor of St. John's Lutheran Church, Charleston*. Charleston SC: Walker, Evans & Cogswell Co., 1888.

Bachman, John. *The Doctrine of the Unity of the Human Race Examined on the Principles of Science*. Charleston SC: C. Canning, 1850.

———. *John Bachman: Selected Writings on Science, Race, and Religion*. Edited by Gene Waddell. Athens: University of Georgia Press, 2011.

Bannon, Lois Elmer. *Handbook of Audubon Prints*. Gretna LA: Pelican Publishing Co., 1985.

Boehme, Sarah. "Omega: John James Audubon's Final Artistic Journey." In *John James Audubon in the West: The Last Expedition*, edited by Sarah E. Boehme, 35–65. New York: Harry N. Abrams.

Butler, Charles T., ed. *Audubon's Last Wilderness Journey: The Viviparous Quadrupeds of North America*. London: Jule Collins Smith Museum of Fine Art, Auburn University, and D. Giles Limited, 2018 [a facsimile of the complete Imperial Folio edition].

Clark, Lois Elmer, and Taylor Clark. *Handbook of Audubon Prints*. Gretna LA: Pelican Publishing Company, 1985.

Cooper, Susan Fenimore. *Rural Hours*. Edited by Rochelle Johnson and Daniel Patterson. Athens: University of Georgia Press, 1998.

Darwin, Charles. *The Correspondence of Charles Darwin*. Edited by Frederick Burkhardt. 30 vols. Cambridge: Cambridge University Press, 1985.

———. *On the Origin of Species*. London: John Murray, 1859.

Dwight, Edward H. *Audubon: Watercolors and Drawings*. Utica NY: Munson Williams-Proctor Institute; New York: Pierpont Morgan Library, 1965.

Ford, Alice. *John James Audubon: A Biography*. New York: Abbeville Press, 1988.

Harris, Edward. *Up the Missouri with Audubon: The Journal of Edward Harris*. Edited by John Francis McDermott. Norman: University of Oklahoma Press, 1951.

Herman, Daniel Justin. *Hunting and the American Imagination*. Washington DC: Smithsonian Institution Press, 2001.

Herrick, Francis Hobart. *Audubon the Naturalist: A History of His Life and Time.* 2 vols. New York: D. Appleton-Century Company, 1917, 1938.

Holbrook, John Edwards. *North American Herpetology.* Philadelphia: J. Dobson, 1836–40, 1842.

Howe, Daniel Walker. *What Hath God Wrought: The Transformation of America, 1815–1848.* New York: Oxford University Press, 2007.

Kheel, Marti. "License to Kill: An Ecofeminist Critique of Hunter's Discourse." In *Animals and Women: Feminist Theoretical Explorations*, edited by Carol J. Adams and Josephine Donovan. Durham NC: Duke University Press, 1995.

Latour, Bruno. *We Have Never Been Modern.* Translated by Catherine Porter. Cambridge MA: Harvard University Press, 1993.

Lewis, Andrew J. *A Democracy of Facts: Natural History in the Early Republic.* Philadelphia: University of Pennsylvania Press, 2011.

Lewis, Meriwether, and William Clark. *History of the Expedition under the Command of Captains Lewis and Clark.* Edited by Nicholas Biddle [without attribution]. 2 vols. Philadelphia: Bradford and Inskeep, 1814.

Lott, Dale F. *American Bison: A Natural History.* Berkeley: University of California Press, 2002.

MacArthur, Robert H., and E. O. Wilson. *The Theory of Island Biogeography.* Princeton NJ: Princeton University Press, 1967.

Montaigne, Michel de. *Essais.* Paris: L'Angelier, 1588.

Moulton, Gary E., ed. *The Definitive Journals of Lewis and Clark.* 8 vols. Lincoln: University of Nebraska Press, 1983–2001.

Numbers, Ronald L., and Lester D. Stephens. "Darwinism in the American South." In *Evolution Education in the American South*, edited by Christopher D. Lynn, 5–32. New York: Palgrave, 2017.

Parrish, Susan Scott. "The Female Opossum and the Nature of the New World." *William and Mary Quarterly* 54, no. 3 (1997): 475–514.

Patterson, Daniel. "*Quadrupeds* Timeline." In *Audubon's Last Wilderness Journey: The Viviparous Quadrupeds of North America*, edited by Charles T. Butler, 253–64. London: Jule Collins Smith Museum of Fine Art, Auburn University, 2018.

Peale, Charles Willson. *Introduction to a Course of Lectures on Natural History: Delivered in the University of Pennsylvania, November 16, 1799.* Philadelphia, 1800.

Regis, Pamela. *Describing Early America: Bartram, Jefferson, Crèvecoeur, and the Rhetoric of Natural History.* DeKalb: Northern Illinois University Press, 1992.

Russell, Eric. "Modern Violence: Animal Studies, Speciesism, and the Writings of John James Audubon." *ESQ: A Journal of Nineteenth-Century American Literature and Culture* 65, no. 3 (2019): 470–510.

Shuler, Jay. *Had I the Wings: The Friendship of Bachman and Audubon*. Athens: University of Georgia Press, 1995.

Stephens, Lester D. *Science, Race, and Religion in the American South: John Bachman and the Charleston Circle of Naturalists, 1815–1895*. Chapel Hill: University of North Carolina Press, 2000.

Stroud, Patricia Tyson. *The Emperor of Nature: Charles-Lucien Bonaparte and His World*. Philadelphia: University of Pennsylvania Press, 2000.

Thoreau, Henry D. *"Walden," "Civil Disobedience," and Other Writings*. Edited by William Rossi. New York: W. W. Norton, 2008.

Tyler, Ron. "The Publication of *The Viviparous Quadrupeds of North America*." In *John James Audubon in the West: The Last Expedition*, edited by Sarah Boehme, 119–82. New York: Harry N. Abrams, 2000.

Webber, Charles Wilkins. *The Hunter-Naturalist: Romance of Sporting; or, Wild Scenes and Wild Hunters*. Philadelphia: J. B. Lippincott and Company, 1856.

———. "The Quadrupeds of North America." *American Review: A Whig Journal of Politics, Literature, Art and Science* 4, no. 6 (December 1846): 625–38.

———. "The Viviparous Quadrupeds of North America." *Southern Quarterly Review* 12, no. 24 (October 1847): 273–306.

Wilson, Alexander. *The American Ornithology*. 9 vols. Philadelphia: Bradford and Inskeep, 1808–14.

EDITORIAL PRINCIPLES

The text of the quadruped essays presented herein reproduces exactly that of the first edition, *The Viviparous Quadrupeds of North America*, 3 vols. (New York: Published by J. J. Audubon and V. G. Audubon, 1846–54), except where a significant change was made for the Octavo edition. Any such change is explained in a textual note. The taxonomic names given in the title block of each essay are those of the original publication. We also retain the original capitalization in the Latin genus and species names. Relatedly, the order of the essays is that of their publication, which—as in *The Birds of America*—is generally that of their creation, not following a settled taxonomic order. We silently emend obvious typesetting errors (e.g., omitted closing quotation marks) and inconsistent spellings and capitalization (e.g., "raccoon" for "racoon," "carcass" for "carcase," "Prong-horned" for "Prong-Horn"). We insert in square brackets conjectural readings where a word seems to have been inadvertently omitted in the original.

TENACIOUS OF LIFE

THE QUADRUPED ESSAYS

No 1

PLATE I.

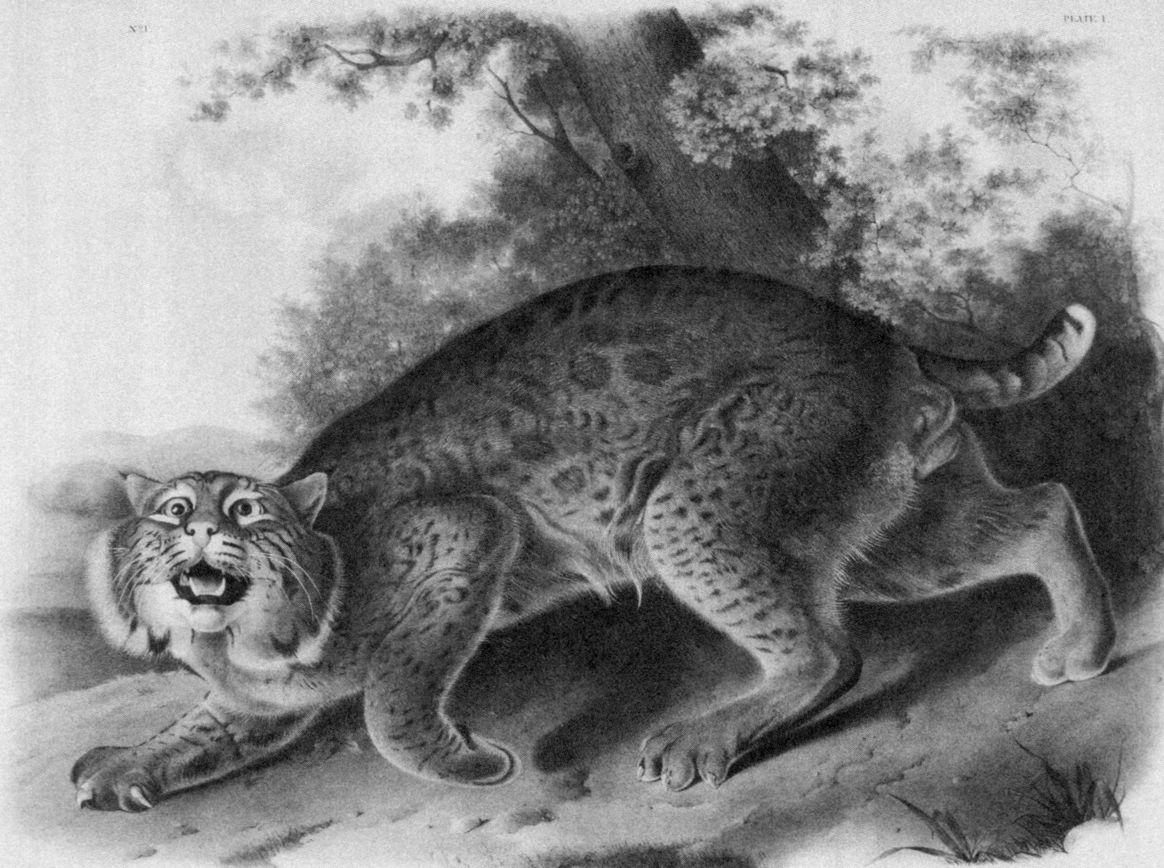

LYNX RUFUS, GULDENSTAED.

Common American Wildcat

IN 1850 BACHMAN CLAIMED IN *THE DOCTRINE OF THE UNITY OF the Human Race Examined on the Principles of Science* that humans were the only animals governed by reason.[1] When "Common American Wild Cat" was published four years prior in the first volume of *The Viviparous Quadrupeds of North America* (1846–54), he didn't seem quite as confident. The bobcat is so clever at diffusing its scent that Bachman was compelled to describe the strategy as "an exercise of instinct, so closely bordering on reason, that we are bewildered in the attempt to separate it from the latter." Seeing both instinct and reason in the same animal behavior indicates that Bachman, despite later claims, conceived of the natural world as a complex realm challenging the efficacy of rigid philosophical labels. This observation, however, is just one piece of commentary on the bobcat's status as an object of human sport. In this, the very first of the quadruped essays, readers are introduced to Audubon and Bachman's acute attention to hunting imagery and the cast of characters involved, thus establishing vivid hunting narratives as a central theme to be found throughout the three volumes.

---

3. John James Audubon, Nagel and Weingartner, lithographers, *Common American Wild Cat*, 1845–46, hand-colored lithograph, plate 1 of *The Viviparous Quadrupeds of North America*. Courtesy of the Jule Collins Smith Museum of Fine Art, Auburn University; Louise Hauss and David Brent Miller Audubon Collection.

LYNX RUFUS.—GULDENSTAED.

COMMON AMERICAN WILD CAT.—BAY LYNX.

PLATE I.—MALE.

HABITS.

The general appearance of this species conveys the idea of a degree of ferocity, which cannot with propriety be considered as belonging to its character, although it will, when at bay, show its sharp teeth, and with outstretched claws and infuriated despair, repel the attacks of either man or dog, sputtering the while, and rolling its eyes like the common cat.

It is, however, generally cowardly when attacked, and always flies from its pursuers, if it can, and although some anecdotes have been related to us of the strength, daring, and fierceness of this animal; such as its having been known to kill at different times a sheep, a full-grown doe, attack a child in the woods, &c.; yet in all the instances that have come under our own notice, we have found it very timid, and always rather inclined to beat a retreat, than to make an attack on any animal larger than a hare or a young pig. In the American Turf register, there is an interesting extract of a letter from Dr. Coleman, U.S.A., written at Fort Armstrong, Prairie du Chien, giving an account of a contest between an eagle and a Wild Cat. After a fierce struggle, in which the eagle was so badly wounded as to be unable to fly, the Cat, scratched and pierced in many places, and having had one eye entirely "gouged out" in the combat, was found lying dead.

In hunting at night for raccoons and opossums, in which sport the negroes on the plantations of Carolina take great delight, a Cat is occasionally "treed" by the dogs; and the negroes, who seldom carry a gun, climb up the tree and shake him off as they would do a raccoon, and although he fights desperately, he is generally killed by the dogs. During a botanical excursion through the swamps of the Edisto river, our attention was attracted by the barking of a small terrier at the foot of a sapling, (young tree.) On looking up, we observed a Wild Cat, about twenty feet from the ground, of at least three times the size of the dog, which he did not appear to be much afraid of. He seemed to have a greater dread of man, however, than of this diminutive specimen of the canine race, and leaped from the tree as we drew near.

The Wild Cat pursues his prey with both activity and cunning, sometimes bounding suddenly upon the object of his rapacity, sometimes with stealthy pace, approaching it in the darkness of night, seizing it with his strong retractile claws and sharp teeth, and bearing it off to his retreat in the forest. The individual from which our figure was drawn had been caught in a steel-trap, and was brought to us alive. We kept it for several weeks; it was a fine male, although not the largest we have seen. Like most of the predacious animals, it grew fat in confinement, being regularly fed on the refuse parts of chickens and raw meat, as well as on the common brown rat.

The Bay Lynx, (as this animal is sometimes called,) is fond of swampy, retired situations, as well as the wooded sides of hills, and is still seen occasionally in that portion of the Alleghany mountains, which traverses the States of Pennsylvania and New York. It is abundant in the *Cane-brakes*, (patches or thickets of the *Miegia macrosperma*, of MICHAUX, which often extend for miles, and are almost impassable,) bordering the lakes, rivers, and lagoons of Carolina, Louisiana, and other Southern and South Western States. This species also inhabits the mountains and the undulating or rolling country of the Southern States, and frequents the thickets that generally spring up on deserted cotton plantations, some of which are two or three miles long, and perhaps a mile wide, and afford, from the quantity of briars, shrubs, and young trees of various kinds which have overgrown them, excellent cover for many quadrupeds and birds. In these bramble-covered old fields, the "Cats" feed chiefly on the rabbits and rats that make their homes in their almost impenetrable and tangled recesses; and seldom does the cautious Wild Cat voluntarily leave so comfortable and secure a lurking place, except in the breeding season, or to follow in very sultry weather, the dry beds of streams or brooks, to pick up the catfish, &c., or cray-fish and frogs that remain in the deep holes of the creeks, during the drought of summer.

The Wild Cat not only makes great havoc among the chickens, turkeys, and ducks of the planter, but destroys many of the smaller quadrupeds, as well as partridges, and such other birds as he can surprise roosting on the ground. The hunters often run down the Wild Cat with packs of fox-hounds. When hard pressed by fast dogs, and in an open country, he ascends a tree with the agility of a squirrel, but the baying of the dogs calling his pursuers to the spot, the unerring rifle brings him to the ground, when, if not mortally wounded, he fights fiercely with the pack until killed. He will, however, when pursued by hunters with hounds, frequently elude both dogs and huntsmen, by an exercise of instinct, so closely bordering on reason, that we are bewildered in the attempt to separate it from the latter. No sooner does he become aware that the enemy is on his track, than, instead of taking a straight course for the deepest forest, he speeds to one of the largest old-fields overgrown with briary thickets, in the neighbourhood; and having reached this tangled maze, he runs in a variety of circles, crossing and re-crossing his path many times, and when he thinks the scent has been diffused sufficiently in different directions by this manœuvre, to puzzle both men and dogs, he creeps slyly forth, and makes for the woods, or for some well known swamp, and if he should be lucky enough to find a half-dried-up pond, or a part of the swamp, on which the clayey bottom is moist and sticky, he seems to know that the adhesive soil, covering his feet and legs, so far destroys the scent, that although the hounds may be in full cry on reaching such a place, and while crossing it, they will lose the track on the opposite side, and perhaps not regain it without some difficulty and delay.

At other times the "Cat," when chased by the dogs, gains some tract of "burnt wood," common especially in the pine lands of Carolina, where fallen and upright trees are alike blackened and scorched, by the fire that has run among them burning before it every blade of grass, every leaf and shrub, and destroying many of the largest trees in its furious course; and here, the charcoal and ashes on the ground, after he has traversed the burnt district a short distance, and made a few leaps along the trunk of a fallen tree, that has been charred in the conflagration, will generally put any hounds at fault. Should no such chance of safety be within his reach, he does not despair, but exerting his powers of flight to the utmost, increases his distance from the pursuing pack, and following as intricate and devious a path as possible, after many a weary mile has been run over, he reaches a long-fallen trunk of a tree, on which he may perchance at some previous time have baffled the hunters as he is now about to do. He leaps on to it, and hastily running to the farther end, doubles and returns to the point from which he gained the tree, and after running backward and forward repeatedly on the fallen trunk, he makes a sudden and vigorous spring, leaping as high up into a tree some feet distant, as he can; he then climbs to its highest forks, (branches,) and closely squatted, watches the movements of his pursuers. The dogs are soon at fault, for he has already led them through many a crooked path; the hunters are dispirited and weary, and perhaps the density of the woods, or the approach of night favours him. The huntsmen call off their dogs from the fruitless search, and give up the chase; and shortly afterwards the escaped marauder, descends leisurely to the earth, and wanders off in search of food, and to begin a new series of adventures.

In some parts of Carolina, Georgia, Mississippi, and Louisiana, the Wild Cat has at times become so great a nuisance as to have aroused the spirit of vengeance in the hearts of the planters, who are constant sufferers from his depredations. They have learned by experience, that one Cat will do as much mischief among the pigs and poultry as a dozen gray foxes. They are now determined to allow their hounds, which they had hitherto kept solely for the favourite amusement of deer hunting, and which had always been whipped-in, from the trail of the Wild Cat, to pursue him, through thicket, briar patch, marsh, and morass, until he is caught or killed.

Arrangements for the Cat-hunt are made over night. Two or three neighbours form the party, each one bringing with him all the hounds he can muster. We have seen thirty of the latter brought together on such occasions, some of which were not inferior to the best we have examined in England, indeed, great numbers of the finest fox-hounds are annually imported into Carolina.

At the earliest dawn, the party is summoned to the spot previously fixed on as the place of meeting. A horn is sounded, not low and with a single blast, as is usual in hunting the deer, lest the timid animal should be startled from its bed among the broom-grass (*Andropagon dissitiflorus*) and bound away out of the drive, beyond the reach of the hunter's double-barrel loaded with buckshot; but with a loud, long, and oft-repeated blast, wakening the echoes that rise from the rice-fields and marshes, and are reverberated from shore to shore of the winding sluggish river, until lost among the fogs and shadows of the distant forest.

An answering horn is heard half a mile off, and anon comes another response from a different quarter. The party is soon collected, they are mounted, not on the fleetest and best-blooded horses, but on the most sure-footed, (sometimes called "Old field Tackies,") which know how to avoid the stump-holes on the burnt grounds of the pine lands, which stand the fire of the gun, and which can not only go with tolerable speed, but are, to use a common expression, "tough as a pine knot." The hunters greet each other in the open-hearted manner characteristic of the Southern planter. Each pack of dogs is under the guidance of a coloured driver, whose business it is to control the hounds and encourage and aid them in the hunt. The drivers ride in most cases the fleetest horses on the ground, in order to be able, whilst on a deer hunt to stop the dogs. These men, who are so important to the success of the chase, are possessed of a good deal of intelligence and shrewdness, are usually much petted, and regarding themselves as belonging to the aristocracy of the plantation, are apt to look down upon their fellow-servants as inferiors, and consider themselves privileged even to crack a joke with their masters. The drivers are ordered to stop the dogs if a deer should be started, a circumstance which often occurs, and which has saved the life of many a Cat, whose fate five minutes before this unlucky occurrence was believed to be sealed. Orders are given to destroy the Cat fairly, by running him down with the hounds, or if this cannot be done, then by shooting him if he ascends a tree or approaches within gun shot of the stand which the hunter has selected as the most likely place for him to pass near. The day is most auspicious—there is not a breath of wind to rustle the falling leaves, nor a cloud to throw its shadows over the wide joyous landscape. The dew-drops are sparkling on the few remaining leaves of the persimmon tree, and the asters and dog-fennel hang drooping beneath their load of moisture. The dogs are gamboling in circles around, and ever and anon, in spite of all restraint, the joyous note breaks forth—the whole pack is impatient for the chase, and the young dogs are almost frantic with excitement.

But we have not time for a farther description of the scene—whilst we are musing and gazing, the word is given, "Go!" and off start the hounds, each pack following its own driver to different parts of the old fields, or along the borders of the swamps and marshes. Much time, labour and patience are usually required, before the "Cat" can be found by the dogs: sometimes there is a sudden burst from one or the other of the packs, awakening expectation in the minds of the huntsmen, but the driver is not to be so easily deceived, as he has some dogs that never open at a rabbit, and the snap of the whip soon silences the riotous young babblers. Again there is a wild burst and an exulting shout, giving assurance that better game than a rabbit is on foot; and now is heard a distant shot, succeeded in a second of time by another, and for an instant all is still: the echoes come roaring up through the woods, and as they gradually subside, the crack of the whip is again heard stopping the dogs. The story is soon told; a deer had been started—the shot was too small—or the distance too great, or any other excuses, (which are always at hand among hunters of fertile imagination,) are made by the unsuccessful sportsman who fired, and the dogs are carried back to the "trail" of the Cat, that has been growing fresher and fresher for the last half hour. At length, "Trimbush," (and a good dog is he,) that has been working on the cold trail for some time, begins to give tongue,[2] in a way that brings the other dogs to his aid. The drivers now advance to each other, encouraging their dogs; the trail becomes a drag;[3] onward it goes through a broad marsh at the head of a rice-field. "He will soon be started now!" "He is up!" What a burst! you might have heard it two miles off—it comes in mingled sounds, roaring like thunder, from the muddy marsh and from the deep swamp. The barred owl, frightened from the monotony of his quiet life among the cypress trees, commences hooting in mockery as it were, of the wide-mouthed hounds. Here they come, sweeping through the resounding swamp like an equinoctial storm—the crackling of a reed, the shaking of a bush, a glimpse of some object that glided past like a shadow, is succeeded by the whole pack, rattling away among the vines and fallen timbers, and leaving a trail in the mud as if a pack of wolves in pursuit of a deer had hurried by. The Cat has gone past. It is now evident that he will not climb a tree. It is almost invariably the case that where he can retreat to low swampy situations, or briar patches, he will not take a tree, but seeks to weary the dogs by making short windings among the almost impassable briar patches. He has now been twisting and turning half a dozen times in a thicket covering only three or four acres—let us go in and take our stand on the very trail where he last passed, and shoot him if we can. A shot is heard on the opposite edge of the thicket, and again all is still; but once more the pack is in full cry. Here he comes, almost brushing our legs as he dashes by and disappears in the bushes, before we can get sight of him and pull trigger. But we see that the dogs are every moment pressing him closer, that the marauder is showing evidences of fatigue and is nearly "done up." He begins to make narrower circles, there are restless flashes in his eye, his back is now curved upwards, his hair is bristled nervously forward, his tongue hangs out—we raise our gun as he is approaching, and scarcely ten yards off—a loud report—the smoke has hardly blown aside, ere we see him lifeless, almost at our very feet—had we waited three minutes longer, the hounds would have saved us the powder and shot!

One fine morning in autumn, when we had crossed the Ohio river at Henderson, in Kentucky, with the view of shooting some wild turkeys, geese, and perhaps a deer, we chanced to seat ourselves about fifty yards from a prostrate tree, and presently saw a Wild Cat leap on to it, and go through the manœuvres we have described in a preceding page. He did not see us, and had scarcely reached one of the higher branches of a tall white-oak, after springing into it from the fallen tree, when we heard the dogs, which soon came up, with the hunters following not far behind. They asked, when they perceived us, whether we had seen the "Cat" that had given them the slip. Always willing to assist the hunter who has lost his game, and having no particular liking towards this species, we answered in the affirmative, and showed them the animal, closely squatted on a large branch some distance from the ground. One of the party immediately put his rifle to his shoulder and pulled the trigger: the Cat leaped from the branch into the air, and fell to the earth quite dead. Whilst residing in Louisiana some twenty years since, we chanced one afternoon to surprise one of these depredators. He had secured a hare, (commonly called rabbit,) and was so eagerly engaged in satisfying his hunger as not to observe us, until we were near the spot where he was partially concealed behind a rotten log. At sight of us, he squatted flat on the ground. As we looked at him, we heard a squirrel close by, and turned our head for an instant, but scarce had we glanced at the squirrel, when looking again for the Wild Cat, he had disappeared, carrying the remains of the hare away with him.

About twenty miles from Charleston, South Carolina, resides a worthy friend of ours, a gentleman well known for his skill in the sports of the field, his hospitality to both friends and strangers, and the excellent manner in which his plantation is managed. The plantation of Dr. DESEL is, in short, the very place for one who likes the sight of several fine bucks hanging on the branches of an old Pecan-nut tree; while turkeys, geese, and poultry of other kinds, are seen in abundance in his well stocked poultry yards, affording certainty of good cheer to his visitors.

The Doctor's geese were nightly lodged near the house, in an enclosure which was rendered apparently safe, by a very high fence. As an additional security, several watch dogs were let loose about the premises; besides an excellent pack of hounds, which by an occasional bark or howl during the night, sounded a note of warning or alarm in case any marauder, whether biped or quadruped approached.

Notwithstanding these precautions, a goose disappeared almost every night, and no trace of the ingress or egress of the robber could be discovered. Slow in attaching suspicion to his servants, the Dr. waited for time and watchfulness to solve the mystery. At length, the feathers, and other remains of his geese, were discovered in a marsh about a quarter of a mile from the house, and strong suspicions were fastened on the Wild Cat; still, as he came at odd hours of the night, all attempts to catch or shoot him proved for a time unavailing.

One morning, however, he came about day-light, and having captured a good fat goose, was traced by the keen noses of the hounds. The chase was kept up for some time through the devious windings of the thickets, when his career of mischief was brought to a close by a shot from the gun of our friend the Doctor, who, in self-defence, became his executioner. Thus ended his career. In this respect he fared worse than he deserved, compared with those beings of a superior nature, who, not understanding that "Honesty is the best policy" outdo our Wild Cat in his destructive habits, until the laws, so just and useful, when mildly, but always, enforced, put an effectual stop to their criminal proceedings.

The Wild Cat is a great destroyer of eggs, and never finds a nest of grouse or partridge, wild turkey or other bird, without sucking every egg in it.

Indeed, it will if practicable, seize on both young and old birds of these and other species. Its "penchant" for a "poulet au naturel" has suggested the following method of capturing it in Georgia, as related to us by our friend MAJOR LECONTE, late of the United States Army.

A large and strong box-trap is constructed, and a chicken-cock, (rooster,) placed at the farthest end of it from the door, is tied by one leg, so that he cannot move. There is a stout wire partition about half way between the fowl and the door, which prevents the Cat when entering the trap, from seizing the bird. The trap is then set so that when the animal enters, the open door closes behind him by a spring, (commonly the branch of some tree bent down for the purpose, and released by a trigger set at the entrance or just within the trap.) These traps are placed in different parts of the plantations, or in the woods, and the Wild Cat is generally attracted by the crowing of the cock at early dawn of day.

MAJOR LECONTE has caught many of them by this artifice, on and about his plantations in the neighbourhood of Savannah, in Georgia; and this method of capturing the Wild Cat is also quite common in South Carolina. Indeed, this species does not seem to possess the suspicion and cunning inherent in the fox, enabling the latter to avoid a trap of almost any kind. We have seen the Wild Cat taken from the common log-traps set for raccoons. We saw one in a cage, that had been caught in a common box-trap, baited with a dead partridge, and have heard intelligent domestics residing on the banks of the Santee river, state, that after setting their steel traps for otters, they frequently found the Wild Cat caught in them instead.

When this animal discovers a flock of wild turkeys, he will generally follow them at a little distance for some time, and after having ascertained the direction in which they are proceeding, make a rapid detour, and concealing himself behind a fallen tree, or in the lower branches of some leafy maple, patiently wait in ambush until the birds approach, when he suddenly springs on one of them, if near enough, and with one bound secures it. We once, while resting on a log in the woods, on the banks of the Wabash river, perceived two wild turkey cocks at some distance below us, under the bank near the water, pluming and picking their feathers; on a sudden, one of them flew across the river, and the other we saw struggling in the grasp of a Wild Cat, which almost instantly dragged it up the bank into the woods, and made off. On another occasion we observed an individual of this species, about nine miles from Charleston, in pursuit of a covey of partridges, (*Ortyx Virginiana*,)—so intent was the Cat upon its prey, that it passed within ten steps of us, as it was making a circle to get in advance and in the path of the birds,—its eyes were constantly fixed on the covey, and it stealthily concealed itself behind a log it expected the birds to pass. In a second attempt the marauder succeeded in capturing one of the partridges, when the rest in great affright flew and scattered in all directions.

An individual that was kept alive at Charleston, and afterwards for a short time at our house in the City of New-York, showed its affinity to the domestic cat, by purring and mewing at times loud enough to be heard at some distance. At the former place its cry was several times mistaken for that of the common house-cat. In the woods, during the winter season, its loud catterwauling can be heard at the distance of a mile.

Although this species may perhaps be designated as nocturnal in its habits, it is, by no means, exclusively so, as is shown by the foregoing account. We have, in fact, in several instances seen this Cat engaged in some predatory expedition in full sun-shine, both in winter and summer.

It is not a very active swimmer, but is not averse to taking the water. We witnessed it on one occasion crossing the Santee river when not pursued, and at another time saw one swimming across some ponds to make its escape from the dogs. It has been observed, however, that when it has taken to the water during a hard chase, it soon after either ascends a tree or is caught by the hounds.

The domicile of the Wild Cat is sometimes under an old log, covered with vines such as the *Smilax*, *Ziziphus volubilus*, *Rubus*, &c., but more commonly in a hollow tree. Sometimes it is found in an opening twenty or thirty feet high, but generally much nearer the ground, frequently in a cavity at the root, and sometimes in the hollow trunk of a fallen tree, where, after collecting a considerable quantity of long moss and dried leaves to make a comfortable lair, it produces from two to four young. These are brought forth in the latter end of March in Carolina; in the Northern States, however, the kittens appear later, as we have heard of an instance in Pennsylvania where two young were found on the 15th day of May, apparently not a week old. Our friend Dr. SAMUEL WILSON, of Charleston, a close observer of nature, has made the following note in our memorandum book. "April 15th, 1839, shot a female Wild Cat as it started from its bed, out of which four young ones were taken, their eyes were not yet open." Our friend Dr. DESEL, whom we have already mentioned, saw three young ones taken out from the hollow of a tree which was thirty feet from the ground. On four occasions, we have had opportunities of counting the young, either in the nest or having been very recently taken from it. In every case there were three young ones. In one instance the nest was composed of long moss, (*Tillandsia usneoides*,) which seemed to have been part of an old, deserted, squirrel's nest.

We once made an attempt at domesticating one of the young of this species, which we obtained when only two weeks old. It was a most spiteful, growling, snappish little wretch, and showed no disposition to improve its habits and manners under our kind tuition. We placed it in a wooden box, from which it was constantly striving to gnaw its way out. It, one night, escaped into our library, where it made sad work among the books, (which gave us some valuable lessons on the philosophy of patience, we could not have so readily found among our folios,) and left the marks of its teeth on the mutilated window-sashes. Finally we fastened it with a light chain, and had a small kennel built for it in the yard. Here it was constantly indulging its carnivorous propensities, and catching the young poultry, which it enticed within reach of its chain by leaving a portion of its food at the door of its house, into which it retreated until an opportunity offered to pounce on its unsuspecting prey. Thus it continued, growing if possible, more wild and vicious every day, growling and spitting at every servant that approached it, until at last, an unlucky blow, as a punishment for its mischievous tricks, put an end to its life, and with it to one source of annoyance.

The Bay Lynx is generally in fine order, and often very fat. The meat is white, and has somewhat the appearance of veal. Although we omitted to taste it, we have seen it cooked, when it appeared savoury, and the persons who partook of it pronounced it delicious.

The muscular powers of this species are very great, and the fore-feet and legs are rather large in proportion to the body.

No. 2

PLATE XI.

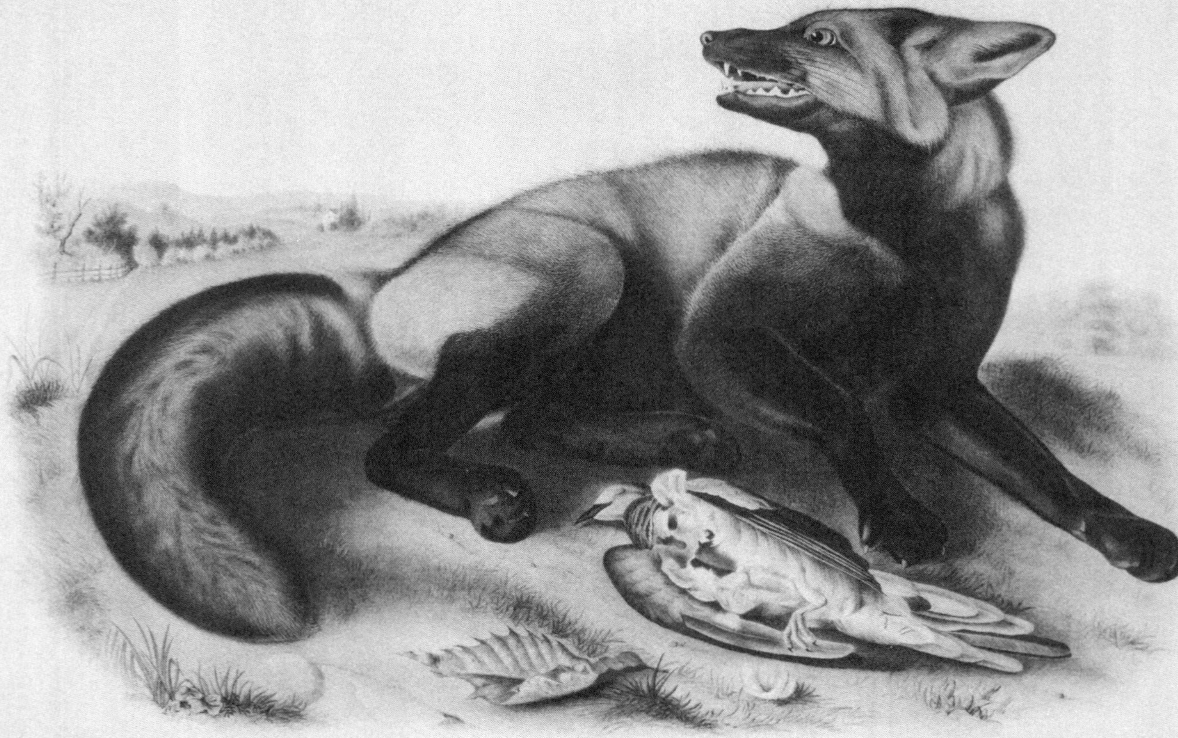

CANIS (VULPES) FULVUS, DESMARET.

VAR. DECUSSATUS.

AMERICAN CROSS FOX.

American Cross Fox

BACHMAN SUGGESTS THAT HUMAN PERSECUTION HAS ACTUALLY benefited this fox by making it cleverer, stating that he would not be surprised if "a constant succession of attempts to capture it together with the instinctive desire for self-preservation possessed by all animals, should sharpen its wits and render it more cautious and wild than those species less frequently molested." It's a reasonable conclusion, perhaps, but nonetheless an ironic one to think that being exhausted and torn apart by hunting hounds or being driven from a den by smoke and then clubbed to death is in some way good for the fox. The larger implication, however, is that Bachman is perceiving in this species an ability to negotiate its environment based on what it has learned from its contentious relationship with humans. In other words, Bachman is coming to recognize nature not simply as a static creation but rather as a dynamic creation populated by adaptive nonhumans.

---

4. John James Audubon, Nagel and Weingartner, lithographers, *American Cross Fox*, 1845–46, hand-colored lithograph, plate 6 of *The Viviparous Quadrupeds of North America*. Courtesy of the Jule Collins Smith Museum of Fine Art, Auburn University; Louise Hauss and David Brent Miller Audubon Collection.

VULPES FULVUS.—DESM.: *VAR. DECUSSATUS.*—PENNANT.

AMERICAN CROSS FOX

PLATE VI.—MALE.

HABITS.

In our youth we had opportunities whilst residing in the northern part of the State of New York, of acquiring some knowledge of the habits of the Fox, and many other animals, which then were abundant around us.

Within a few miles dwelt several neighbours who vied with each other in destroying foxes, and other predacious animals, and who kept a strict account of the number they captured or killed each season. As trappers, most of our neighbours were rather unsuccessful—the wary foxes, especially, seemed very soon, as our western hunters would say, to be "up to trap." Shooting them by star-light, from behind a hay-stack in the fields, when they had for some time been baited and the snow covered the ground so that food was eagerly sought after by them, answered pretty well at first, but after a few had been shot at, the whole tribe of foxes—red, gray, cross, and black—appeared to be aware that safety was no longer to be expected in the vicinity of hay-stacks, and they all gave the latter a wide berth.

With the assistance of dogs, pick-axes, and spades, our friends were far more successful, and we think might have been considered adepts. We were invited to join them, which we did on a few occasions, but finding that our ideas of sport did not accord precisely with theirs, we gradually withdrew from this club of primitive fox-hunters. Each of these sportsmen was guided by his own "rules and regulations" in the "chase;" the horse was not brought into the field, nor do we remember any scarlet coats. Each hunter proceeded in the direction that to him seemed best—what he killed he kept—and he always took the shortest possible method he could devise, to obtain the fox's skin. He seldom carried a gun, but in lieu of it, on his shoulder was a pick-axe and a spade, and in his pocket a tinder box and steel.

A half-hound, being a stronger and swifter dog than the thorough-bred, accompanied him, the true foxhound being too slow, and too noisy for his purpose; we remember one of these half-bred dogs, which was of great size and extraordinary fleetness; it was said to have a cross of the grey-hound.

In the fresh-fallen and deep snows of mid-winter, the hunters were most successful. During these severe snow storms, the ruffed grouse, (*Tetrao umbellus*,) called in our Eastern States the partridge, is often snowed up and covered over; or sometimes plunges from on wing into the soft snow, where it remains concealed for a day or two. The fox occasionally surprises these birds, and as he is usually stimulated at this inclement season by the gnawings of hunger, he is compelled to seek for food by day as well as by night; his fresh tracks may be seen in the fields, along the fences, and on the skirts of the farm-yard, as well as in the deep forest.

Nothing is easier than to track the Fox under these favourable circumstances, and the trail having been discovered, it is followed up, until Reynard is started. Now the chase begins; the half-hound yells out, in tones far removed from the mellow notes of the thorough-bred dog, but equally inspiriting perhaps, through the clear frosty air, as the solitary hunter eagerly follows, as fast as his limited powers of locomotion will admit. At intervals of three or four minutes, the sharp cry of the dog resounds, the Fox has no time to double and shuffle, the dog is at his heels almost, and speed, speed, is his only hope for life. Now the shrill baying of the hound becomes irregular; we may fancy he is at the throat of his victim; the hunter is far in the rear, toiling along the track which marks the course so well contested, but occasionally the voice of his dog softened by the distance, is borne on the wind to his ear. For a mile or two the Fox keeps ahead of his pursuer, but the latter has the longest legs, and the snow impedes him less than it does poor Reynard; every bound and plunge into the snow, diminishes the distance between the fox and his relentless foe. Onward they rush through field, fence, brushwood, and open forest, the snow flying from bush and briar as they dart through the copse, or speed across the newly-cleared field. But this desperate race cannot last longer. The fox must gain his burrow, or some cavernous rock, or he dies. Alas! he has been lured too far away from his customary haunts and from his secure retreat, in search of prey, he is unable to reach his home; the dog is even now within a foot of his brush. One more desperate leap, and with a sudden snappish growl he turns upon his pursuer, and endeavours to defend himself with his sharp teeth. For a moment he resists the dog, but is almost instantly overcome. He is not killed, however, in the first onset; both dog and fox, are so fatigued that they now sit on their haunches facing each other, resting, panting, their tongues hanging out, and the foam from their lips dropping on the snow. After fiercely eyeing each other for a while, both become impatient—the former to seize his prey, and the latter to escape. At the first leap of the fox, the dog is upon him; with renewed vigour he seizes him by the throat, and does not loose his hold until the snow is stained with his blood, and he lies rumpled, draggled, with blood-shot eye, and frothy open mouth, a mangled carcass on the ground.

The hunter soon comes up: he has made several short cuts, guided by the baying of his hound; and striking the deep trail in the snow again, at a point much nearer to the scene of the death-struggle, he hurries toward the place where the last cry was heard, and pushes forward in a half run until he meets his dog, which on hearing his master approach, generally advances towards him, and leads the way to the place where he has achieved his victory.

We will now have another hunt, and pursue a Fox, that is within reach of his burrow when we let loose our dog upon him. We will suppose him "started;" with loud shouts we encourage our half-hound; he dashes away on the Fox's track, whilst the latter, with every muscle strained to the utmost, is shortening the distance between himself and his stronghold; increasing his speed with his renewed hopes of safety, he gains the entrance to his retreat, and throws himself headlong into it, rejoicing at his escape. Whilst yet panting for breath, he hears his foe barking at the entrance of his burrow, and flatters himself he is now beyond a peradventure safe [distance]. But perhaps we do injustice to his sagacity; he may have taken refuge in his hole, well aware of the possibility of his being attacked there—yet what better could he do? However this may be, he has escaped one enemy, by means of a swift pair of heels, and has only to dread the skill, perseverance, and invention of the hunter; who in time comes up, rigged out pretty much as we have already described him, with spade, pick-axe, flint and steel.

On arriving at the spot where the Fox has been (in select phrase) "holed," the sportsman surveys the place, and if it is on level ground, where he can use the spade, he throws off his coat, and prepares for his work with a determination to have "that" fox, and no mistake! He now cuts a long slender stick, which he inserts in the hole, to ascertain in what direction he shall dig the first pit. The edge or mouth of the burrow, is generally elevated a little above the adjacent surface of the ground, by the earth which the Fox has brought from within; and this slight embankment, serves to keep out the rain water, that might otherwise flow in from the vicinity in stormy weather.

The burrow at first inclines downward, for four or five feet, at an angle of about twenty-five degrees, it then inclines upward a little, which is an additional security against inundations, and is continued, at a depth of about three or four feet from the surface, until it reaches a point where it is divided into two or three galleries.

This dividing point the hunter discovers after sinking three or four pits—it is generally twenty or thirty feet from the entrance of the burrow. The excavation is now made larger, the earth and rubbish thrown out, the dog is placed in the hole thus laid open, and his aid is sought, to ascertain into which branch of the gallery the Fox has retreated. There are seldom any tortuous windings beyond the spot whence the galleries diverge—the Fox is not far off. The stick is again inserted, and either reaches him, and the hunter is made aware of his whereabouts, by his snapping at it and growling, which calls forth a yelp of fierce anxiety from the dog; or, as frequently happens, the Fox is heard digging for life, and making no contemptible progress through the earth. Should no rocks or large roots interfere, he is easily unearthed, and caught by the dog.

It however very frequently occurs, that the den of the Fox is situated on the mountain side; and that its winding galleries run beneath the enormous roots of some stately pine or oak; or it may be amongst huge masses of broken rock, in some fissure of too great depth to be sounded, and too contracted to be entered by man or dog. What is then to be done? Should a "dead-fall" be set at the mouth of the hole, the Fox will (unless the ground be frozen too hard,) dig another opening, and not go out by the old place of egress; place a steel-trap before it, and he will spring it without being caught. He will remain for days in his retreat, without once exposing himself to the danger of having a dog snapping at his nose, or a load of duck-shot whistling round his ears. Our hunter, however, is not much worried with such reflections as we have just made; he has already gathered an armful or two of dry wood, and perhaps some resinous knots, or bits of the bark of the pine-tree; he cuts up a portion into small pieces, pulls out his tinder-box, flint, and steel, and in a few moments a smart fire is lighted within the burrow; more wood is thrown on, the mass pushed further down the hole, and as soon as it begins to roar and blaze freely, the mouth is stopped with brush-wood covered with a few spadefuls of earth, and the den is speedily exhausted of pure air, and filled with smoke and noxious gases.

There is no escape for the Fox—an enemy worse than the dog or the gun, is destroying him; he dies a protracted, painful death by suffocation! In about an hour the entrance is uncovered, large volumes of smoke issue into the pure air, and when the hunter's eye can pierce through the dense smoky darkness of the interior, he may perhaps discern the poor Fox extended lifeless in the burrow, and may reach him with a stick. If not quite dead, the Fox is at least exhausted and insensible; this is sometimes the case, and the animal is then knocked on the head.

The number of Foxes taken by our neighbours, in the primitive mode of hunting them we have attempted to describe, was, as nearly as we can now recollect, about sixty every winter, or an average of nearly twenty killed by each hunter. After one or two seasons, the number of Foxes in that part of the country was sensibly diminished, although the settlements had not increased materially and the neighbourhood was at that time very wild.

At this time Pennant's marten (*Mustela Canadensis*) was not very scarce in Rensselaer county, and we had three different specimens brought to us to examine.

These, the people called Black Foxes. They were obtained by cutting down hollow trees, in which they were concealed, and to which their tracks on the snow directed the hunters.

We cannot now find any note, in regard to the number of Cross Foxes taken, as compared to the Red, Gray, and Black Foxes; about one-fourth of the whole number captured, however, were Gray Foxes, and we recollect but a single one that was perfectly black with the exception of a white tip at the end of its tail, like the specimen figured in our work.[4]

On examining several packages of Fox skins at Montreal, we saw about four specimens only of the Cross Fox, and three of the Black Fox, in some three hundred skins. We were informed during our recent visit to the Upper Missouri country, that from fifty to one hundred skins of the Cross Fox were annually procured by the American Fur Company, from the hunters and Indians.

The specimen from which our drawing was made, was caught in a steel-trap, by one of its fore-feet, not far from the falls of Niagara, and was purchased by J. W. AUDUBON of the proprietor of the "Museum" kept there to gratify the curiosity of the travellers who visit the great Cataract.

Dr. RICHARDSON (Fauna Boreali Americana, p. 93) adheres to the opinion of the Indians, who regard the Cross Fox of the fur traders as a mere variety of the Red Fox. He says, "I found on inquiry that the gradations of colour between characteristic specimens of the Cross and Red Fox, are so small that the hunters are often in doubt with respect to the proper denomination of a skin; and I was frequently told, 'This is not a Cross Fox yet, but it is becoming so.'" It is worthy of remark, moreover, that the European Fox (*Vulpes vulgaris*) is subject to similar varieties, and that the "*Canis crucigera* of GESNER, differs from the latter animal in the same way that the American Cross Fox does from the red one."

We have had several opportunities of examining *C. crucigera* in the museums of Europe, and regard it as a variety of the common European Fox, but it differs in many particulars from any variety of the American Red Fox that we have seen.

The Cross Fox is generally regarded as being more wary and swift of foot than the Red Fox; with regard to its greater swiftness, we doubt the fact. We witnessed a trial of speed between the mongrel greyhound already referred to in this article, and a Red Fox, in the morning, and another between the same dog and a Cross Fox, about noon, on the same day. The former was taken after an hour's hard run in the snow, and the latter in half that time, which we accounted for from the fact that the Cross Fox was considerably the fattest, and from this circumstance became tired out very soon. We purchased from a country lad a specimen of the Cross Fox, in the flesh, which he told us he had caught with a common cur dog, in the snow, which was then a foot in depth.

In regard to the cunning of this variety, there may be some truth in the general opinion, but this can be accounted for on natural principles: the skin is considered very valuable, and the animal is always regarded as a curiosity; hence the hunters make every endeavour to obtain one when seen, and it would not be surprising if a constant succession of attempts to capture it, together with the instinctive desire for self-preservation possessed by all animals, should sharpen its wits, and render it more cautious and wild than those species that are less frequently molested. We remember an instance of this kind, which we will here relate.

A Cross Fox, nearly black, was frequently seen in a particular cover. We offered what was in those days considered a high premium, for the animal in the flesh. The fox was accordingly chased and shot at by the farmers' boys in the neighbourhood. The autumn and winter passed away, nay, a whole year, and still the fox was going at large. It was at last regarded by some of the more credulous as possessing a charmed life, and it was thought that nothing but a silver ball could kill it. In the spring, we induced one of our servants to dig for the young Foxes that had been seen at the burrow which was known to be frequented by the Cross Fox. With an immense deal of labour and fatigue the young were dug out from the side of a hill; there were seven. Unfortunately we were obliged to leave home and did not return until after they had been given away and were distributed about the neighbourhood.

Three were said to have been black, the rest were red. The blackest of the young whelps, was retained for us, and we frequently saw at the house of a neighbour, another of the litter, that was red, and differed in no respect from the common Red Fox. The older our little pet became, the less it grew like the Black, and the more like the Cross Fox. It was, very much to our regret, killed by a dog when about six months old, and as far as we can now recollect, was nearly of the colour of the specimen figured in our work.

The following autumn, we determined to try our hand at procuring the enchanted fox which was the parent of these young varieties, as it could always be started in the same vicinity. We obtained a pair of fine fox-hounds and gave chase. The dogs were young, and proved no match for the fox, which generally took a straight direction through several cleared fields, for five or six miles, after which it began winding and twisting among the hills, where the hounds on two occasions lost the scent and returned home.

On a third hunt, we took our stand near the corner of an old field, at a spot we had twice observed it to pass. It came at last, swinging its brush from side to side, and running with great rapidity, three-quarters of a mile ahead of the dogs, which were yet out of hearing.—A good aim removed the mysterious charm—We killed it with squirrel-shot, without the aid of a silver bullet. It was nearly jet-black, with the tip of the tail white. This fox was the female which had produced the young of the previous spring that we have just spoken of; and as some of them, as we have already said, were Cross Foxes and others Red Foxes, this has settled the question in our minds, that both the Cross Fox and the Black Fox are mere varieties of the Red.

J. W. Audubon brought the specimen he obtained at Niagara, alive to New-York, where it was kept for six or seven weeks. It fed on meat of various kinds: it was easily exasperated, having been much teased on its way from the Falls. It usually laid down in the box in which it was confined, with its head toward the front, and its bright eyes constantly looking upward, and forward, at all intruders. Sometimes during the night it would bark like a dog, and frequently, during the day, its movements corresponded with those of the latter animal. It could not bear the sun-light shining into its prison, and continued shy and snappish to the last.

The fur of the Cross Fox was formerly in great demand; a single skin sometimes selling for twenty-five dollars; at present, however, it is said not to be worth more than about three times the price of that of the Red Fox.

No. 2. PLATE VII.

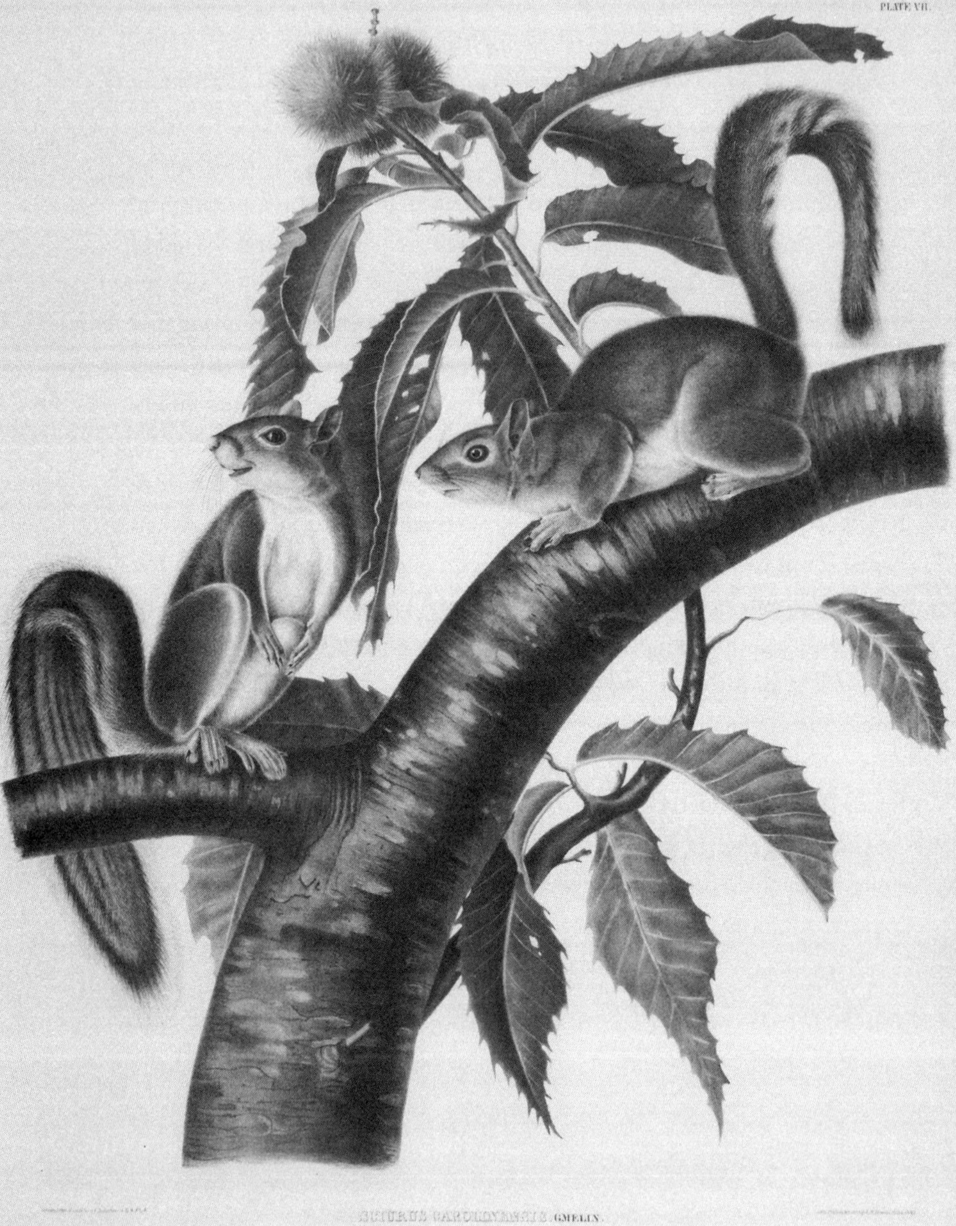

SCIURUS CAROLINENSIS, GMELIN.

CAROLINA GREY SQUIRREL.

Carolina Gray Squirrel

UP IN NEW ENGLAND IN THE EARLY 1840S, SERIOUS INTELLECTUALS like Margaret Fuller had explored the possibilities of mesmerism as part of their Transcendentalist project. Edgar Allan Poe was working mesmerism and related peculiarities into his stories in often troubling ways. As the first volume of *Quadrupeds* was being published in 1846, the mesmerism craze was sweeping into Charleston, and Bachman was perturbed that many Charlestonians were willing to embrace what he considered "humbug."[5] As Bachman turns to a discussion of squirrels' predators in this essay, then, it might not be surprising that he devotes considerable effort to discounting snakes' supposed ability to "charm" their prey, asserting that since this phenomenon can be explained "by natural deductions [. . .] it would be absurd to imagine that anything mysterious or supernatural is connected with the subject." As he tells us in *The Doctrine of the Unity of the Human Race Examined on the Principles of Science*, he believed in two books: the book of nature and the book of God.[6] Clearly, mesmerism and predation via "fascination" had places in neither.

---

5. John James Audubon, Nagel and Weingartner, lithographers, *Carolina Gray Squirrel*, 1845–46, hand-colored lithograph, plate 7 of *The Viviparous Quadrupeds of North America*. Courtesy of the Jule Collins Smith Museum of Fine Art, Auburn University; Louise Hauss and David Brent Miller Audubon Collection.

SCIURUS CAROLINENSIS.—GMEL.

CAROLINA GRAY SQUIRREL.

PLATE VII.—MALE AND FEMALE.

HABITS.

This species differs as much in its habits from the Northern Gray Squirrel, as it does in form and colour. From an intimate acquaintance with the habits of the latter, we are particularly impressed with the peculiarities of the present species. Its bark has not the depth of tone of that of the Northern species, and is more shrill and querulous. Instead of mounting high on the tree when alarmed, which the latter always does, the *Sc. Carolinensis* generally plays round the trunk, and on the side opposite to the observer, at a height of some twenty or thirty feet, often concealing itself beneath the Spanish moss (*Tillandsia Usneoides*) which hangs about the tree. When a person who has alarmed one of these Squirrels remains quiet for a few moments, it descends a few feet, and seats itself on the first convenient branch, in order the better to observe his movements.

It is, however, capable of climbing to the extremity of the branches, and leaping from tree to tree with great agility, but is less wild than the Northern species, and is almost as easily approached as the chickaree, (*Sc. Hudsonius.*) One who is desirous of obtaining a specimen, has only to take a seat for half an hour in any of the swamps of Carolina, and he will be surprised at the immense number of these squirrels that may be seen running along the logs or leaping among the surrounding trees. A great many are killed, and their flesh is both juicy and tender.

The Carolina Gray Squirrel is sometimes seen on high grounds, among the oak and hickory trees, although its usual haunts are low swampy places, or trees overhanging streams, or growing near the margin of some river. In deep cypress swamps, covered in many places with several feet of water during the whole year, it takes up its constant residence, moving among the entwined branches of the dense forest with great facility. Its hole in such situations may sometimes be found in the trunk of a decayed cypress. On the large tupelo trees, (*Nyssa aquatica,*) which are found in the swamps, many nests of this species, composed principally of Spanish moss and leaves, are every where to be seen. In these nests, or in some woodpecker's hole, they produce their young. These are five or six in number, and are brought forth in March; it is well ascertained also that the female litters a second time in the season, probably about mid-summer.

This species has one peculiarity which we have not observed in any other. It is in some degree nocturnal, or at least crepuscular, in its habits. In riding along by-paths through the woods, long after sunset, we are often startled by the barking of this little Squirrel, as it scratches among the leaves, or leaps from tree to tree, scattering over the earth the seeds of the maple, &c., which are shaken off from the uppermost branches, as it passes over them.

This species is seldom, if ever, seen in company with the Fox Squirrel, (*Sc. Capistratus,*) or even found in the same neighbourhood; this arises, probably, not so much from any antipathy to each other, as from the fact that very different localities are congenial to the peculiar habits of each.

We have observed the Carolina Gray Squirrel on several occasions by moonlight, as actively engaged as the Flying Squirrel usually is in the evening, and this propensity to prolong its search after food, or its playful gambols, until the light of day is succeeded by the moon's pale gleams, causes it frequently to fall a prey to the Virginian owl, or the barred owl; which last especially, is very abundant in the swamps of Carolina, where, gliding on noiseless pinions between the leafy branches, it seizes the luckless Squirrel ere it is aware of its danger, or can make the slightest attempt to escape. The gray fox and the wild cat often surprise this and other species by stratagem or stealth. We have beheld the prowling lynx, concealed in a heap of brushwood near an old log, or near the foot of a tree frequented by the Squirrel he hopes to capture. For hours together will he lie thus in ambush, and should the unsuspicious creature pass within a few feet of him, he pounces on it with a sudden spring, and rarely fails to secure it.

Several species of snakes, the rattle-snake, (*Crotalus durassus,*) black snake, (*Coluber constrictor,*) and the chicken snake, (*coluber quadrivittatus,*) for instance, have been found, on being killed, to have a Squirrel in their stomach, and the fact that Squirrels, birds, &c., although possessing great activity and agility, constitute a portion of the food of these reptiles, being well established, the manner in which the sluggish serpent catches animals so far exceeding him in speed, and some of them endowed with the power of rising from the earth, and skimming away with a few flaps of their wings, has been the subject of much speculation. Some persons have attributed a mysterious power, more especially to the rattle-snake and black snake—we mean the power of fascinating, or as it is commonly called, *charming*.

This supposed faculty of the serpent has, however, not been accounted for. The basilisk of the ancients killed by a look; the eye of the rattle-snake is supposed so to paralyze, and at the same time attract, its intended prey, that the animal slowly approaches, going through an infinite variety of motions, alternately advancing and retreating, until it finally falls powerless into the open jaws of its devourer.

As long as we are able to explain by natural deductions, the very singular manœuvres of birds and squirrels, when "fascinated" by a snake, it would be absurd to imagine that anything mysterious or supernatural is connected with the subject; and we consider that there are many ways of accounting for all the appearances described on these occasions. Fear and surprise cause an instinctive horror, when we find ourselves unexpectedly within a foot or two of a rattle-snake; the shrill, startling, noise proceeding from the rattles of its tail, as it vibrates rapidly, and its hideous aspect, no doubt produce a much greater effect on birds and small quadrupeds. It is said that the distant roar of the African lion causes the oxen to tremble, and stand paralyzed in the fields; and Humboldt relates that in the forests of South America, the mingled cries of monkeys and other animals resound through the whole night, but as soon as the roar of the jaguar, the American tiger, is heard, terror seizes on all the other animals, and their voices are suddenly hushed. Birds and quadrupeds are very curious, also, and this feeling prompts them to draw near to strange objects. "Tolling" wild ducks and loons, as it is called, by waving a red handkerchief or a small flag, or by causing a little dog to bound backward and forward on the beach, has long been successfully practised by sportsmen on the Chesapeake Bay and elsewhere.

The Indians attract the reindeer, the antelope, and other animals, until they are within bow-shot, by waving a stick to which a piece of red cloth is attached, or by throwing themselves on their backs, and kicking their heels up in the air. If any strange object is thrown into the poultry-yard, such as a stuffed specimen of a quadruped or bird, &c., all the fowls will crowd near it, and scrutinize it for a long time. Every body almost may have observed at some time or other dozens of birds collected around a common cat in a shrubbery, a tortoise, or particularly a snake. The Squirrel is remarkable for its fondness for "sights," and will sometimes come down from the highest branch of a tree to within three feet of the ground, to take a view of a small scarlet snake, (*Rhinostoma coccinea*,) not much larger than a pipe-stem, and which, having no poisonous fangs, could scarcely master a grasshopper. This might be regarded by believers in the fascinating powers of snakes, as a decided case in favour of their theories, but they would find it somewhat difficult to explain the following circumstances which happened to ourselves. After observing a Squirrel come down to inspect one of the beautiful little snakes we have just been speaking of, the reptile being a rare species, was captured and secured in our carriage box. After we had driven off, we recollected that in our anxiety to secure the snake, we had left our box of botanical specimens at the place where we had first seen the latter, and on returning for it, we once more saw the Squirrel, darting backward and forward, and skipping round the root of the tree, eyeing with equal curiosity the article we had left behind, and we could not help making the reflection, that if the little snake had "charmed" the Squirrel, the same "fascinating" influence was exercised by our tin box!

Quadrupeds and birds have certain antipathies, they are capable of experiencing many of the feelings that appertain to mankind; they are susceptible of passion, are sometimes spiteful and revengeful, and are wise enough to know their "natural enemies" without a formal introduction. The blue jay, brown thrush, white-eyed fly-catcher, and other little birds are often to be heard scolding, and fluttering about a thicket, in which some animal is concealed; and on going to examine into the cause of their unwonted excitement, you will probably see a wild cat or fox spring forth from the covert. Every one familiar with the habits of our feathered tribes must have seen at times the owl or buzzard chased by the smallest birds, which unite on such occasions for the purpose of driving off a common enemy; in these cases, the birds sometimes approach too near, and are seized by the owl. We once observed some night-hawks (*Chordeiles Virginianus*) darting round a tree upon which an owl was perched. Whilst looking on, we perceived the owl make a sudden movement, and found that he had caught one of them in his sharp claws, and notwithstanding the cries and menaces of the others, he instantly devoured it.

Birds dart in the same manner at snakes, and no doubt are often caught by passing too near—shall we, therefore, conclude that they are fascinated?

One of the most powerful "attractions" which remain to be considered, is the love of offspring. This feeling, which is so deeply rooted in the system of nature, as to be a rule, almost without an exception, is manifested strongly by birds and quadrupeds; and snakes are among the most to be dreaded destroyers of eggs and young birds, and of the young of small species of viviparous animals; is it not likely therefore, that many of the (supposed) cases of fascination that are related, may be referred to the intrepidity of the animals or birds, manifested in trying to defend their young, or drive away their enemy from their vicinity? In our work, the "Birds of America," we represented a mocking-bird's nest attacked by a rattle-snake, and the nest of a red thrush invaded by a black snake; these two plates each exhibit several birds assisting the pair whose nest has been robbed by the snake, and also show the mocking-bird and thrush courageously advancing to the jaws even of their enemy. These pictures were drawn after the actual occurrence before our eyes, of the scenes which we endeavoured to represent in them, and supposing a person but little acquainted with natural history, to have seen the birds, as we did, he might readily have fancied that some of them at least were fascinated, as he could not probably have been near enough to mark the angry expression of their eyes, and see their well concealed nest.

Our readers will, we trust, excuse us for detaining them yet a little longer on this subject, as we have more to say of the habits of the rattle-snake, in connexion with the subject we are upon.

This snake, the most venomous known in North America, subsists wholly on animal food; it digests its food slowly, and is able to exist without any sustenance for months, or even years, in confinement; during this time it often increases in size, and the number of its rattles is augmented. In its natural state it feeds on rabbits, squirrels, rats, birds, or any other small animals that may come in its way. It captures its prey by lying in wait for it, and we have heard of an instance, in which one of these snakes remained coiled up for two days before the mouth of the burrow of the Florida rat, (*Neotoma Floridana,*) and on its being killed it was found to have swallowed one of these quadrupeds.

As far as we have been able to ascertain, it always strikes its intended prey with its fangs, and thus kills it, before swallowing it. The bite is sudden, and although the victim may run a few yards after it is struck, the serpent easily finds it when dead. Generally the common species of rattle-snake refuses all food when in a cage, but occasionally one is found that does not refuse to eat whilst in captivity. When a rat is turned loose in a cage with one of these snakes, it does not immediately kill it, but often leaves it unmolested for days and weeks together. When, however, the reptile, prompted either by irritation or hunger, designs to kill the animal, it lies in wait for it, cat-like, or gently crawls up to it, and suddenly gives it the mortal blow, after which, it very slowly and deliberately turns it over into a proper position, and finally swallows it.

We have seen a rattle-snake, in a very large cage, using every means within its power, and exerting its cunning, for a whole month, before it could succeed in capturing a brown thrush that was imprisoned with it. At night the bird roosted beyond the reach of the snake, and during the day-time it was too cautious in its movements, and too agile, snatching up its food at intervals, and flying instantly back to its perch, to be struck by the unwieldy serpent. We now added a mouse to the number of the inmates of the cage; the affrighted animal retreated to a corner, where the snake, slowly crawling up to it, with a sudden blow darted his fangs into and killed it; soon after which he swallowed it. About a week after this adventure, the snake again resumed his attempts to capture the thrush, and pursued it all round the cage.

This experiment offered a fair opportunity for the rattle-snake to exert its powers of fascination, had it possessed any; but as it did not exhibit them, we do not hesitate to say that it was entirely destitute of any faculty of the kind.

After some hours' fruitless manœuvring, the snake coiled itself up, near the cup of water from which the bird drank. For two days the thrush avoided the water; on the third, having become very thirsty, it showed a constant desire to approach the cup; the snake waited for it to come within reach, and in the course of the day struck at it two or three times, the bird darted out of its way, however, and was not killed until the next day.

If, notwithstanding these facts, it is argued, that the mysterious and inexplicable power of fascination is possessed by the snake, because birds have been seen to approach it, and with open wings and plaintive voice, seemed to wait upon its appetite, we must be prepared to admit that the same faculty is possessed by other animals. On a certain day, we saw a mocking-bird exhibiting every appearance, usually, according to descriptions, witnessed when birds are under the influence of fascination. It approached a hog, which was occupied in munching something at the foot of a small cedar. The bird fluttered before the grunter with open wings, uttered a low and plaintive note, alighted on his back, and finally began to peck at his snout. On examining into the cause of these strange proceedings, we ascertained that the mocking-bird had a nest in the tree, from which several of her younglings had fallen, which the hog was eating! Our friend, the late Dr. WRIGHT, of Troy, informed us that he witnessed a nearly similar scene between a cat-bird and a dog which had disturbed her brood, on which occasion the cat-bird went through many of the movements generally ascribed to the effect of fascination.

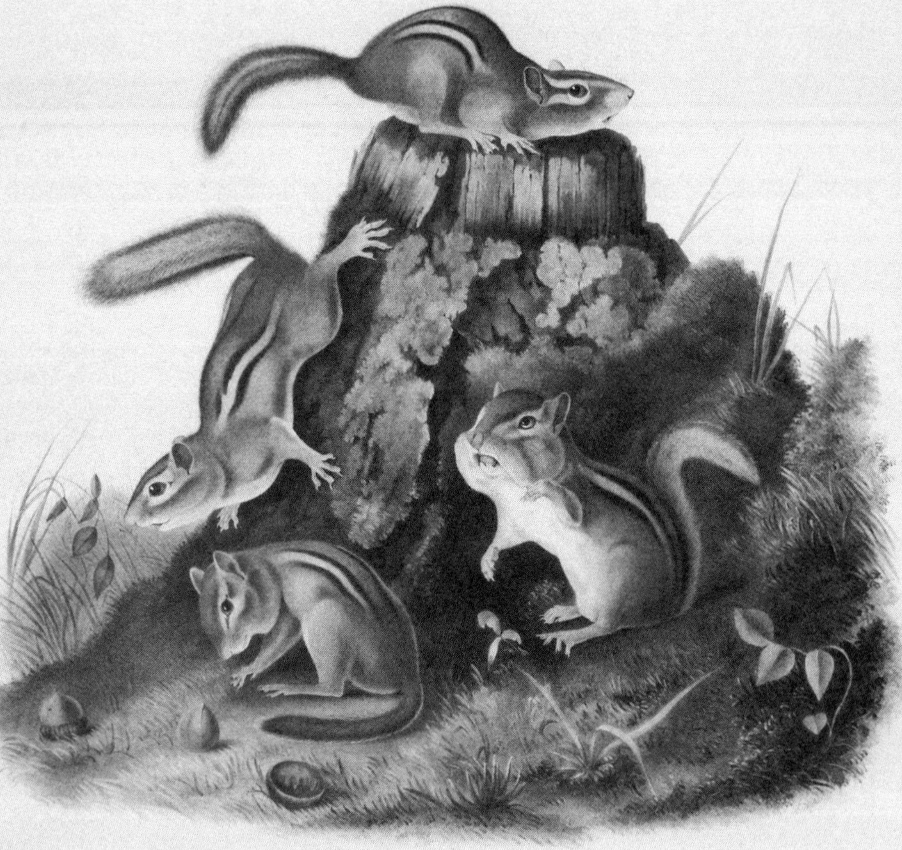

TAMIAS LYSTERI, RAY.

CHIPPING SQUIRREL, HACKEE &c.

Natural Size

Chipping Squirrel

A QUAINT SENSE OF NOSTALGIA PUNCTUATES THIS ESSAY: the ground squirrel calls to Bachman's mind "pleasing reminiscences" of "little sports of boyhood." Employing a sort of comic Americana mode, Bachman describes the excitement and challenges experienced by boys as they team up to hunt chipmunks. As one boy strikes a chipmunk on the nose and captures it, Bachman anticipates the reader's disapproval: "Perhaps the little animal is not dead, only stunned, and is carried home to be made a pet." Dead or alive, it is placed in "a calabash, a stocking, or a small bag prepared for the occasion by some fond little sister, who whilst sewing for her brother half longed to enjoy the romp and the sport herself." Bachman's memories are so pleasing that he even envisions a little girl wishing for activity beyond the domestic sphere assigned to her, but he ultimately believes she is better off where she is: "Reader, don't smile at this group of juvenile sportsmen; older and bigger 'boys' are often engaged in amusements not more rational, and not half so innocent." As Lester Stephens has noted, Bachman vehemently objected to Victor Audubon inserting writerly clichés like the above "Reader" into his essay manuscripts.[7] In any case, the quaint nature of the entire episode is eclipsed as Bachman tells us about the weasel, which kills every chipmunk it finds in a burrow seemingly in "bloodthirsty" sport: "It seldom devours one fifth of the animals it so wantonly kills."

---

6. John James Audubon, Nagel and Weingartner, lithographers, *Chipping Squirrel, Hackee*, 1845–46, hand-colored lithograph, plate 8 of *The Viviparous Quadrupeds of North America*. Courtesy of the Jule Collins Smith Museum of Fine Art, Auburn University; Louise Hauss and David Brent Miller Audubon Collection.

TAMIAS LISTERI.—RAY.

CHIPPING SQUIRREL, HACKEE, &C.

PLATE VIII.—MALE, FEMALE, AND YOUNG (FIRST AUTUMN).

HABITS.

The Chipping Squirrel, as this little animal is usually called, or Ground Squirrel, as it is named almost as frequently, is probably, with the exception of the common flying squirrel, (*Pteromys volucella*,) one of the most interesting of our small quadrupeds. It is found in most parts of the United States, and being beautifully marked in its colouring, is known to every body. From its lively and busy habits, one might consider it among the quadrupeds as occupying the place of the *wren* among the feathered tribes. Like the latter, the Ground Squirrel, full of vivacity, plays with the utmost grace and agility among the broken rocks or uprooted stumps of trees about the farm or wood pasture; its clucking resembles the chip, chip, chip, of a young chicken, and although not musical, like the song of the little winter wren, excites agreeable thoughts as it comes on the air. We fancy we see one of these sprightly Chipping Squirrels as he runs before us with the speed of a bird, skimming along a log or fence, his chops distended by the nuts he has gathered in the woods; he makes no pause till he reaches the entrance of his subterranean retreat and store-house. Now he stands upright, and his chattering cry is heard, but at the first step we make towards him, he disappears. Stone after stone we remove from the aperture leading to his deep and circuitous burrow; but in vain is all our labour—with our hatchets we cut the tangled roots, and as we follow the animal, patiently digging into his innermost retreat, we hear his angry, querulous tones. We get within a few inches of him now, and can already see his large dark eyes; but at this moment out he rushes, and ere we can "grab" him, has passed us, and finds security in some other hiding place, of which there are always plenty at hand, that he is well accustomed to fly to; and we willingly leave him unmolested, to congratulate himself on his escape.

The Chipping Squirrel makes his burrow generally near the roots of trees, in the centre of a decayed stump, along fences or old walls, or in some bank, near the woods whence he obtains the greater portion of his food.

Some of these retreats have two or three openings, at a little distance from each other. It rarely happens that this animal is caught by digging out its burrow. When hard pressed and closely pursued, it will betake itself to a tree, the trunk of which it ascends for a little distance with considerable rapidity, occasionally concealing itself behind a large branch, but generally stopping within twelve or fifteen feet of the ground, where it often clings, with its body so closely pressed to the trunk, that it is difficult to detect it; and it remains so immoveable that it appears like a piece of bark or some excrescence, till the enemy has retired from the vicinity, when it once more descends, and by its renewed clucking, seems to chuckle over its escape.

We are doubtful whether this species can at any time be perfectly tamed. We have preserved it in cages from time to time, and generally found it wild and sullen. Those we had, however, were not young when captured.

At a subsequent period we obtained in the State of New York, five or six young ones almost half grown. We removed them to Carolina, where they were kept during winter and spring. They were somewhat more gentle than those we had formerly possessed, occasionally took a filbert or a ground-nut from the fingers, but never became tame enough to be handled with safety, as they on more than one occasion were disposed to test the sharpness of their teeth on our hand.

The skin which covered the vertebrae of their tails was so brittle that nearly all of them soon had mutilated them. They appeared to have some aversion to playing in a wheel, which is so favourite an amusement of the true squirrels. During the whole winter they only left their nest to carry into it the rice, nuts, Indian corn, &c., placed in their cage as food.

Late in the following spring, having carried on our experiments as far as we cared to pursue them, we released our pets, which were occasionally seen in the vicinity for several months afterward, when they disappeared.

We were once informed of a strange carnivorous propensity in this species. A lady in the vicinity of Boston said to us, "We had in our garden a nest of young robins, (*Turdus migratorius*,) and one afternoon as I was walking in the garden, I happened to pass very close to the tree on which this nest was placed; my attention was attracted by a noise which I thought proceeded from it, and on looking up I saw a Ground Squirrel tearing at the nest, and actually devouring one of the young ones. I called to the gardener, who came accompanied by a dog, and shook the tree violently, when the animal fell to the earth, and was in an instant secured by the dog." We do not conceive that the unnatural propensity in the individual here referred to, is indicative of the genuine habit of this species, but think that it may be regarded as an exception to a general rule, and referred to a morbid depravity of taste sometimes to be observed in other genera, leading an individual to feed upon that which the rest of the species would loathe and reject. Thus we have known a horse which preferred a string of fish to a mess of oats; and mocking-birds, in confinement, kill and devour jays, black-birds, or sparrows.

We saw and caught a specimen of this beautiful TAMIAS, in Louisiana, that had no less than sixteen chinquapin nuts (*Castanea pumila*) stowed away in its cheek-pouches. We have a specimen now lying before us, sent from Pennsylvania in alcohol, which contains at least one and a half table-spoonfuls of Bush trefoil (*Hedysarum cannabinum*) in its widely-distended sacks. We have represented one of our figures in the plate, with its pouches thus filled out.

This species is to a certain extent gregarious in its habits. We had marked one of its burrows in autumn, which we conceived well adapted to our purpose, which was to dig it out. It was in the woods, on a sandy piece of ground, and the earth was strewed with leaves to the depth of eight inches, which we believed would prevent the frost from penetrating to any considerable depth. We had the place opened in January, when the ground was covered with snow about five inches deep. The entrance of the burrow had been closed from within. We followed the course of the small winding gallery with considerable difficulty. The hole descended at first almost perpendicularly for about three feet. It then continued with one or two windings, rising a little nearer the surface, until it had advanced about eight feet, when we came to a large nest made of oak leaves and dried grasses. Here lay, snugly covered, three Chipping Squirrels. Another was subsequently dug from one of the small lateral galleries, to which it had evidently retreated to avoid us. They were not dormant, and seemed ready to bite when taken in the hand; but they were not very active, and appeared somewhat sluggish and benumbed, which we conjectured was owing to their being exposed to sudden cold, from our having opened their burrow.

There was about a gill of wheat and buckwheat in the nest; but in the galleries we afterwards dug out, we obtained about a quart of the beaked hazel nuts, (*Corylus rostrata,*) nearly a peck of acorns, some grains of Indian corn, about two quarts of buckwheat, and a very small quantity of grass seeds. The late Dr. JOHN WRIGHT, of Troy, in an interesting communication on the habits of several of our quadrupeds, informs us, in reference to this species, that "It is a most provident little creature, continuing to add to its winter store, if food is abundant, until driven in by the severity of the frost. Indeed, it seems not to know when it has enough, if we may judge by the surplus left in the spring, being sometimes a peck of corn or nuts for a single Squirrel. Some years ago I watched one of these animals whilst laying up its winter store. As there were no nuts to be found near, I furnished a supply. After scattering some hickory nuts on the ground near the burrow, the work of carrying in was immediately commenced. It soon became aware that I was a friend, and approached almost to my feet for my gifts. It would take a nut from its paws, and dexterously bite off the sharp point from each end, and then pass it to its cheek-pouch, using its paws to shove it in, then one would be placed on the opposite side, then again one along with the first, and finally, having taken one between its front teeth, it would go into the burrow. After remaining there for five or ten minutes it would reappear for another load. This was repeated in my presence a great number of times, the animal always carrying four nuts at a time, and always biting off the asperities."

We perceive from hence, that the Chipping Squirrels retire to winter quarters, in small families, in the early part of November, sooner or later, according to the coldness or mildness of the season, after providing a store of food in their subterranean winter residence. When the snows are melted from the earth in early spring, they leave the retreat to which they had resorted during the first severe frosts in autumn. We have seen them sunning themselves on a stump during warm days about the last of February, when the snows were still on the earth here and there in patches a foot deep; we remarked, however, that they remained only for half an hour, when they again retreated to their burrows.

The young are produced in May, to the number of four or five at a birth, and we have sometimes supposed from the circumstance of seeing a young brood in August, that they breed twice a year.

The Chipping Squirrel does but little injury to the farmer. It seldom disturbs the grain before it is ripe, and is scarcely more than a gleaner of the fields, coming in for a small pittance, when the harvest is nearly gathered. It prefers wheat to rye, seems fond of buckwheat, but gives the preference to nuts, cherry-stones, the seeds of the red gum, or pepperidge, (*Nyssa Multiflora*,) and those of several annual plants and grasses.

This species is easily captured. It enters almost any kind of trap without suspicion. We have seen a beautiful muff and tippet made of a host of little skins of this TAMIAS ingeniously joined together so as to give the appearance of a regular series of stripes around the muff, and longitudinally along the sides of the tippet. The animals had in most cases been captured in rat-traps.

There is, besides, a simple, rustic, but effectual mode of hunting the Ground Squirrel, to which we are tempted to devote a paragraph.

Man has his hours of recreation, and so has the school-boy; while the former is fond of the chase, and keeps his horses, dogs, and guns, the latter, when released from school, gets up a little hunt, agreeable to his own taste and limited resources. The boys have not yet been allowed to carry fire-arms, and have been obliged to adhere to the command of a careful mother—"don't meddle with that gun, Billy, it may go off and kill you." But the Chip Muck can be hunted without a gun, and Saturday, the glorious weekly return of their freedom and independence from the crabbed schoolmaster and the puzzling spelling-book, is selected for the important event.

There are some very pleasing reminiscences associated with these little sports of boyhood. The lads, full of delightful anticipations, usually meet half an hour before the time appointed. They come with their "shining morning faces," full of glee and talking of their anticipated success.[8] In lieu of fire-arms they each carry a stick, about eight feet long. They go along the old fashioned worm-fences that skirt the woods, a crop of wheat or of buckwheat has just been gathered, and the little Hackee is busily engaged in collecting its winter store.

In every direction its lively chirrup is heard, with answering calls from adjacent parts of the woods, and here and there you may observe one mounted on the top of a fence-stake, and chipping away as it were in exultation at his elevated seat. One of the tiny huntsmen now places his pole on a fence rail, the second or third from the bottom, along which the Ground Squirrel is expected to pass; a few yards behind him is another youngster, ready with his stick on another rail, in case the Chip Muck escapes the first enemy. One of the juveniles now makes a circuit, gets behind the little Hackee, and gives a blow on the fence to drive him toward the others, who are eagerly expecting him. The unsuspecting little creature, with a sweep of his half-erected tail, quickly descends from the top of the fence, along a stake, and betaking himself to some of the lower rails, makes a rapid retreat. If no stone-heaps or burrows are at hand, it runs along the winding fence, and as it is passing the place where the young sportsmen are lying in wait; they brush the stick along the rail with the celerity of thought, hitting the little creature on the nose, and knocking it six yards off. "He is ours," is the exulting shout, and the whole party now hurry to the spot. Perhaps the little animal is not dead, only stunned; and is carried home to be made a pet. It is put into a calabash, a stocking, or a small bag, prepared for the occasion by some fond little sister, who whilst sewing it for her brother, half longed to enjoy the romp and the sport herself. Reader, don't smile at this group of juvenile sportsmen; older and bigger "boys" are often engaged in amusements not more rational, and not half so innocent.

Several species of hawks are successful in capturing the Chipping Squirrel. It furnishes also many a meal for the hungry fox, the wild cat, and the mink; but it possesses an enemy in the common weasel or ermine, (*mustela erminea*) more formidable than all the rest combined. This bloodthirsty little animal pursues it into its dwelling, and following it to the farthest extremity, strikes its teeth into its skull, and like a cruel savage of the wilderness, does not satiate its thirst for blood, until it has destroyed every inhabitant of the burrow, old and young, although it seldom devours one fifth of the animals it so wantonly kills. We once observed one pursue a Chipping Squirrel into its burrow. After an interval of ten minutes it reappeared, licking its mouth, and stroking its fur with its head, by the aid of its long neck. We watched it as it pursued its way through a buckwheat field, in which many roots and stumps were yet remaining, evidently in quest of additional victims. On the following day we were impelled by curiosity to open the burrow we had seen it enter. There we found an old female ground squirrel, and five young, half-grown, lying dead, with the marks of the weasel's teeth in their skulls.

GENERAL REMARKS.

We have, at the head of this article, endeavoured to preserve TAMIAS as a valuable genus distinct from SCIURUS. We hope we have offered such reasons as will induce naturalists to separate this interesting and increasing little group, mostly of American species, from the squirrels, to which they bear about the same affinity, as do the marmot squirrels (SPERMOPHILUS) to the true marmots (ARCTOMYS.) We will now inquire whether the present species, (*Tamias Lysteri*,) is a foreigner from Siberia, naturalized in our Western world; or whether it is one of the aborigines of our country, as much entitled to a name as the grizzly bear or the cougar.

Two of our American naturalists, HARLAN and GODMAN, supposed that it was the Asiatic species, the *S. striatus* of KLEIN, PALLAS, SCHREBER, and other authors; Dr. RICHARDSON (1829) believed that the descriptions given of *Sciurus striatus*, did not exactly correspond with American specimens, and as he had no opportunity of instituting a comparison, he adopted the specific name of RAY, *Sciurus* (TAMIAS) *Lysteri*, for our species; and quoted what PALLAS had written in regard to the habits of the Asiatic animal, as applying to those of our little Chipping Squirrel. Very recently (1842) Dr. DEKAY, in the work on American quadrupeds, published by order of the State of New York, has again referred it to *S. striatus* of LINNÆUS, and endeavoured to prove the identity of the two species, from European writers. We suspect he had no opportunity of making a comparison from actual specimens.

Reasoning from analogy in regard to the species of birds, or quadrupeds, found to be identical on both continents, we should be compelled to admit that if our species is the *S. striatus* of Asia, it presents a solitary exception to a long-established general rule. That many species of water-birds, such as geese, ducks, gulls, auks, and guillemots, which during the long days of summer, crowd toward the polar regions to engage in the duties and pleasures of reproduction, should be found on both continents, cannot be a matter of surprise; and that the ptarmigan, the white snow-bird, Lapland long-spur, &c., which resort annually to them, should, at that season, take wing and stray to either continent, is so probable a case, that we might think it strange if it were otherwise. Neither need we regard it as singular if a few quadrupeds, with peculiar constitutions and habits suited to the polar regions, should be inhabitants of the northern portions of both continents. Thus, the polar bear which delights in the snow and ice, and which is indifferent as to whether it is on the land or on an iceberg at sea; the reindeer, which exists only in cold regions, and which, by alternately swimming and walking, can make its way over the icy waters in winter, and over rivers and arms of the sea in summer, and which migrates for thousands of miles; the beaver, which is found all over our continent; on the banks of the Mackenzie river, leading into the polar sea, in latitude 68°, and in the Russian settlements near Behring's Straits; the ermine, which riots in the snow-drifts, and has been found as far to the north as man has ever travelled; and the common wolf, which is a cosmopolite, exhibits itself in all colours, and strays from the tropics to the north pole, may be found on both continents, without surprising us: but if this little land animal, the Chipping Squirrel, which is unable to swim, and retires to the earth in cold weather, should be found both in Asia and America, it would oppose all our past experience in regard to American quadrupeds, and be the only exception to a long and universally admitted theory. The highest northern range in which this species has ever been seen is above Lake Huron, as far as latitude 50°; from thence there is a distance of more than 90° of longitude and 18° of latitude, before we reach its Asiatic range, and in its migrations either way it would have to cross Behring's Straits, and traverse regions, which even in summer are covered with snow and ice. From the above facts, and from our knowledge of the adaptation of various animals for extensive migrations, we must conclude that this species cannot possibly exist on both continents, even admitting the correctness of the supposition, that these continents had in some former age been united.

Dr. Richardson says, (p. 181,) "I am not aware that the identity of the species on the two continents has been established by actual comparison."

In this he was quite correct. At the period at which his valuable work on American quadrupeds was published, nearly all the figures, and many of the descriptions of *Tamias striatus* of the Eastern continent, were taken from American specimens of *Tamias Lysteri*; and the authors supposing them to be identical, were not sufficiently cautious to note this important fact.

In 1838 we carried to Europe, American specimens of nearly all those species which had their congeners on the Eastern continent. We were surprised at finding no specimen of the *T. striatus* in the museums of either England or France. At Berlin, however, an excellent opportunity was afforded us for instituting a comparison. Through the kindness of Dr. LICHTENSTEIN, the superintendent of the museum, we were permitted to open the cases, examine several specimens in a fine state of preservation, and compare them with our American species, which we placed beside them. The differences, at first sight were so striking that we could only account for their ever having been considered identical, from the fact that the descriptions of the old authors were so loose and unsatisfactory that many minute but important characteristics had not been noted. The following memorandum was made by us on the occasion:—"The *Tamias striatus* differs so widely from our American Chipping Squirrel or Hackee, that it is unnecessary to be very minute in making the comparison. The two species can always be distinguished from each other by one remarkable characteristic, which I have observed running through all the specimens. The stripes on the Asiatic, (*T. striatus*) running over the back, extend to the root of the tail; whilst those on the American, (*T. Lysteri*) do not reach so far by a full inch. There are many other differences which may as well be noticed. *T. striatus* is a little the largest, the stripes on the back are situated nearer each other, and are broader than in the other species; the stripes on each side of the back are nearly black, instead of yellowish-brown; on each side of the black stripe on the centre of the back of *Tamias Lysteri*, there is a broad space of reddish-gray. In *T. striatus* this part of the animal is yellowish; being an alternate stripe of black and yellowish-white. The tail of the latter is black towards the extremity, and tipped with white; its tail and ears also are larger than those of *T. Lysteri*: in short, these two species differ as widely from each other, as *Tamias Lysteri* differs from the four-lined ground squirrel of SAY, (*T. quadrivittatus.*)"

No. 2

PLATE X.

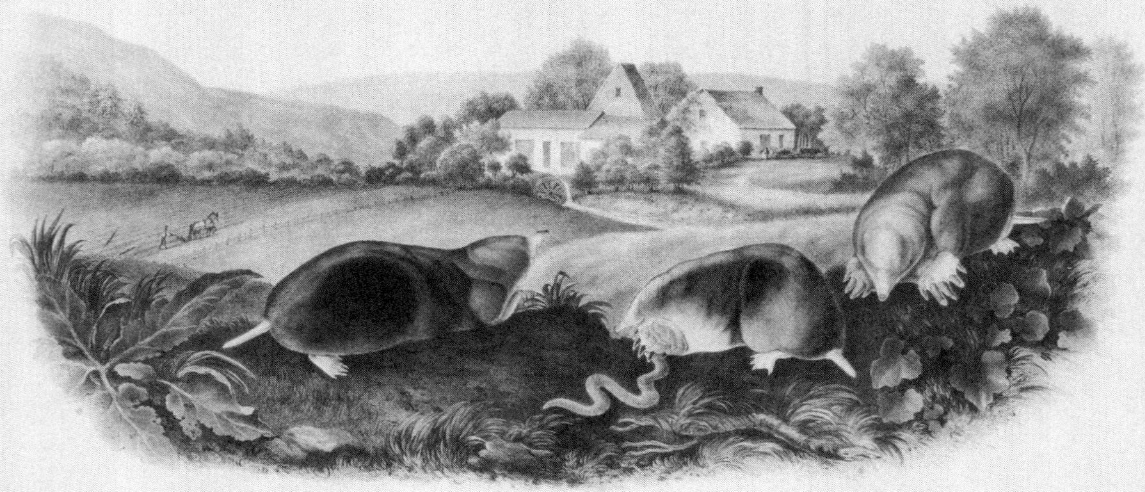

Scalops Aquaticus, Linn.

Common American Shrew Mole

CONTAINING THE FIRST USE OF THE PHRASE "TENACIOUS OF life" in *Quadrupeds*, this essay emphasizes not only the mole's physical fortitude but also its position as a largely misunderstood creature. As a comedic mood colors much of the essay, Bachman teaches readers to view this mole with more benevolence by dispelling misconceptions about its impact on crops and gardens. Ultimately, Bachman wants readers to see that even though moles can detrimentally affect human property, these animals are worthy of our tolerance and interest, thus reflecting another of *Quadrupeds'* themes—that of setting the mammalogical record straight, sometimes through compassionate understanding. This essay also contains rare political commentary: two moles are busy fighting and are shot in their distraction, "falling victims to their own passions; and if we would read aright, affording us an instructive lesson, either as individuals, or in a national point of view." Bachman is hinting at the political tensions of the 1840s that would find their fullest expression in the Civil War.

---

7. John James Audubon, J. T. Bowen, lithographer, *Common American Shrew Mole*, 1845–46, hand-colored lithograph, plate 10 of *The Viviparous Quadrupeds of North America*. Courtesy of the Jule Collins Smith Museum of Fine Art, Auburn University; Louise Hauss and David Brent Miller Audubon Collection.

SCALOPS AQUATICUS.—LINN.

COMMON AMERICAN SHREW MOLE.

PLATE X.—MALE AND FEMALE.

HABITS.

Whilst almost every farmer or gardener throughout the Northern and Eastern States is well acquainted with this curious animal, as far as the mere observation of its meandering course through his fields and meadows, his beds of green peas or other vegetables, is concerned, but few have arrived at proper conclusions in regard to the habits of the Shrew Mole; and it is generally caught and killed whenever practicable; the common idea being, that the Mole feeds on the roots of tender plants, grasses, &c.; while the fact that the animal devours great quantities of earth-worms, slugs, and grubs, all hurtful to the fruit trees, to the grasses, and the peas and other vegetables, seems to be unknown, or overlooked.

In justice to the farmer and gardener, however, we must say, that the course taken occasionally by this species, directly along a row of tender plants, throwing them out of the earth, as it does, or zig-zag across a valuable bed or beautiful lawn, is rather provoking, and we have ourselves caused traps to be set for moles, being greatly annoyed by their digging long galleries under the grass on our sloping banks, which during a heavy shower soon filled with water, and presently increased to large gutters, or deep holes, requiring repairs forthwith. At such times also, a Mole-track through loose soil where there is any descent, will be found by the gardener, perchance, to have become a miniature ravine some twenty or thirty yards in length, and a few (anticipated) bushels of carrots are destroyed. In neglected or sandy soils, one of these gutters becomes deep and wide in a short time, and we may perhaps not err in hazarding the opinion that some of the unsightly ravines which run almost through large estates, occasionally might be traced to no higher origin than the wandering of an unlucky mole!

We kept one of this species alive for some days, feeding it altogether upon earth-worms, but we soon found it difficult to procure a sufficient supply; forty or fifty worms of moderate size did not appear too much for its seemingly insatiable appetite. At the expiration of four days, another of this species which we had in confinement would not touch any vegetable substances, although the cage was filled with clods covered with fine clover, pieces of sweet apples, bread, &c.

We were much interested in observing, that no matter how soiled its coat might have become in the cage, it would resume its beauty and glossiness after the mole had passed and re-passed through the earth eight or ten times, which it always accomplished in a few minutes. We frequently remarked with surprise the great strength of this animal, which enabled it to lift the lid or top of a box in which it was kept, although it was large and heavy; the box-top was not however fastened down. Seating ourselves quietly in the room, after putting back the mole into the box, the animal supposing itself no longer watched, very soon raised its body against the side of the box, which was partly filled with earth, and presently its snout was protruded through the small space between the box and the cover; and after a few efforts the creature got his fore-feet on to the edge of the box, raised itself over the latter, and fell upon a table on which we had placed the box. It immediately ran to the edge of the table, and thence tumbled on to the floor; this, however, did not at all incommode it, for it made off to a dark corner of the room at once, and remained there until again replaced in its prison.

When this Mole was fed on earth-worms, (*Lumbricus terrenus*,) as we have just related, we heard the worms crushed in the strong jaws of the animal, with a noise somewhat like the grating of broken glass, which was probably caused by its strong teeth gnashing on the sand or grit contained in the bodies of the worms. These were placed singly on the ground near the animal, which after smelling around for a moment turned about in every direction with the greatest activity, until he felt a worm, when he seized it between the outer surface of his hands or fore-paws, and pushed it into his mouth with a continually repeated forward movement of the paws, cramming it downward until all was in his jaws.

Small-sized earth-worms were despatched in a very short time; the animal never failing to begin with the anterior end of the worm, and apparently cutting it as he ate, into small pieces, until the whole was devoured. On the contrary, when the earth-worm was of a large size, the Mole seemed to find some difficulty in managing it, and munched the worm sideways, moving it from one side of its mouth to the other. On these occasions the gritting of its teeth, which we have already spoken of, can be heard at the distance of several feet.

We afterwards put the Mole into a large wire rat-trap, and to our surprise saw him insert his fore-paws or hands between the wires, and force them apart sufficiently to give him room to pass out through them at once, and this without any great apparent effort. It is this extraordinary muscular power in the fore-paws and arms, that enables the Shrew Moles to traverse the galleries they excavate with so much rapidity, in doing which they turn the backs of their palms or hands toward each other, push them forward as far as the end of their snout, and then open and bring them round backward, in the manner of a person moving his hands and arms when swimming. When running along on the surface of the ground, they extend the fore-legs as far forward as they will reach, turning the backs of the hands or paws (as just mentioned) towards each other, and placing them edge-wise, instead of flat on the earth as might be supposed, and in this manner they run briskly and without any awkward movement, crossing beaten-roads or paved walks, and sometimes running swiftly twenty or thirty feet before they can get into the ground.

The Shrew Mole varies somewhat in its habits, according to our observations: for while a solitary individual will occasionally for some weeks occupy and root up a large plot of grass or a considerable portion of a garden, and on his being caught in a trap, the place will remain free from fresh Mole-tracks for a long period, proving that all the mischief was the work of a single Mole, at other times we have caught several out of one gallery on the same day; and while excavating a root-house, the lower part of which was rock, four of these animals came during the night through one gallery and tumbled down into the pit, where, the rock preventing their digging a way out, they were found in the morning. No others ever came through that gallery while the cellar was in progress, and those thus caught may probably have been one family.

Although generally known to run through the same galleries often, so much so that the most common method of capturing them is to set a trap anywhere in one of these tracks to intercept them when again passing through it, we have known a trap to remain set in a fresh track for eleven days before the animal passed that way, when it was caught; and we are of opinion that many of their tracks are only passed through once, as this animal is known to travel from one field or wood to another, and probably the only galleries they regularly traverse are those adjacent to the spot they have selected for rearing their young. In relation to this subject, Dr. Godman says—

"It is remarkable how unwilling they are to relinquish a long frequented burrow; I have frequently broken down or torn off the surface of the same burrow for several days in succession, but would always find it repaired at the next visit. This was especially the case with one individual whose nest I discovered, which was always repaired within a short time, as often as destroyed. It was an oval cavity, about five or seven inches in length by three in breadth, and was placed at about eight inches from the surface in a stiff clay. The entrance to it sloped obliquely downwards from the gallery about two inches from the surface; three times I entirely exposed this cell, by cutting out the whole superincumbent clay with a knife, and three times a similar one was made a little beyond the situation of the former, the excavation having been continued from its back part. I paid a visit to the same spot two months after capturing its occupant, and breaking up the cell, all the injuries were found to be repaired, and another excavated within a few inches of the old one. Most probably numerous individuals, composing a whole family, reside together in these extensive galleries. In the winter they burrow closer to the streams, where the ground is not so deeply frozen."

This species whilst beneath the earth's surface seems to search for food with the same activity and untiring perseverance that are observable in animals that seek for their provender above ground. It works through the earth not only in a straight-forward direction, but loosens it to the right and left, beneath and above, so that no worm or insect can escape it. When in contact with any one of the objects of which it has been in search, it seizes it with remarkable quickness both with its fore-feet and its sharp teeth, drawing itself immediately backward with its prize, upon which it begins to prey at once. The Shrew Mole passes through loose soil with nearly the same ease and speed that it displays in running, or "scrabbling" along above ground. It moves backward almost as rapidly as it goes forward. The nose is often seen protruded above the surface of the ground.

The snout of this species, although apparently delicate, is most powerfully muscular, as well as flexible; the animal can turn it to the right or left, upward or downward, and at times inserts it in its mouth, as if for the purpose of cleansing it, and then suddenly withdraws it with a kind of smack of its lips; this habit we observed three times in the course of a few minutes. The Shrew Mole is exceedingly tenacious of life; it cannot easily be put to death, either by heavy pressure or strangling, and a severe blow on the head seems to be the quickest mode of despatching it.

Although this species, as we have seen, feeds principally on worms, grubs, &c., we have the authority of our friend OGDEN HAMMOND, Esq., for the following example either of a most singular perversity of taste, or of habits hitherto totally unknown as appertaining to animals of this genus, and meriting a farther inquiry. While at his estate near Throg's Neck, on Long Island Sound, his son, who is an intelligent young lad, and fond of Natural History, observed in company with an old servant of the family, a Shrew Mole in the act of swallowing, or devouring, a common toad—this was accomplished by the Mole, and he was then killed, being unable to escape after such a meal, and was taken to the house, when Mr. HAMMOND saw and examined the animal, with the toad partially protruding from its throat. This gentleman also related to us some time ago, that he once witnessed an engagement between two Moles, that happened to encounter each other in one of the noon-day excursions this species is so much in the habit of making. The combatants sidled up to one another like two little pigs, and each tried to root the other over, in attempting which their efforts so much resembled the manner of two boars fighting, that the whole affair was supremely ridiculous to the beholder, although no doubt to either of the bold warriors the consequences of an overthrow would have been very serious; for the conqueror would vent his rage upon the fallen hero, and punish him severely with his sharp teeth. We have no doubt these conflicts generally take place in the love season, and are caused by rivalry, and that some "fair Mole" probably rewards the victor. When approached, the Moles attempted to escape, but were both shot on the spot, thus falling victims to their own passions; and if we would read aright, affording us an instructive lesson, either as individuals, or in a national point of view.[9]

The Shrew Moles are able to work their way so rapidly, that in soft or loamy soil it is almost impossible for the most active man to overtake and turn them out with a spade, unless he can see the spot where they are working by the movement of the earth, in which case they can be thrown out easily by sticking the spade in front of them or at one side of their gallery, and with a quick movement tossing them on to the surface.

They have been known to make a fresh track after rain, during one night, several hundred yards in length; oftentimes they proceed for a considerable distance in nearly a straight or direct line, then suddenly begin to excavate around and across a small space of not more than a few feet in diameter, until you could hardly place your foot on a spot within this subterranean labyrinth without sinking through into their track; at this time they are most probably in pursuit of worms, or other food, which may be there imbedded.

Although cold weather appears to us to put a stop to the movements of the Mole, we do not feel by any means certain that such is the case; and very probably the hardness of the ground when frozen, and the depth at which the Mole is then obliged to seek his food, may be a sufficient reason for our seeing no traces of this busy creature's movements during cold winter weather. We have, however, often perceived their tracks after a day or two of warm weather in January, and have repeatedly observed them about during a thaw, after the first autumnal frosts had occurred. In Carolina there are not many weeks in a winter in which we are not able to find here and there traces of the activity of the Mole. We admit, however, that even in this comparatively mild climate, they appear to be far less active in winter than at other seasons.

From the foregoing facts we are inclined to think the Mole does not become torpid at any time; and in corroboration of this idea, we find that the animal is not at any season found in high Northern latitudes. Dr. RICHARDSON thinks "the absence of the Shrew Mole from these countries is owing to the fact that the earth-worm on which the Scalops, like the common Mole, principally feeds, is unknown in the Hudson's Bay countries."[10]

The idea commonly entertained by uninformed persons, that Moles have no eyes, is an error; although our own experience confirms the opinion of others, that they appear to possess the power of seeing only in a very limited degree. We must not forget, however, that a wise Providence has adapted their organs of vision to the subterraneous life they lead. Shut out from the light of the sun by a law of nature requiring them to search for food beneath the earth's surface, these animals would find a large pair of eyes one of the greatest of evils, inasmuch as they would be constantly liable to be filled with sand; thus causing inflammation, blindness, and eventually death.

It is not, however, beyond the reach of possibility, nor contrary to the economy of Nature, to suppose that during the night, when this species is seen occasionally above ground, or when engaged in running or fighting, or for purposes we have not yet discovered, this animal may have the power of expanding its minute orbs, and drawing back the hair that entirely conceals its eyes. This, however, is a mere conjecture, which we have thrown out for the consideration of those who are fond of investigating Nature in her minutest operations.

The inquiry has often been made, if the Shrew Mole does not feed upon the grains or roots of the corn, peas, potatoes, &c., planted in rows or in hills, why is it that this pest so ingeniously and so mischievously follows the rows, and as effectually destroys the young plants as if it had consumed them? We answer, it is not the spirit of mischief by which the Mole is actuated; it is the law of self-preservation. In the rows where these seeds have been sown, or these vegetables planted, the ground has been manured; this, and the consequent moisture around the roots of the plants, attracts worms and other insects that are invariably found in rich moist earth. To the accusations made against the Shrew Mole as a destroyer of potatoes, and other vegetables, he might often with great truth plead an alibi. LECONTE's pine mouse, (*Arvicola pinetorum,*) is usually the author of the mischief, whilst all the blame is thrown upon the innocent Shrew Mole. We are, moreover, inclined to think that whilst the earth-worm is the general, it is by no means the only food of the latter, and we had an opportunity of discovering to our cost, that when in captivity, this species relishes other fare. We preserved one in a cage in Carolina, during a winter, for the purpose of ascertaining on what kind of food it was sustained, and whether it became dormant. It at no time touched grains or vegetables; the lower part of the cage was filled with a foot of moist earth, in which we occasionally placed a pint of earth-worms. It devoured pieces of beef, and for a week was engaged in demolishing a dead pigeon. Until the middle of January we found it every day actively running through the earth in search of worms. Suddenly, however, it seemed to have gone to winter quarters, as we could see no more traces of its customary burrowing. We now carefully searched for it in the box, to ascertain its appearance in a dormant state. But the little creature had forced itself through the wooden bars, and was gone. We examined every part of the room without success, and finally supposed it had escaped through the door. The cage of the Mole had been set on a box, full of earth, in which the chrysolides of some sixty or seventy species of rare butterflies, moths, and sphinges, had been carefully deposited. In this box we a few days afterwards heard a noise, and on looking, discovered our little fugitive. On searching for our choice insects we found not one left; they had all been devoured by the Shrew Mole. This greatly disappointed us, and put an end to all our hopes of reading the following spring a better lesson on entomology than ever could have been taught us—either by FABRICIUS, SPENCE, or KIRBY.[11]

We had an opportunity on two different occasions of examining the nests and young of the Shrew Mole. The nests were about eight inches below the surface, the excavation was rather large and contained a quantity of oak leaves on the outer surface, lined with soft dried leaves of the crab-grass, (*Digitaria sanguinalis.*) There were galleries leading to this nest, in two or three directions. The young numbered in one case, five, and in another, nine.

Our kind friend, J. S. HAINES, Esq., of Germantown, near Philadelphia, informed us that he once kept several Shrew Moles in confinement for the purpose of investigating their habits, and that having been neglected for a few days, the strongest of them killed and ate up the others; they also devoured raw meat, especially beef, with great avidity.

Nº 5

PLATE XXII

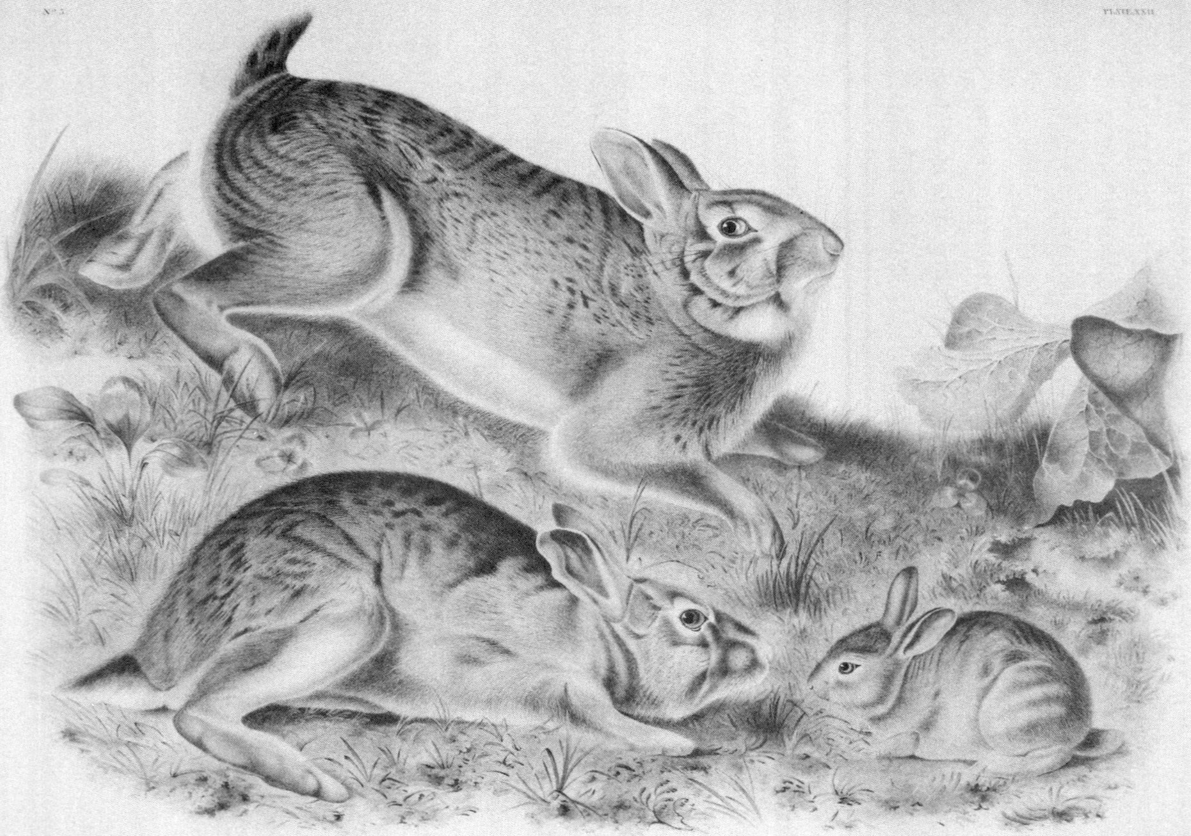

LEPUS SYLVATICUS, BACHMAN

GREY RABBIT

Gray Rabbit

PERHAPS THE MOST STRIKING QUALITY OF THIS ESSAY IS ITS extensive "General Remarks" section, only part of which is included here, hence the abrupt conclusion. Bachman goes to great investigative lengths to explain why identification of this species had been so troublesome for naturalists, showing how James DeKay, in his *Natural History of New York* (1842), misidentified the northern hare as the gray rabbit. He is able to locate DeKay's error in that naturalist's citing of another naturalist's work, that of Johann von Schreber. What we now call the cottontail rabbit thus provided Bachman an opportunity to demonstrate that only informed personal observation could prevent errors in species identification.

---

8. John James Audubon, Nagel and Weingartner, lithographers, *Gray Rabbit*, 1845–46, hand-colored lithograph, plate 22 of *The Viviparous Quadrupeds of North America*. Courtesy of the Jule Collins Smith Museum of Fine Art, Auburn University; Louise Hauss and David Brent Miller Audubon Collection.

LEPUS SYLVATICUS.—BACHMAN.

GRAY RABBIT.

PLATE XXII.—OLD MALE, FEMALE, AND YOUNG.

HABITS.

This species abounds in our woods and forests, even in their densest coverts; it is fond of places overgrown with young pines thickly crowded together, or thickets of the high bush-blackberry, (*Rubus villosus*;) and is also fond of frequenting farms and plantations, and occupying the coppices and grassy spots in the neighbourhood of cultivation, remaining in its form by day, concealed by a brush-heap, a tuft of grass, or some hedge-row on the side of an old fence; from which retreat it issues at night, to regale itself on the clover, turnip, or corn-fields of the farmer. It not unfrequently divests the young trees in the nursery of their bark; it makes frequent inroads upon the kitchen-garden, feasting on the young green peas, lettuces, cabbages, &c., and doing a great deal of mischief; and when it has once had an opportunity of tasting these dainties, it becomes difficult to prevent its making a nightly visit to them; although the place it enters at may be carefully closed, the Rabbit is sure to dig a fresh hole every night in its immediate vicinity; and snares, traps, or guns, are the best auxiliaries in such cases, soon putting an end to farther depredations.

This animal, when first started, runs with greater swiftness, and makes fewer doublings than the Northern hare, (*L. Americanus*;) having advanced a hundred yards or more it stops to listen; finding itself pursued by dogs, should the woods be open and free from swamps or thickets, it runs directly toward some hole in the root of a tree or hollow log. In the lower parts of Carolina, where it finds protection in briar patches, and places thickly overgrown with smilax and other vines, it continues much longer on foot, and by winding and turning in places inaccessible to larger animals, frequently makes its escape from its pursuers, without the necessity of resorting for shelter to a hollow tree.

The Gray Rabbit possesses the habit of all the other species of this genus, with which we are acquainted, of stamping with its hind feet on the earth when alarmed at night, and when the males are engaged in combat. It is also seen during the spring season, in wood-paths and along the edges of fields, seeking food late in the mornings and early in the afternoons, and during the breeding season even at mid-day: on such occasions, it may be approached and shot with great ease. This species, like all the true hares, has no note of recognition, and its voice is never heard, except when wounded or at the moment of its capture, when it utters a shrill, plaintive cry, like that of a young child in pain; in the Northern hare this cry is louder, shriller, and of longer continuance. The common domesticated European rabbit seems more easily made to cry out in this way than any other of the genus.

Dr. RICHARDSON, in his work on the American quadrupeds, expresses an opinion from a careful examination of many specimens in different States that the change to the winter dress in the Northern hare, is effected not by a shedding of its hair, but by a lengthening and blanching of the summer fur. Having watched the progress of this change, in the present species in a state of confinement, and having also examined many specimens at all seasons of the year, we have arrived at the opposite conclusion as far as regards the Gray Rabbit. In autumn, the greater portion, if not all, the summer fur drops off in spots, and is gradually replaced by the winter coat. In this state, as there are shades of difference between the summer and winter colours, the animal presents a somewhat singular appearance, exhibiting at the same time, (as in the Northern hare, although far less conspicuously,) patches of different colours. The Gray Rabbit, although it breeds freely in enclosed warrens, seldom becomes tame, and will probably never be domesticated. When captive, it seems to be constantly engaged in trying to find some means of escape, and though it digs no burrows in a state of nature, yet, when confined, it is capable of digging to the depth of a foot or more under a wall, in order to effect its object. We, however, at the house of Dr. DE BENNEVILLE at Milestown, near Philadelphia, saw five or six that were taken from the nest when very young and brought up by hand, so completely tamed that they came at the call and leapt on to the lap of their feeder; they lived sociably and without restraint in the yard, among the dogs and poultry. The former, although accustomed to chase the wild rabbit, never molesting those which had, in this manner, grown up with them, and now made a part of the motley tenants of the poultry-yard. We have not only observed dogs peacefully associating with the hare when thus tamed; but have seen hounds, accustomed to the chase of the deer, eating from the same platter with one of those animals that was domesticated and loose in the yard, refraining from molesting it, and even defending it from the attacks of strangers of their own species, that happened to come into the premises; and when this tame deer, which occasionally visited the woods, was started by the pack of hounds here referred to, they refused to pursue it.

The Gray Rabbit is one of the most prolific of all our species of this genus; in the Northern States it produces young about three times in the season, from five to seven at a litter, whilst in Carolina, its young are frequently brought forth as early as the twentieth of February, and as late as the middle of October, and in all the intermediate months. Nature seems thus to have made a wise provision for the preservation of the species, since no animal is more defenceless or possesses more numerous enemies. Although it can run with considerable swiftness for some distance, its strength in a short time is exhausted, and an active dog would soon overtake it if it did not take shelter in some hole in the earth, heap of logs, or stones, or in a tree with a hollow near its root; in these retreats it is often captured by young hunters.

In the Northern and Middle States, where the burrows of the Maryland marmot (*Arctomys monax*) and the holes resorted to by the common skunk, (*Mephitis chinga*,) are numerous, the Gray Rabbit, in order to effect its escape when pursued, betakes itself to them, and as they are generally deep, or placed among rocks or roots, it would require more labour to unearth it when it has taken possession of either of these animals' retreats than it is worth, and it is generally left unmolested. It is not always safe in these cases, however, for the skunk occasionally is "at home" when the Rabbit runs into his hole, and often catches and devours the astonished fugitive before it can retrace its steps and reach the mouth of the burrow.

This species is also captured occasionally by the skunk and other carnivorous animals when in its form. Its most formidable enemy, however, is the ermine, which follows its tracks until it retires to a hole in the earth or to a hollow tree, which the little but ferocious creature, although not one-fourth as large as the timid Rabbit, quickly enters and kills it—eating off the head, and leaving the body until a want of food compels it to return for more.

Whilst residing in the State of New-York many years ago, we were desirous of preserving a number of Rabbits during the winter from the excessive cold and from the hands of the hunters, who killed so many that we feared the race would be nearly extirpated in our neighbourhood; our design being to set them at liberty in the spring. At this period we had in confinement several weasels of two species existing in that part of the country, (*Putorius erminea* and *P. fusca*,) in order to ascertain in what manner their change of colour from brown in summer to white in winter, and vice versa, was effected.

We bethought ourselves of using one of each species of these weasels instead of a ferret, to aid in taking the Rabbits we wanted, and having provided ourselves with a man and a dog to hunt the Rabbits to their holes, we took the weasels in a small tin box with us, having first tied a small cord around their necks in such a manner as to prevent them from escaping, or remaining in the holes to eat the Rabbits, whilst it could not slip and choke them.

We soon raced a Rabbit to its hole, and our first experiment was made with the little brown weasel, (*P. fusca*;) it appeared to be frightened, and refused to enter the hole; the common species, (*P. erminea*,) although we had captured the individual but a few days before, entered readily; but having its jaws at liberty, it killed the Rabbit. Relinquishing the weasel to our man, he afterwards filed its teeth down to prevent it from destroying the Rabbits; and when thus rendered harmless, the ermine pursued the Rabbits to the bottom of their holes, and terrified them so that they instantly fled to the entrance and were taken alive in the hand; and although they sometimes scrambled up some distance in a hollow tree, their active and persevering little foe followed them, and instantly forced them down. In this manner the man procured twelve Rabbits alive in the course of one morning, and more than fifty in about three weeks, when we requested him to desist.

On more than one occasion we have seen the tracks of this species on the snow, giving evidence, by their distance from each other, that the animal had passed rapidly, running under the influence of fear. Examining the surface of the snow carefully, we observed the foot-prints of the weasel, as if in pursuit, and following up the double trail, we found, at the mouth of a hole a short distance beyond, the mutilated remains of the luckless Rabbit.

The Canada lynx, the Bay lynx, (wild cat,) the red and the gray fox, &c., capture this species by stratagem or stealth; various species of hawks and owls prey upon them, and the rattle-snake, chicken-snake, and other serpents have been killed with the Gray Rabbit in their stomach. These reptiles probably caught their victims by stratagem, or by stealing upon them when in their form, and enclosing them in their twining folds, as the boa constrictor captures larger animals.

In order to catch or kill the Gray Rabbit, different means are resorted to according to the fancy of the hunter or the nature of the locality in which the animal may be. In the northern parts of the United States it is pursued with dogs, and either shot or taken from the hole or other retreat to which it may have been driven. It is also frequently captured in box-traps, or snares placed in the gaps of some brush-fence made in the woods for the purpose. In the Southern States it is generally hunted with pointer dogs and shot at the moment when it leaps from its form.

GENERAL REMARKS.

This being the most common of our hares in the Atlantic States of America, it has been longest and most familiarly known. HERRIOTT, who gave an account of the third voyage of the English to Virginia in 1586, in enumerating the natural productions of that country, under the head of Conies, says, "Those that we have seen, and all that we can hear of, are of a gray colour like unto hares; in some places there are such plenty that all the people, of some towns, make them mantles of the fur, or fleece of the skins of those which they usually take." It is subsequently mentioned by the intrepid GOVERNOR SMITH of Virginia, by LAWSON and by CATESBY. KALM, in the 1st vol. of his Travels in America, gave a correct description, not only of the animal, but of its habits. The following is an extract from his Journal, the entry was made either at Philadelphia or his favourite retreat "Raccoon," in the vicinity of that city, on the 6th Jan. 1749. "There are a great number of hares in this country, but they differ from our Swedish ones in their size, which is very small, and but little bigger than that of a rabbit; they keep almost the same gray colour both in summer and winter, which our Northern hares have in summer only; the tip of their ears is always gray, and not black; the tail is likewise gray on the upper side, at all seasons; they breed several times a year. In spring they lodge their young ones in hollow trees, and in summer, in the months of June and July, they breed in the grass. When they are surprised they commonly take refuge in hollow trees, out of which they are taken by means of a crooked stick, or by cutting a hole into the tree, opposite to the place where they lie; or by smoke which is occasioned by making a fire on the outside of the tree. On all these occasions the greyhounds must be at hand. These hares never bite, and can be touched without any danger. In the day-time they usually lie in hollow trees, and hardly ever stir from thence, unless they be disturbed by men or dogs; but in the night they come out and seek their food. In bad weather, or when it snows, they lie close for a day or two, and do not venture to leave their retreats. They do a great deal of mischief in the cabbage-fields, but apple-trees suffer infinitely more from them, for they peel off all the bark next to the ground. The people here are agreed that the hares are fatter in a cold and severe winter, than in a mild and wet one, for which they could give me several reasons from their own conjectures. The skin is useless because it is so loose that it can be drawn off; for when you would separate it from the flesh, you need only pull at the fur and the skin follows. These hares cannot be tamed. They were at all times, even in the midst of winter, plagued with a number of common fleas."

In 1820 (as we have observed in our article on *L. Americanus*) DESMAREST mistaking the species, gave a pretty good description of the Gray Rabbit, and unfortunately referred it to *L. Americanus*. He had evidently been misled by FORSTER, SCHŒPFF, who having confounded these two species, induced him to believe that as he was describing an American hare, only one American species at that time being known, it must be the one referred to by previous authors. Hence he quoted GMELIN, SCHŒPFF, ERXLEBEN, PALLAS and BODD, and gave to the species the extravagant geographical range, from Churchill, Hudson's Bay, to California, and assigned it a habitation in New-Albion, Louisiana, Florida, the two Carolinas, &c. HARLAN, in giving an account of the American quadrupeds in 1825, finding the Gray Rabbit described by DESMAREST, translated the article very literally, even to its faults, from the French of that author, (See Encyclopédie de Mammalogie, p. 351.) HARLAN's translation represents the fur as "becoming whiter during winter, but the ears and tail remaining always of the same gray." In the following year GODMAN (Amer. Nat. Hist., vol. ii., p. 157) once more described this species under the (wrong) name of *Lepus Americanus*. In speaking of its colour, he says, "in winter the pelage is nearly or altogether white," and he gives it the extraordinary weight of seven pounds. This is rather surprising, as we know no city in the union where the market in winter is better supplied with this species of hare than Philadelphia.

In this singular manner the Gray Rabbit, the most common and best known of all the species of quadrupeds in America, had never received a specific name that was not pre-occupied. In 1827, we proposed the name of *Lepus sylvaticus*, and assigned our reasons for so doing in a subsequent paper, (See Journ. Acad. Nat. Sc, vol. viii., part 1, p. 75.) In 1840, Dr. EMMONS also, (Report on Quadrupeds of Massachusetts,) described it under the (wrong) name of *L. Americanus*, giving as synonymous, *L. Hudsonius*, PALLAS; American hare, FORSTER, PENNANT, Arct. Zool. HEARNE's Journey, SABINE, PARRY and RICHARDSON; who each described the Northern hare, and not this species. He, however, quoted HARLAN and GODMAN correctly, with the exception of the name which they had misapplied.

In 1842 Dr. DEKAY (See Nat. Hist. N. York, part 1st, p. 93,) refers this species to *Lepus nanus* of SCHREBER, supposing the description of that author, (which is contained in an old work that is so scarce in America that our naturalists have seldom had an opportunity of referring to it,) to have escaped the notice of modern authors. After giving a translation from Schreber, he remarks, "The whole history of the habits of this species, and its abundance, sufficiently confirm the fact, that SCHREBER had our Rabbit in view, although he was misled by SCHŒPFF and PENNANT, and confounded two species."

We regret that we are obliged to differ from an author who is generally accurate, and who is always courteous in his language towards other naturalists, but in this case we must do so.

In order to save the student of natural history the labour of searching for SCHREBER's work, to refer to his description, we have concluded to insert it here, together with our translation of the article, adding the references to authors, &c., which were omitted by DEKAY, and which we conceive very important in pursuing our inquiries.

N°8

PLATE XXXIX.

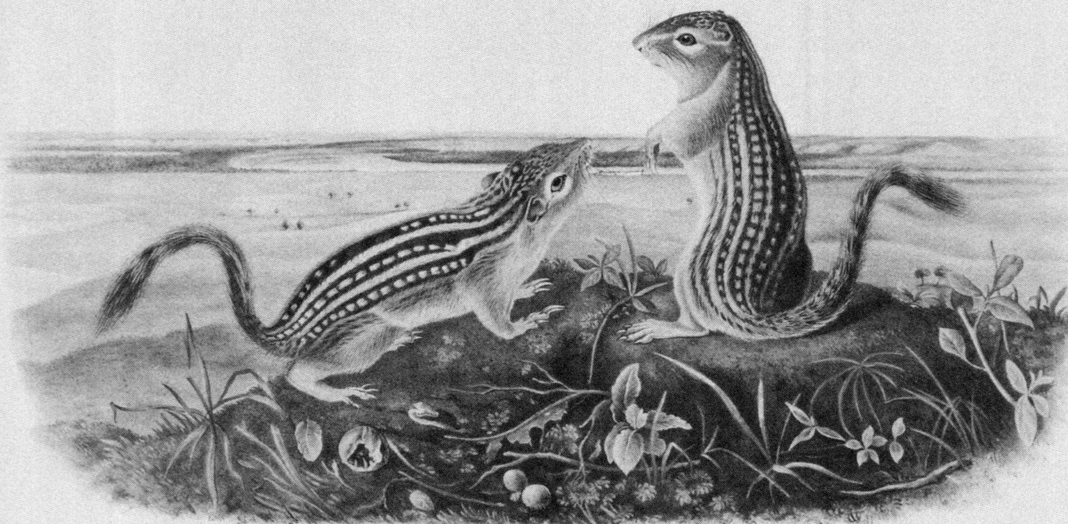

SPERMOPHILUS TRIDECEMLINEATUS, MITCHELL.

LEOPARD SPERMOPHILE.

Leopard Spermophile

NEARLY THREE HUNDRED PAGES INTO THE FIRST VOLUME, Bachman is compelled to explain why quadrupeds are worth studying. He acknowledges that birds "present greater attractions to the student of nature" in his culture, but he reminds readers that mammals possess no shortage of beauty and interesting habits; they too reflect the "infinite [. . .] wisdom" of nature, and they are the creatures most like us. This likeness, however, was not always considered a categorically good thing. In his lengthy albeit positive review of the first volume, Charles Wilkins Webber asserts that, unlike birds, quadrupeds "appeal to our material, lower and purely animal instincts"—they even kill their prey in messier, uglier ways—but despite this ugliness, which has its analogs in human society, it is all ordained by God and therefore noble.[12] From this vantage point, *Quadrupeds* can be viewed as a study of God's darker side.

---

9. John James Audubon, J. T. Bowen, lithographer, *Leopard Spermophile*, 1845–46, hand-colored lithograph, plate 39 of *The Viviparous Quadrupeds of North America*. Courtesy of the Jule Collins Smith Museum of Fine Art, Auburn University; Louise Hauss and David Brent Miller Audubon Collection.

SPERMOPHILUS TRIDECEM LINEATUS.—MITCHILL.

LEOPARD SPERMOPHILE

PLATE XXXIX.—MALE AND FEMALE.

HABITS.

We believe it is generally supposed that "birds," with their varied and pleasing forms, their gay and beautiful plumage, their tuneful throats, and their graceful movements through the air, present greater attractions to the student of nature than "quadrupeds," and awaken in him a stronger desire to acquire a knowledge of their natures and characters than he may entertain to study the habits of the mammalia.

In addition, however, to the fact that the latter are like ourselves viviparous, and approach our own organization, it should be remembered that all the productions of nature are the work of so infinite a wisdom, that they must, in every department of the physical world, excite our greatest interest and our admiration, even when examined superficially.

Among the quadrupeds, there are innumerable varieties of form and character, and although most animals are nocturnal, and therefore their habits cannot be studied with the same facility with which the manners and customs of the lively diurnal species of birds may be observed; yet when we follow them in their nightly wanderings, penetrate into their retreats, and observe the sagacity and extraordinary instincts with which they are endowed, we find in them matter to interest us greatly, and arouse our curiosity and astonishment.

Owls seem to us a dull and stupid race, principally because we only notice them during the day, which nature requires them to spend in sleep, the structure of their eyes compelling them to avoid the light, and seek concealment in hollow trees, in caves, and obscure retreats. But we should recollect that the diurnal birds are, during night, the time for their repose, as dull and stupid as owls are during the day. We should therefore not judge the habits of quadrupeds by the same standard. In regard to their fur, and external markings, there are many that will strike even the most careless observer as eminently beautiful. The little animal which is here presented to you is one of this description. In the distribution of the tints that compose its gaudy dress, in the regularity of its lines and spots, and in the soft blendings of its various shades of colour, we have evidence that even species whose habitations are under ground, may present to the eye as rich and beautiful a vesture as is found in the garb of a majority of the lively songsters of the woods.

In the warm days of spring the traveller on our Western prairies is often diverted from the contemplation of larger animals, to watch the movements of this lively little species. He withdraws his attention for a moment from the bellowing buffalo herd that is scampering over the prairies, to fix his eyes on a lively little creature of exquisite beauty seated on a diminutive mound at the mouth of its burrow, which seems by its chirrupings and scoldings to warn away the intruder on its peaceful domains. On a nearer approach it darts into its hole; but although concealed from view, and out of the reach of danger, its tongue, like that of other scolds of a more intelligent race, is not idle; it still continues to vent its threats of resentment against its unwelcome visitor by a shrill and harsh repetition of the word "seek—seek."

There is a great similarity in the habits of the various spermophili that compose the interesting group to which the present species belongs.

They live principally on the open prairies, make their burrows in the earth, and feed on roots and seeds of various kinds, which they carry in their pouches to their dark retreats under ground.

The holes of this species, according to RICHARDSON, run nearly perpendicularly, and are so straight, that they will admit a stick to be inserted to the depth of four or five feet. He supposes that owing to the depth of their burrows, which the sun does not penetrate very early in spring, they do not make their appearance as early as some others, especially *S. Richardsonii.*

As soon as they feel the warmth of spring they come forth and go in quest of their mates; at this period they seem fearless of danger, and are easily captured by the beasts and birds of prey that frequent the plains. The males are said to be very pugnacious at this season.

This is believed to be the most active and lively of all our known species of marmot-squirrels; we recently observed one in New York that played in a wheel in the manner of the squirrel. We saw in Charleston a pair in a cage, that were brought from Missouri by an officer of the army. They were adults, had but recently been captured, and were rather wild. They seemed to keep up a constant angry querulous chattering; they were fed on various kinds of nuts and grains, but principally on corn-meal and pea or ground-nuts, (*Arachis hypogæa.*) They would come to the bars of the cage and take a nut from the hand, but would then make a hasty retreat to a little box in the corner of their domicile. On our placing a handful of filberts in front of the cage, they at first came out and carried off one by one to their store-house, but after we had retired so as not to be observed, they filled their pouches by the aid of their paws, and seemed to prefer this mode of transporting their provisions. As we were desirous of taking measurements and descriptions, we endeavoured to hold one in the hand by the aid of a glove, but it struggled so lustily and used its teeth so savagely that we were compelled to let it go.

This species frequently takes up its residence near the fields and gardens of the settlers, and in the neighbourhood of Fort Union and other places, was represented as particularly destructive to the gardens.

We found the Leopard-Spermophile quite abundant near Fort Union, on the Upper Missouri. Their burrows were made in a sandy gravelly soil, they were never deep or inclined downwards, but ran horizontally within about a foot of the surface of the earth. This difference in habit from those observed by RICHARDSON may be owing to the nature of the different soils. We dug some of their burrows and discovered that the holes ran in all directions, containing many furcations.

RICHARDSON states that "the males fight when they meet, and in their contests their tails are often mutilated." All the specimens, however, that we obtained were perfect and in good order.

The Leopard-Spermophile has two more teats than are found in the majority of the species of this genus, and hence it may be expected to produce an additional number of young. RICHARDSON informs us that ten young were taken from a female killed at Carlton House. This was on the 17th May, and we from hence presume that they produce their young soon after this period.

N° 9

PLATE XLI

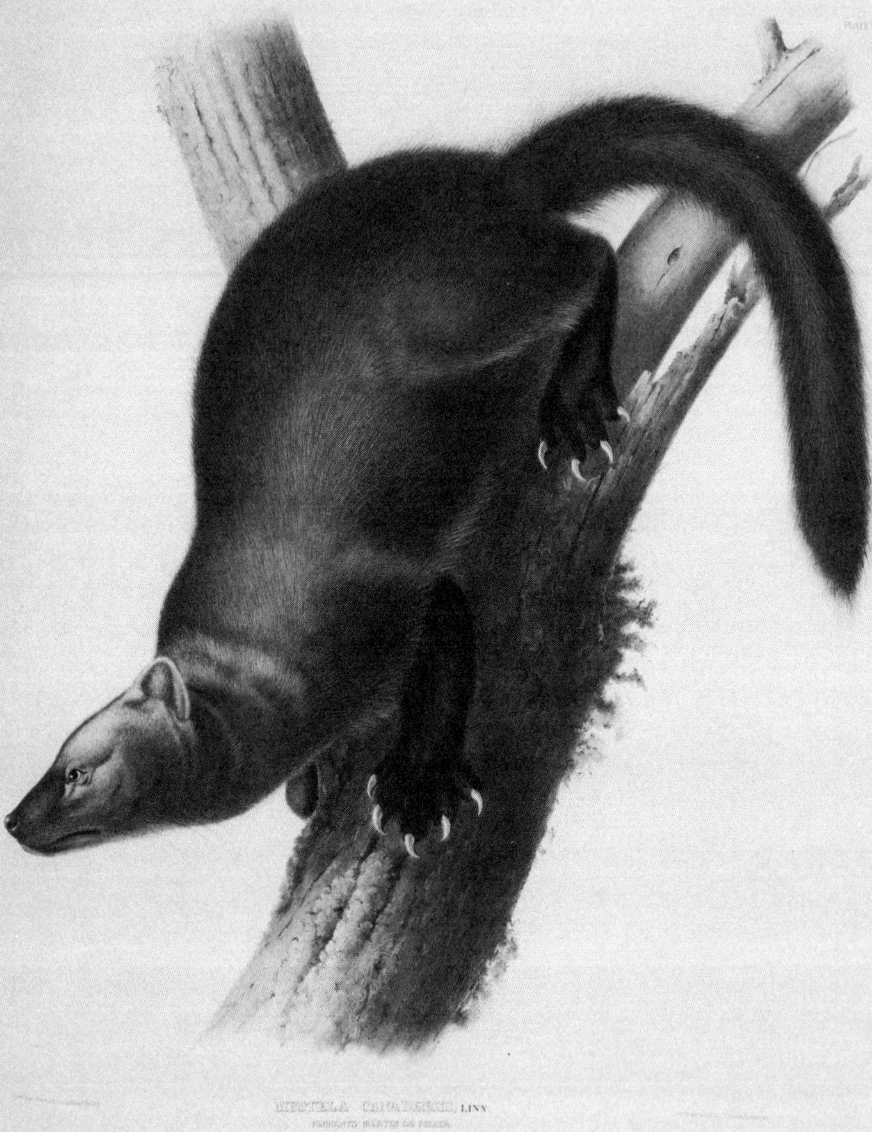

MUSTELA CANADENSIS, LINN.

PENNANTS MARTEN OR FISHER.

Pennant's Marten or Fisher

THE "HABITS" SECTION OF THIS ESSAY IS UNIQUE BECAUSE IT begins with language that resonates with our current environmental crisis. Bachman points out that although this species can still be found in many places in the 1840s, it is "met with in diminished numbers"; since the second decade of the nineteenth century, it has "become still more rare, and in places where it was then known, scarcely any vestige of the knowledge of its former existence can now be traced"; in New York in 1827 specifically, "it was every where represented as a species that was fast disappearing." Today, the fisher's range is reduced, but encouragingly it is no longer endangered.

---

10. John James Audubon, J. T. Bowen, lithographer, *Pennant's Marten or Fisher*, 1845–46, hand-colored lithograph, plate 41 of *The Viviparous Quadrupeds of North America*. Courtesy of the Jule Collins Smith Museum of Fine Art, Auburn University; Louise Hauss and David Brent Miller Audubon Collection.

MUSTELA CANADENSIS.—SCHREBER.

PENNANT'S MARTEN OR FISHER.

BLACK FOX OR BLACK CAT OF THE NORTHERN HUNTERS.

PLATE XLI.—MALE.

HABITS.

Although this species is represented as having been rather common in every part of the Northern and Middle States, in the early periods of our history, and is still met with in diminished numbers, in the thinly settled portions of our country; very little of its history or habits has been written, and much is still unknown. We have occasionally met with it, but it has been to us far from a common species. Even in the mountainous portions of the Northern and Eastern States, the Fisher, thirty years ago, was as difficult to procure as the Bay lynx. It has since become still more rare, and in places where it was then known, scarcely any vestige of the knowledge of its former existence can now be traced.

Dr. DEKAY (Nat. Hist. N.Y., p. 32, 1843,) states that, "in Hamilton county, (N.Y.,) it is still numerous and troublesome." On an excursion we made in the State of New York, in 1827, we heard of it occasionally near the head waters of Lake Champlain, along the St. Lawrence, and to the west as far as Lake Erie, but it was every where represented as a species that was fast disappearing.

Whilst residing in the northern part of our native State, (New York,) thirty-five years ago, the hunters were in the habit of bringing us two or three specimens of this Marten in the course of a winter. They obtained them by following their tracks in the snow, when the animals had been out in quest of food on the previous night, thus tracing them to the hollow trees in which they were concealed, which they chopped down. They informed us that as a tree was falling, the Fisher would dart from the hollow, which was often fifty feet from the ground, and leap into the snow, when the dogs usually seized and killed him, although not without a hard struggle, as the

Fisher was infinitely more dangerous to their hounds than either the gray or the red fox. They usually called this species the Black Fox.

A servant, on one occasion, came to us before daylight, asking us to shoot a raccoon for him, which, after having been chased by his dogs the previous night, had taken to so large a tree that he neither felt disposed to climb it nor to cut it down. On our arrival at the place, it was already light, and the dogs were barking furiously at the foot of the tree. We soon perceived that instead of being a raccoon, the animal was a far more rare and interesting species, a Fisher. As we were anxious to study its habits we did not immediately shoot, but teased it by shaking some grape vines that had crept up nearly to the top of the tree. The animal not only became thoroughly frightened but seemed furious; he leaped from branch to branch, showing his teeth and growling at the same time; now and then he ran half way down the trunk of the tree, elevating his back in the manner of an angry cat, and we every moment expected to see him leap off and fall among the dogs. He was brought down after several discharges from the gun. He seemed extremely tenacious of life, and was game to the last, holding on to the nose of a dog with a dying grasp. This animal proved to be a male, the body measured twenty-five inches, and the tail, including the fur, fifteen. The servant who had traced him, informed us that he appeared to have far less speed than a fox, that he ran for ten minutes through a swamp in a straight direction, and then took to a tree.

The only opportunity that was ever afforded us of judging of the speed of the Fisher occurred near the Virginia Gray-Sulphur Springs, in 1839. We had ascended Peter's Mountain in search of rare plants for our herbarium; out of health and fatigued, we had for some time been seated on a rock to rest, when we observed a gray squirrel pass within ten feet of us, seemingly in a great fright, and with all the speed it could command, with a Fisher in full pursuit. They were both too much occupied with their own affairs to take any notice of us. The Fisher seemed to make more rapid progress than the squirrel, and we feel confident that if the latter had not mounted a tree it would have been overtaken before it could have advanced many feet farther; it ran rapidly up the sides of a cucumber tree, (*Magnolia acuminati*,) still pursued by its hungry foe. The squirrel leaped lightly among the smaller branches, on which its heavier pursuer seemed unwilling to trust himself. At length the affrighted animal pitched from one of the topmost boughs and landed on its feet unhurt among the rocks beneath. We expected every moment to see the Fisher give us a specimen also of his talent at lofty tumbling, but he seemed to think that the "better part of valour was discretion," and began to run down the stem of the tree. At this point we interfered. Had he imitated the squirrel in its flying leap, he might have been entitled to the prey, provided he could overtake it, but as he chose to exercise some strategy and jockeying in the race, when the chances were so much in his favour, we resolved to end the chase by running to the foot of the tree which the Fisher was descending. He paused on the opposite side as if trying to ascertain whether he had been observed; we were without a gun, but rattled away with a knife on our botanizing box, which seemed to frighten the Marten in his turn, most effectually;—the more noise we created the greater appeared to be his terror; after ascending to the top of the tree he sprang to another, which he rapidly descended, till within twenty feet of the earth, when he jumped to the ground, and with long leaps ran rapidly down the side of the mountain, and was out of sight in a few moments.

This scene occurred in the morning of a warm day in the month of July, a proof that this species is not altogether nocturnal in its habits. We are, however, inclined to believe that the above was only an exception to the general character of the animal.

Species that are decidedly nocturnal in their habits, frequently may be seen moving about by day during the period when they are engaged in providing for their young. Thus the raccoon, the opossum, and all our hares, are constantly met with in spring, and early summer, in the morning and afternoon, whilst in the autumn and winter they only move about by night.

In the many fox-hunts, in which our neighbours were from time to time engaged, not far from our residence at the north, during the period when we obtained the information concerning their primitive mode of enjoying that amusement, which we have laid before our readers, in pages 49 and 50,[13] (where we also spoke of Pennant's Marten as not being very scarce at that time in Rensselaer county, N.Y.,) we never heard of their having encountered a single Fisher in the day-time; but when they traversed the same grounds at night, in search of raccoons, it was not unusual for them to discover and capture this species. We were informed by the trappers that they caught the Fisher in their traps only by night.

The specimen, from which the figure in our plate was drawn, was taken alive in some part of the Alleghany Mountains, in the State of Pennsylvania, and we soon afterwards received a letter from our esteemed friend, SPENCER F. BAIRD, Esq., of Carlisle, in that State, informing us of its having been captured, which enabled us, through that gentleman, to purchase it. We received it at New-York, in good condition, in a case tinned inside, with iron bars in front, to prevent the animal from making its escape, as it was so strong and so well supplied with sharp teeth that it could easily have eaten its way out of a common wooden box. In Mr. BAIRD's note he says, "all the account I was able to procure respecting this species was the following:—It was found in company with an older one, in Peter's Mountain, six miles above Harrisburgh, about five weeks ago. (His letter is dated Carlisle, March 16th, 1844.) After a most desperate resistance the old one was killed, after having beaten off the dogs, to whose assistance the hunters were obliged to come. This individual ran up a tree, and being stoned by the hunters, jumped off from a height of about forty feet! when being a little stunned by the leap, the men ran up quickly, threw their coats over it, and thus secured it. The old one was said to have been about the size of a pointer dog. The young one is very savage, and emits a rather strong musky odour."

We kept this individual alive for some days, feeding it on raw meat, pieces of chicken, and now and then a bird. It was voracious, and very spiteful, growling snarling and spitting when approached, but it did not appear to suffer much uneasiness from being held in captivity, as, like many other predacious quadrupeds it grew fat, being better supplied with food than when it had been obliged to cater for itself in the woods.

The older one, which, as Mr. Baird mentions, was killed by the dogs and the hunters, was a female, and no doubt was the mother of the one that was captured, and probably died in the hope of saving her young.

On several occasions we have seen the tracks of the Fisher in the snow; they resemble those of the pine marten, but are double their size. To judge by them, the animal advances by short leaps in the manner of a mink.

Pennant's Marten appears to prefer low swampy grounds. We traced one which had followed a trout stream for some distance, and ascertained that it had not gone into the water. Marks were quite visible in different places where it had scratched up the snow by the side of logs and piles of timber, to seek for mice, or other small quadrupeds, and we have no doubt it preys upon the Northern hare, gray rabbit, and ruffed grouse, as we observed a great many tracks of those species in the vicinity. It further appears, that this animal makes an occasional meal on species which are much more closely allied to it than those just mentioned. In a letter we received from Mr. Fothergill, in which he furnished us with notes on the habits of some of the animals existing near Lake Ontario, he informs us that "a Fisher was shot by a hunter named Marsh, near Port Hope, who said it was up in a tree, in close pursuit of a pine marten, which he also brought with it." Mr. Fothergill stuffed them both at the time.

LEWIS and CLARK state that in Oregon the Fisher captures not only the squirrel, but the raccoon, and that in pursuing them it leaps from tree to tree.

RICHARDSON remarks that, "the Fisher is said to prey much on frogs in the summer season; but I have been informed that its favourite food is the Canada porcupine, which it kills by biting in the belly." He says also, "it will feed on the hoards of frozen fish laid up by the residents."

We can scarcely conceive in what manner it is able to overturn the porcupine, so as to bite it on the belly, as it is large and heavy, and is armed with bristles at all points.

It is stated by Dr. DEKAY, on the authority of a person who resided many years near Lake Oneida, New York, that the name (Fisher) "was derived from its singular fondness for the fish used to bait traps."

An individual of this species, which had been caught in a steel-trap, was brought to Charleston and exhibited in a menagerie. It had been taken only a few months, and was sullen and spiteful; when fed, it gulped down a moderate quantity of meat in great haste, swallowing it nearly whole, and then retired in a growling humour to a dark corner of its cage.

The Fisher is represented as following the line of traps set by the trappers, and in the manner of the wolverene, robbing them of their bait. The season for hunting this species is stated, by Dr. DEKAY, to commence in the western part of New York, about the 10th of October, and to last till the middle of May; and he says the ordinary price paid for each skin is a dollar and a half.

This species brings forth once a year, depositing its young in the trunk of a large tree, usually some thirty or forty feet from the ground. Dr. RICHARDSON observes that it produces from two to four young at a litter; DEKAY confines the number to two. We once saw three extracted from the body of a female on the 20th of April, in the northern part of New York.

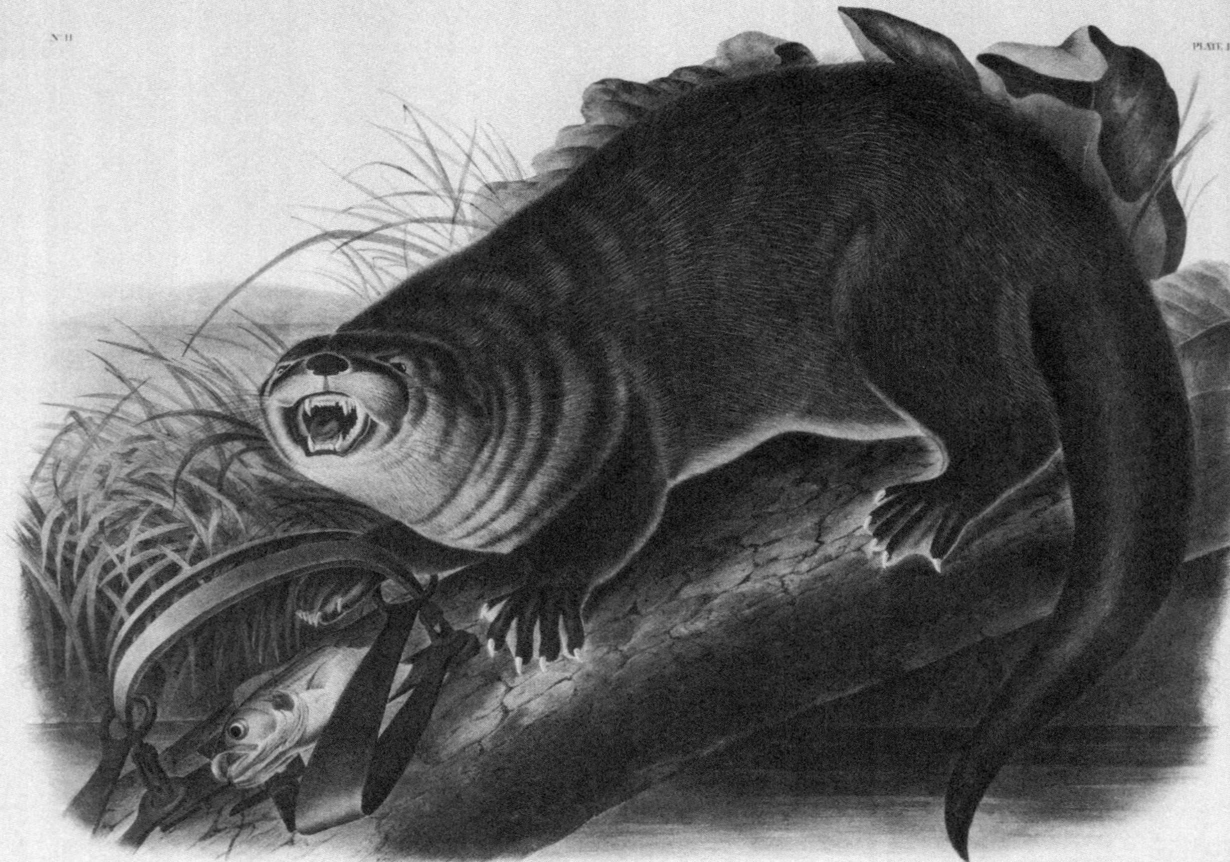

LUTRA CANADENSIS, SABINE.

CANADA OTTER.

MALE.

Canada Otter

"CANADA OTTER" IS A STRIKING EXAMPLE OF THE RECURRING juxtaposition of romantic, nostalgic art and the violence humans impose on and depict in the lives of animals—as well as of the unselfconscious suddenness of switching from peacefulness to violence.[14] The tone of Audubon's opening nostalgic reminiscence is quiet and peaceful, not preparing the reader for the violence that follows. Then he makes clear that in his painting of this beautiful mammal, he intended to "represent the pain and terror felt by the creature" caught in the trap. This is followed immediately by a pastoral scene of otters enjoying a perfectly secure, happy, and playful existence untouched by humankind. In addition to testimony of the former abundance of the species, the essay displays the insouciant denial of ethical considerability of mammals prevalent at the time. Nevertheless, lovely images of affectionate pet otters foreshadow future improved relations.

---

11. John James Audubon, J. T. Bowen, lithographer, *Canada Otter*, 1845–46, hand-colored lithograph, plate 51 of *The Viviparous Quadrupeds of North America*. Courtesy of the Jule Collins Smith Museum of Fine Art, Auburn University; Louise Hauss and David Brent Miller Audubon Collection.

LUTRA CANADENSIS.—SABINE.

CANADA OTTER.

PLATE LI.—MALE.

HABITS.

We concluded our first volume with a brief account of *Spermophilus Richardsonii*, the last animal figured in plates 1 to 50 inclusive, of our illustrations of the Quadrupeds of North America. Having, since that volume was written, published about 60 more plates, we now take up our pen to portray the habits and describe the forms and colours of the species figured in plates 51 to 100 inclusive, and shall, we hope, be able to give our readers tolerably good accounts of them; although, alas! the days of our youth are gone, when, full of enthusiasm, and anxious to examine every object in nature within our reach, the rising sun never found us slumbering away the fresh hours of the morning, but beamed upon our path through the deep forest, or lighted up to joy and gladness the hill side or mountain top, which we had already gained in quest of the birds or the beasts that were to be met with; and where we often prolonged our rambles until the shades of evening found us yet at a distance from our camp, loaded with wild turkeys, ducks, geese, and perchance an Otter.

Fresh and pleasant in our mind is the recollection of our early expeditions among the wild woods, and along the unvisited shores of our new country; and although more than forty years of varied and busy life have passed since the Otter was shot and drawn, whose figure we have given, we will try to take you with us to a spot on the eastern banks of the fair Ohio. It is a cold wintry morning: the earth concealed by a slight covering of snow, and the landscape in all its original wildness. Here let us proceed cautiously, followed by that constant companion, our faithful dog. Whilst we are surveying the quiet waters as they roll onward toward the great Mississippi, in whose muddy current they will lose their clear and limpid character, and become as opaque and impetuous as the waves of that mighty river of the West, we see a dark object making its way towards the spot on which we stand, through the swiftly dividing element. It has not observed us: we remain perfectly still, and presently it is distinctly visible; it is an Otter, and now within the range of our old gun "Tear Jacket," we take but one moment to raise our piece and fire; the water is agitated by a violent convulsive movement of the animal, our dog plunges into the river, and swimming eagerly to the Otter, seizes it, but the latter dives, dragging the dog with it beneath the surface, and when they reappear, the Otter has caught the dog by the nose and is struggling violently. The brave dog, however, does not give up, but in a few moments drags the wounded Otter to the shore, and we immediately despatch it. Being anxious to figure the animal, we smooth its disordered fur and proceed homewards with it, where, although at that time we had not drawn many quadrupeds, we soon select a position in which to figure the Otter, and accordingly draw it with one foot in a steel-trap, and endeavour to represent the pain and terror felt by the creature when its foot is caught by the sharp saw-like teeth of the trap.

Not far from the town of Henderson, (Kentucky), but on the opposite side of the Ohio river, in the State of Indiana, there is a pond nearly one mile in length, with a depth of water varying from twelve to fifteen feet. Its shores are thickly lined with cane, and on the edge of the water stand many large and lofty cypress trees. We often used to seat ourselves on a fallen trunk, and watch in this secluded spot the actions of the birds and animals which resorted to it, and here we several times observed Otters engaged in catching fishes and devouring them. When pursuing a fish, they dived expertly and occasionally remained for more than a minute below the surface. They generally held their prey when they came to the top of the water, by the head, and almost invariably swam with it to a half-sunken log, or to the margin of the pond, to eat the fish at their ease, having done which, they returned again to the deep water to obtain more.

One morning we observed that some of these animals resorted to the neighbourhood of the root of a large tree which stood on the side of the pond opposite to us, and with its overhanging branches shaded the water. After a fatiguing walk through the tangled cane-brake and thick underwood which bordered the sides of this lonely place, we reached the opposite side of the pond near the large tree, and moved cautiously through the mud and water towards its roots: but the hearing or sight of the Otters was attracted to us, and we saw several of them hastily make off at our approach. On sounding the tree with the butt of our gun, we discovered that it was hollow, and then having placed a large stick in a slanting position against the trunk, we succeeded in reaching the lowest bough, and thence climbed up to a broken branch from which an aperture into the upper part of the hollow enabled us to examine the interior. At the bottom there was quite a large space or chamber to which the Otters retired, but whether for security or to sleep we could not decide.

Next morning we returned to the spot, accompanied by one of our neighbours, and having approached, and stopped up the entrance under water as noiselessly as possible, we cut a hole in the side of the tree four or five feet from the ground, and as soon as it was large enough to admit our heads, we peeped in and discovered three Otters on a sort of bed composed of the inner bark of trees and other soft substances, such as water grasses. We continued cutting the hole we had made, larger, and when sufficiently widened, took some green saplings, split them at the butt-end, and managed to fix the head of each animal firmly to the ground by passing one of these split pieces over his neck, and then pressing the stick forcibly downwards. Our companion then crept into the hollow, and soon killed the Otters, with which we returned home.

The American Otter frequents running streams, large ponds, and more sparingly the shores of some of our great lakes. It prefers those waters which are clear, and makes a hole or burrow in the banks, the entrance to which is under water.

This species has a singular habit of sliding off the wet sloping banks into the water, and the trappers take advantage of this habit to catch the animal by placing a steel-trap near the bottom of their sliding places, so that the Otters occasionally put their foot into it as they are swiftly gliding toward the water.

In Carolina, a very common mode of capturing the Otter is by tying a pretty large fish on the pan of a steel-trap, which is sunk in the water where it is from five to ten feet deep. The Otter dives to the bottom to seize the fish, is caught either by the nose or foot, and is generally found drowned. At other times the trap is set under the water, without bait, on a log, one end of which projects into the water, whilst the other rests on the banks of a pond or river; the Otter, in endeavouring to mount the log, is caught in the trap.

Mr. GODMAN, in his account of these singular quadrupeds, states that "their favourite sport is sliding, and for this purpose in winter the highest ridge of snow is selected, to the top of which the Otters scramble, where, lying on the belly with the fore-feet bent backwards, they give themselves an impulse with their hind legs and swiftly glide head-foremost down the declivity, sometimes for the distance of twenty yards. This sport they continue apparently with the keenest enjoyment until fatigue or hunger induces them to desist."

This statement is confirmed by CARTWRIGHT, HEARNE, RICHARDSON, and more recent writers who have given the history of this species, and is in accordance with our own personal observations.

The Otters ascend the bank at a place suitable for their diversion, and sometimes where it is very steep, so that they are obliged to make quite an effort to gain the top; they slide down in rapid succession where there are many at a sliding place. On one occasion we were resting ourself on the bank of Canoe Creek, a small stream near Henderson, which empties into the Ohio, when a pair of Otters made their appearance, and not observing our proximity, began to enjoy their sliding pastime. They glided down the soap-like muddy surface of the slide with the rapidity of an arrow from a bow, and we counted each one making twenty-two slides before we disturbed their sportive occupation.

This habit of the Otter of sliding down from elevated places to the borders of streams, is not confined to cold countries, or to slides on the snow or ice, but is pursued in the Southern States, where the earth is seldom covered with snow, or the waters frozen over. Along the reserve-dams of the rice fields of Carolina and Georgia, these slides are very common. From the fact that this occurs in most cases during winter, about the period of the rutting season, we are inclined to the belief that this propensity may be traced to those instincts which lead the sexes to their periodical associations.

RICHARDSON says that this species has the habit of travelling to a great distance through the snow in search of some rapid that has resisted the severity of the winter frosts, and that if seen and pursued by hunters on these journeys, it will throw itself forward on its belly and slide through the snow for several yards, leaving a deep furrow behind it, which movement is repeated with so much rapidity, that even a swift runner on snow shoes has some difficulty in overtaking it. He also remarks that it doubles on its track with much cunning, and dives under the snow to elude its pursuers.

The Otter is a very expert swimmer, and can overtake almost any fish, and as it is a voracious animal, it doubtless destroys a great number of fresh water fishes annually. We are not aware of its having a preference for any particular species, although it is highly probable that it has. About twenty-five years ago we went early one autumnal morning to study the habits of the Otter at Gordon and Spring's Ferry, on the Cooper River, six miles above Charleston, where they were represented as being quite abundant. They came down with the receding tide in groups or families of five or six together. In the space of two hours we counted forty-six. They soon separated, ascended the different creeks in the salt marshes, and engaged in capturing mullets (*Mugil*). In most cases they came to the bank with a fish in their mouth, despatching it in a minute, and then hastened again after more prey. They returned up the river to their more secure retreats with the rising tide. In the small lakes and ponds of the interior of Carolina, there is found a favourite fish with the Otter, called the fresh-water trout (*Grystes salmoides*).

Although the food of the Otter in general is fish, yet when hard pressed by hunger, it will not reject animal food of any kind. Those we had in confinement, when no fish could be obtained were fed on beef, which they always preferred boiled. During the last winter we ascertained that the skeleton and feathers of a wild duck were taken from an Otter's nest on the banks of a rice field reserve-dam. It was conjectured that the duck had either been killed or wounded by the hunters, and was in this state seized by the Otter. This species can be kept in confinement easily in a pond surrounded by a proper fence where a good supply of fish is procurable.

On throwing some live fishes into a small pond in the Zoological Gardens in London, where an Otter was kept alive, it immediately plunged off the bank after them, and soon securing one, rose to the surface holding its prize in its teeth, and ascending the bank, rapidly ate it by large mouthfuls, and dived into the water again for another. This it repeated until it had caught and eaten all the fish which had been thrown into the water for its use. When thus engaged in devouring the luckless fishes the Otter bit through them, crushing the bones, which we could hear snapping under the pressure of its powerful jaws.

When an Otter is shot and killed in the water, it sinks from the weight of its skeleton, the bones being nearly solid and therefore heavy, and the hunter consequently is apt to lose the game if the water be deep; this animal is, however, usually caught in strong steel-traps placed and baited in its haunts; if caught by one of the fore-feet, it will sometimes gnaw the foot off, in order to make its escape.

Otters when caught young are easily tamed, and although their gait is ungainly, will follow their owner about, and at times are quite playful. We have on two occasions domesticated the Otter. The individuals had been captured when quite young, and in the space of two or three days became as tame and gentle as the young of the domestic dog. They preferred milk and boiled corn meal, and refused to eat fish or meat of any kind, until they were several months old. They became so attached to us, that at the moment of their entrance into our study they commenced crawling into our lap—mounting our table, romping among our books and writing materials, and not unfrequently upsetting our ink-stand and deranging our papers.

The American Otter has one litter annually, and the young, usually two and occasionally three in number, are brought forth about the middle of April, according to Dr. RICHARDSON, in high northern latitudes. In the Middle and Southern States they are about a month earlier, and probably litter in Texas and Mexico about the end of February.

The nest, in which the Otter spends a great portion of the day and in which the young are deposited, we have had opportunities of examining on several occasions. One we observed in an excavation three feet in diameter, in the bank of a rice field; one in the hollow of a fallen tree, and a third under the root of a cypress, on the banks of Cooper river, in South Carolina; the materials—sticks, grasses and leaves—were abundant; the nest was large, in all cases protected from the rains, and above and beyond the influence of high water or freshets.

J. W. AUDUBON procured a fine specimen of the Otter, near Lagrange in Texas, on the twenty-third of February, 1846. It was shot whilst playing or sporting in a piece of swampy and partially flooded ground, about sunset,—its dimensions we have already given.

Early writers have told us that the common Otter of Europe had long been taught to catch fish for its owners, and that in the houses of the great in Sweden, these animals were kept for that purpose, and would go out at a signal from the cook, catch fish and bring it into the kitchen in order to be dressed for dinner.

This, however improbable it may at first appear, is by no means unlikely, except that we doubt the fact of the animal's going by itself for the fish.

BEWICK relates some anecdotes of Otters which captured salmon and other fish for their owners, for particulars of which we must refer our readers to his History of Quadrupeds.

Our late relative and friend, N. BERTHOUD, Esq., of St. Louis, told us some time since, that while travelling through the interior of the State of Ohio, he stopped at a house where the landlord had four Otters alive which were so gentle that they never failed to come when he whistled for them, and that when they approached their master they crawled along slowly and with much apparent humility towards him, and looked somewhat like enormous thick and short snakes.

No 11

PLATE LV.

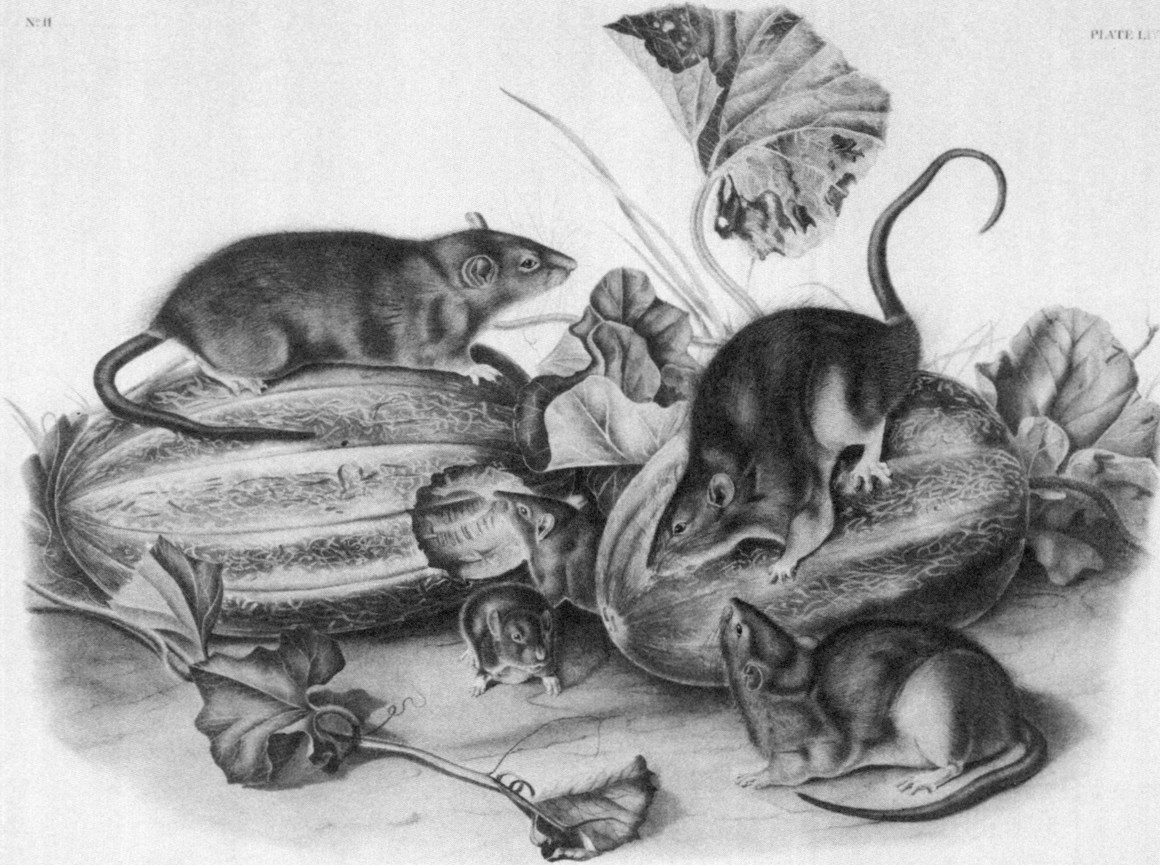

MUS DECUMANUS, LINN.

BROWN OR NORWAY RAT.

Brown or Norway Rat

DESPITE THE DISFAVOR HUMANS HOLD THIS SPECIES IN AND the antagonistic relationship depicted between humans and rats, the species inspired some of the best writing in all of *Quadrupeds*, a striking phenomenon that occurs in many of these essays. The description of late winter floods along the Ohio River provides a rare window onto how humans, mammals, and even arachnids behaved during those times. The vivid images of rats in peach trees and of young people plying about on wooden boards become the literary equivalent of genre painting, the depiction of ordinary lives generally not considered worthy of literary treatment. But our authors argue that these floods are worth spending your vacation there to see. The Norway rat is also shown to be responsible for a slightly darker aspect of ordinary human life: an occasional missing bit of ear, nose, or toe. The mammal essays frequently provide information relevant to environmental history, and here we see how invasive species were introduced to new areas, twice by Audubon himself, if inadvertently.

---

12. John James Audubon, J. T. Bowen, lithographer, *Brown or Norway Rat*, 1845–46, hand-colored lithograph, plate 54 of *The Viviparous Quadrupeds of North America*. Courtesy of the Jule Collins Smith Museum of Fine Art, Auburn University; Louise Hauss and David Brent Miller Audubon Collection.

MUS DECUMANUS.—PALL.

BROWN OR NORWAY RAT.

PLATE LIV.—MALES, FEMALE, AND YOUNG.

HABITS.

The brown rat is unfortunately but too well known almost in every portion of our country, and in fact throughout the world, to require an elaborate account of its habits, but we will give such particulars as may we hope be interesting. It is one of the most prolific and destructive little quadrupeds about the residences of man, and is as fierce as voracious. Some cases are on record where this rat has attacked a man when he was asleep, and we have seen both adults and children who, by their wanting a piece of the ear, or a bit of the end of the nose, bore painful testimony to its having attacked them while they were in bed; it has been known to nibble at an exposed toe or finger, and sometimes to have bitten even the remains of the shrouded dead who may have been exposed to its attacks.

The Norway Rat is very pugnacious, and several individuals may often be seen fighting together, squealing, biting, and inflicting severe wounds on each other. On one occasion, we saw two of these rats in furious combat, and so enraged were they, that one of them whose tail was turned towards us, allowed us to seize him, which we did, giving him at the same time such a swing against a gate post which was near, that the blow killed him instantly—his antagonist making his escape.

During the great floods or freshets which almost annually submerge the flat bottom-lands on the Ohio river at various places, the rats are driven out from their holes and seek shelter under the barns, stables, and houses in the vicinity, and as the increasing waters cover the low grounds, may be seen taking to pieces of drift wood and floating logs, &c., on which they sometimes remain driving along with the currents for some distance. They also at such times climb up into the lofts of barns, smokehouses, &c., or betake themselves to the trees in the orchards or gardens. We once, at Shippingport, near the foot of the falls of the Ohio river, whilst residing with our brother-in-law, the late N. Berthoud, went out in a skiff, during a freshet which had exceeded those of many previous years in its altitude, and after rowing about over the tops of fences that were secured from rising with the waters by being anchored by large cross-timbers placed when they were put up, under the ground, to which the posts were dove-tailed, and occasionally rowing through floating worm-fences which had broken away from their proper locations and were lying flat upon the surface of the flowing tide, we came to the orchard attached to the garden, and found the peach and apple trees full of rats, which seemed almost as active in running among the branches as squirrels. We had our gun with us and tried to shoot some of them, but the cunning rogues dived into the water whenever we approached, and swam off in various directions, some to one tree and some to another, so that we were puzzled which to follow. The rats swam and dived with equal facility and made rapid progress through the water. Many of them remained in the orchard until the freshet subsided, which was in the course of a few days. Whether they caught any fish or not during this time we cannot say, but most of them found food enough to keep them alive until they were able once more to occupy their customary holes and burrows. During these occasional floods on our western rivers, immense numbers of spiders and other insects take refuge in the upper stories of the houses, and the inhabitants find themselves much incommoded by them as well as by the turbulent waters around their dwellings. Such times are, however, quite holidays to the young folks, and skiffs and batteaux of every description are in requisition, while some go about on a couple of boards, or paddle from street to street on large square pine logs. When the flats are thus covered, there is generally but little current running on them, although the main channel of the river flows majestically onward, covered with floating logs and the fragments of sheds, haystacks, &c., which have left their quiet homes on the sides of the river many miles above, to float on a voyage of discovery down to the great Mississippi, unless stopped by the way by the exertions of some fortunate discoverer of their value, who rowing out among the drifting logs, roots and branches, ties a rope to the frail floating tenement, and tows it to the trunk of a tree, where he makes it fast, for the water to leave it ready for his service, when the river has again returned to its quiet and customary channel. Stray flat boats loaded with produce, flour, corn and tobacco, &c., are often thus taken up, and are generally found and claimed afterwards by their owners. The sight of the beautiful Ohio thus swelling proudly along, and sometimes embracing the country with its watery margin extended for miles beyond its ordinary limits, is well worth a trip to the West in February or March. But these high freshets do not occur every year, and depend on the melting of the snows, which are generally dissolved so gradually that the channel of the river is sufficient to carry them off.

In a former work, (Ornithological Biography, vol. 1, p. 155,) we have given a more detailed account of one of the booming floods of the Ohio and Mississippi rivers, to which we beg now to refer such of our readers as have never witnessed one of those remarkable periodical inundations.

Mr. Ogden Hammond, formerly of Throg's Neck, near New-York, furnished us with the following account of the mode in which the Norway Rat captures and feeds upon the small sand clams which abound on the sandy places along the East river below high water mark. He repaired to a wharf on his farm with one of his men at low water: in a few moments a rat was seen issuing from the lower part of the wharf, peeping cautiously around before he ventured from his hiding place. Presently one of the small clams buried in the soft mud and sand which they inhabit, threw up a thin jet of water about a foot above the surface of the ground, upon seeing which, the rat leaped quickly to the spot, and digging with its fore-paws, in a few moments was seen bringing the clam towards his retreat, where he immediately devoured it.

When any of these clams lie too deep to be dug up by the rats, they continue on the watch and dig after the next which may make known its whereabouts by the customary jet of water. These clams are about ¾ of an inch long and not more than ⅝ of an inch wide; their shells are slight, and they are sometimes used as bait by fishermen.

The Brown or Norway Rat was first introduced in the neighbourhood of Henderson, Kentucky, our old and happy residence for several years, within our recollection.

One day a barge arrived from New-Orleans (we think in 1811) loaded with sugar and other goods; some of the cargo belonged to us. During the landing of the packages we saw several of these rats make their escape from the vessel to the shore, and run off in different directions. In a year from this time they had become quite a nuisance; whether they had been reinforced by other importations, or had multiplied to an incredible extent, we know not. Shortly after this period we had our smokehouse floor taken up on account of their having burrowed under it in nearly every direction. We killed at that time a great many of them with the aid of our dogs, but they continued to annoy us, and the readers of our Ornithological Biography are aware, that ere we left Henderson some rats destroyed many of our valued drawings.[15]

This species migrates either in troops or singly, and for this purpose takes passage in any conveyance that may offer, or it plods along on foot. It swims and dives well, as we have already remarked, so that rivers or water-courses do not obstruct its progress. We once knew a female to secrete herself in a wagon, loaded with bale rope, sent from Lexington, (Ky.) to Louisville, and on the wagon reaching its destination, when the coils of rope were turned out, it was discovered that the animal had a litter of several young ones: she darted into the warehouse through the iron bars which were placed like a grating in front of the cellar windows. Some of the young escaped also, but several of them were killed by the wagoner. How this rat was fed during the journey we do not know, but as the wagons stop every evening at some tavern, the probability is that she procured food for herself by getting out during the night and picking up corn, &c.

The Norway Rat frequently deserts a locality in which it has for some time remained and proved a great pest. When this is the case, the whole tribe journey to other quarters, keeping together and generally appearing in numbers in their new locality without any previous warning to the unlucky farmer or housekeeper to whose premises they have taken a fancy.

When we first moved to our retreat, nine miles above the city of New-York, we had no rats to annoy us, and we hoped it would be some time before they discovered the spot where we had located ourselves. But in the course of a few months a great many of them appeared, and we have occasionally had eggs, chickens and ducklings carried off by them to the number of six or more in a night. We have never been able to get rid of this colony of rats, and they have even made large burrows in the banks on the water side, where they can hardly be extirpated.

The Norway Rat is quite abundant in New-York and most other maritime cities, along the wharves and docks, and becomes very large. These animals are frequently destroyed in great numbers, while a ship is in port, after her cargo has been discharged, by smoking them; the fumes of sulphur and other suffocating materials, being confined to the hold by closing all parts, windows and hatches. After a thorough cleaning out, a large ship has been known to have had many thousands on board. Our old friend, Capt. CUMINGS, who in early life made many voyages to the East Indies, relates to us, that one of his captains used to have rats caught, when on long voyages, and had them cooked and served up at his table as a luxury. He allowed his sailors a glass of grog for every rat they caught, and as the supply was generally ample, he used to invite his mates and passengers to partake of them with due hospitality. Our friend, who was a mate, had a great horror of the captain's invitations, for it was sometimes difficult to ascertain in what form the delicate animals would appear, and to avoid eating them. Not having ourselves eaten rats, (as far as we know,) we cannot say whether the old India captain's fondness for them was justified by their possessing a fine flavour, but we do think prejudices are entertained against many animals and reptiles that are, after all, pretty good eating.

In the account of the black rat in our first volume, (*Mus rattus*,) pp. 190, 191, and 192, we gave some details of the habits of the present species, and stated our opinion in regard to its destroying the black rat. Dr. GODMAN considered the Norway Rat so thorough an enemy of the black rat, that he says, (vol. 2, p. 83,) in speaking of the latter, that it is now found only in situations to which the Brown Rat has not extended its migrations. According to the same author, who quotes R. SMITH, Rat Catcher, p. 5, 1768, (see GODMAN, vol. 2, p. 77,) the Brown Rat was not known even in Europe prior to the year 1750. RICHARDSON says, (probably quoting from HARLAN, Fauna, p. 149,) that it was brought from Asia to Europe, according to the accounts of historians of the seventeenth century, and was unknown in England before 1730. PENNANT, writing in 1785, says he has no authority for considering it an inhabitant of the new continent (America).

HARLAN states that the Norwegian rat did not, as he was credibly informed, make its appearance in the United States any length of time previous to the year 1775. HARLAN does not give the Brown Rat as an American species, giving only what he considered indigenous species.

The Brown Rat brings forth from 10 to 15 young at a litter, and breeds several times in a year. Fortunately for mankind, it has many enemies: weasels, skunks, owls, hawks, &c., as well as cats and dogs. We have never known the latter to eat them, but they may at times do so. Rats are also killed by each other, and the weak ones devoured by the stronger.

This species becomes very fat and clumsy when living a long time in mills or warehouses. We have often seen old ones so fat and inactive that they would fall back when attempting to ascend a staircase.

We will take our leave of this disagreeable pest, by saying, that it is omnivorous, devouring with equal voracity meat of all kinds, eggs, poultry, fish, reptiles, vegetables, &c. &c. It prefers eels to other kinds of fish, having been known to select an eel out of a large bucket of fresh fish, and drag it off to its hole. In vegetable gardens it devours melons, cucumbers, &c., and will eat into a melon, entering through a hole large enough to admit its body, consuming the tender sweet fruit, seeds and all, and leaving the rind almost perfect. Where rats have gained access to a field or vegetable garden, they generally dig holes near the fruits or vegetables, into which they can make an easy retreat at the approach of an enemy.

We have represented several of these rats in our plate about to devour muskmelons, for which they have a strong predilection.

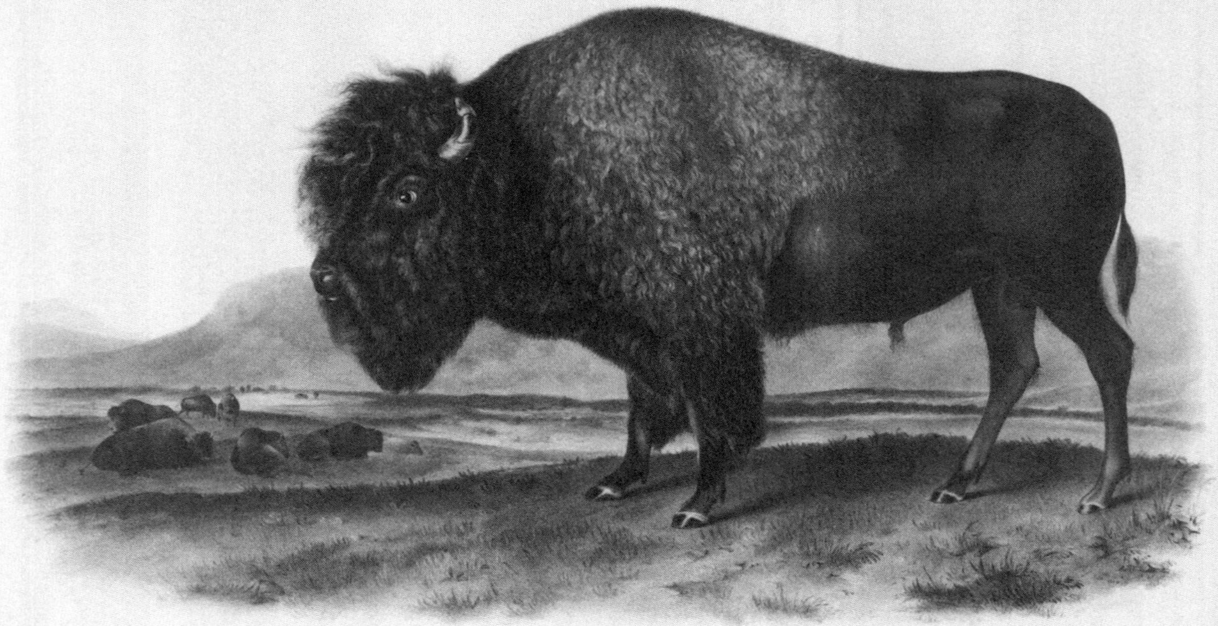

BOS AMERICANUS, GMEL.

AMERICAN BISON OR BUFFALO.

MALE.

American Bison

THIS ESSAY HELPED ESTABLISH THE BISON, "THE MOST important of all our contemporary American quadrupeds," as the prominent and noble symbol of the remote and rugged northern prairies. The opening account of the relentless displacement of bison populations in eastern regions by the westward push of human populations, together with the vivid descriptions of the extremely harsh environment on the northern prairies that are home to the "buffalo," build considerable pathos for the species, as does the account of the many natural causes of death. A third of the way through, however, natural history gives way to action-packed hunting stories, "the mode in which we killed them." Bachman recognized, when he finally read Audubon's 1843 journals, that his friend had described "some Buffalo hunts that were first rate," including the slightly manufactured tale of John G. Bell's saving of Audubon from a wounded and angry bison bull. The journals of Bell and his friend Edward Harris, however, show that Bell saved Harris, not Audubon, and that the famous "American Woodsman" appropriated the story and substituted himself for Harris, apparently to enhance his brand. Later historians and biographers have repeated this version of the story as true.[16]

---

13. John James Audubon, J. T. Bowen, lithographer, *American Bison or Buffalo*, 1845–46, hand-colored lithograph, plate 56 of *The Viviparous Quadrupeds of North America*. Courtesy of the Jule Collins Smith Museum of Fine Art, Auburn University; Louise Hauss and David Brent Miller Audubon Collection.

No. 12. PLATE LVII

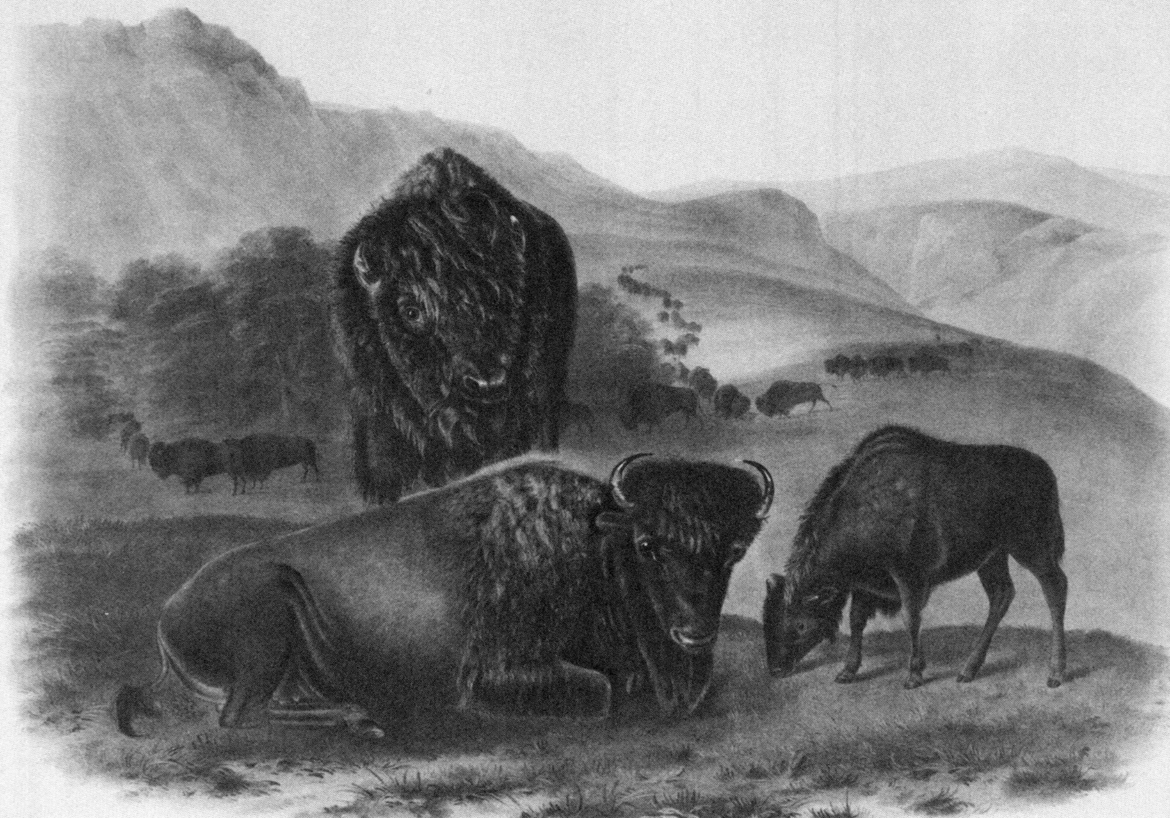

BOS AMERICANUS, GMEL.
AMERICAN BISON OR BUFFALO.

BISON AMERICANUS.—GMEL.

AMERICAN BISON.—BUFFALO.

PLATE LVI.—MALE.

PLATE LVII.—FEMALE, MALE AND YOUNG.

HABITS.

Whether we consider this noble animal as an object of the chase, or as an article of food for man, it is decidedly the most important of all our contemporary American quadrupeds; and as we can no longer see the gigantic mastodon passing over the broad savannas, or laving his enormous sides in the deep rivers of our wide-spread land, we will consider the Buffalo as a link, (perhaps sooner to be forever lost than is generally supposed,) which to a slight degree yet connects us with larger American animals, belonging to extinct creations.

But ere we endeavour to place before you the living and breathing herds of Buffaloes, you must journey with us in imagination to the vast western prairies, the secluded and almost inaccessible valleys of the Rocky Mountain chain, and the arid and nearly impassable deserts of the western table lands of our country; and here we may be allowed to express our deep, though unavailing regret, that the world now contains only few and imperfect remains of the lost races, of which we have our sole knowledge through the researches and profound deductions of geologists; and even though our knowledge of the osteology of the more recently exterminated species be sufficient to place them before our "mind's eye," we have no description and no figures of the once living and moving, but now departed possessors of these woods, plains, mountains and waters, in which, ages ago, they are supposed to have dwelt. Let us however hope, that our humble efforts may

14. John James Audubon, J. T. Bowen, lithographer, *American Bison or Buffalo*, 1845–46, hand-colored lithograph, plate 57 of *The Viviparous Quadrupeds of North America*. Courtesy of the Jule Collins Smith Museum of Fine Art, Auburn University; Louise Hauss and David Brent Miller Audubon Collection.

at least enable us to perpetuate a knowledge of such species as the Giver of all good has allowed to remain with us to the present day. And now we will endeavour to give a good account of the majestic Bison.

In the days of our boyhood and youth, Buffaloes roamed over the small and beautiful prairies of Indiana and Illinois, and herds of them stalked through the open woods of Kentucky and Tennessee; but they had dwindled down to a few stragglers, which resorted chiefly to the "Barrens," towards the years 1808 and 1809, and soon after entirely disappeared. Their range has since that period gradually tended westward, and now you must direct your steps "to the Indian country," and travel many hundred miles beyond the fair valleys of the Ohio, towards the great rocky chain of mountains which forms the backbone of North-America, before you can reach the Buffalo, and see him roving in his sturdy independence upon the vast elevated plains, which extend to the base of the Rocky Mountains.

Hie with us then to the West! let us quit the busy streets of St. Louis, once considered the outpost of civilization, but now a flourishing city, in the midst of a fertile and rapidly growing country, with towns and villages scattered for hundreds of miles beyond it; let us leave the busy haunts of men, and on good horses take the course that will lead us into the Buffalo region, and when we have arrived at the sterile and extended plains which we desire to reach, we shall be recompensed for our toilsome and tedious journey: for there we may find thousands of these noble animals, and be enabled to study their habits, as they graze and ramble over the prairies, or migrate from one range of country to another, crossing on their route water-courses, or swimming rivers at places where they often plunge from the muddy bank into the stream, to gain a sand-bar or shoal, midway in the river, that affords them a resting place, from which, after a little time, they can direct their course to the opposite shore, when, having reached it, they must scramble up the bank, ere they can gain the open prairie beyond.

There we may also witness severe combats between the valiant bulls, in the rutting season, hear their angry bellowing, and observe their sagacity, as well as courage, when disturbed by the approach of man.

The American Bison is much addicted to wandering, and the various herds annually remove from the North, at the approach of winter, although many may be found, during that season, remaining in high latitudes, their thick woolly coats enabling them to resist a low temperature, without suffering greatly. During a severe winter, however, numbers of them perish, especially the old, and the very young ones. The breeding season is generally the months of June and July, and the calves are brought forth in April and May; although occasionally they are produced as early as March or as late as July. The Buffalo most frequently has but one calf at a time, but instances occur of their having two. The females usually retire from the herd either singly or several in company, select as solitary a spot as can be found, remote from the haunt of wolves, bears, or other enemies that would be most likely to molest them, and there produce their young.

Occasionally, however, they bring forth their offspring when the herd is migrating, and at such times they are left by the main body, which they rejoin as soon as possible. The young usually follow the mother until she is nearly ready to have a calf again. The Buffalo seldom produces young until the third year, but will continue breeding until very old. When a cow and her very young calf are attacked by wolves, the cow bellows and sometimes runs at the enemy, and not unfrequently frightens him away; this, however, is more generally the case when several cows are together, as the wolf, ever on the watch, is sometimes able to secure a calf when it is only protected by its mother.

The Buffalo begins to shed its hair as early as February. This falling of the winter coat shows first between the fore-legs and around the udder in the female on the inner surface of the thighs, &c. Next, the entire pelage of long hairs drop gradually but irregularly, leaving almost naked patches in some places, whilst other portions are covered with loosely hanging wool and hair. At this period these animals have an extremely ragged and miserable appearance. The last part of the shedding process takes place on the hump. During the time of shedding, the Bison searches for trees, bushes, &c., against which to rub himself, and thereby facilitate the speedy falling off of his old hair. It is not until the end of September, or later, that he gains his new coat of hair. The skin of a Buffalo, killed in October, the hunters generally consider, makes a good Buffalo robe; and who is there, that has driven in an open sleigh or wagon, that will not be ready to admit this covering to be the cheapest and the best, as a protection from the cold, rain, sleet, and the drifting snows of winter? for it is not only a warm covering, but impervious to water.

The Bison bulls generally select a mate from among a herd of cows and do not leave their chosen one until she is about to calve.

When two or more males fancy the same female, furious battles ensue and the conqueror leads off the fair cause of the contest in triumph. Should the cow be alone, the defeated lovers follow the happy pair at such a respectful distance, as will ensure to them a chance to make their escape, if they should again become obnoxious to the victor, and at the same time enable them to take advantage of any accident that might happen in their favour. But should the fight have been caused by a female who is in a large herd of cows, the discomfited bull soon finds a substitute for his first passion. It frequently happens, that a bull leads off a cow, and remains with her separated during the season from all others, either male or female.

When the Buffalo bull is working himself up to a belligerent state, he paws the ground, bellows loudly, and goes through nearly all the actions we may see performed by the domesticated bull under similar circumstances, and finally rushes at his foe head foremost, with all his speed and strength. Notwithstanding the violent shock with which two bulls thus meet in mad career, these encounters have never been known to result fatally, probably owing to the strength of the spinous process commonly called the hump, the shortness of their horns, and the quantity of hair about all their fore-parts.

When congregated together in fair weather, calm or nearly so, the bellowing of a large herd (which sometimes contains a thousand) may be heard at the extraordinary distance of ten miles at least.

During the rutting season, or while fighting, (we are not sure which,) the bulls scrape or paw up the grass in a circle, sometimes ten feet in diameter, and these places being resorted to, from time to time, by other fighting bulls, become larger and deeper, and are easily recognised even after rains have filled them with water.

In winter, when the ice has become strong enough to bear the weight of many tons, Buffaloes are often drowned in great numbers, for they are in the habit of crossing rivers on the ice, and should any alarm occur, rush in a dense crowd to one place; the ice gives way beneath the pressure of hundreds of these huge animals, they are precipitated into the water, and if it is deep enough to reach over their backs, soon perish. Should the water, however, be shallow, they scuffle through the broken and breaking ice, in the greatest disorder, to the shore.

From time to time small herds, crossing rivers on the ice in the spring, are set adrift, in consequence of the sudden breaking of the ice after a rise in the river. They have been seen floating on such occasions in groups of three, four, and sometimes eight or ten together, although on separate cakes of ice. A few stragglers have been known to reach the shore in an almost exhausted state, but the majority perish from cold and want of food rather than trust themselves boldly to the turbulent waters.

Buffalo calves are often drowned, from being unable to ascend the steep banks of the rivers across which they have just swam, as the cows cannot help them, although they stand near the bank, and will not leave them to their fate unless something alarms them.

On one occasion Mr. Kipp, of the American Fur Company, caught eleven calves, their dams all the time standing near the top of the bank. Frequently, however, the cows leave the young to their fate, when most of them perish. In connection with this part of the subject, we may add, that we were informed when on the Upper Missouri river, that when the banks of that river were practicable for cows, and their calves could not follow them, they went down again, after having gained the top, and would remain by them until forced away by the cravings of hunger. When thus forced by the necessity of saving themselves to quit their young, they seldom, if ever, returned to them.

When a large herd of these wild animals are crossing a river, the calves or yearlings manage to get on the backs of the cows, and are thus conveyed safely over; but when the heavy animals, old and young, reach the shore, they sometimes find it muddy or even deeply miry; the strength of the old ones struggling in such cases to gain a solid footing, enables them to work their way out of danger in a wonderfully short time. Old bulls, indeed, have been known to extricate themselves when they had got into the mire so deep that but little more than their heads and backs could be seen. On one occasion we saw an unfortunate cow that had fallen into, or rather sank into a quicksand only seven or eight feet wide; she was quite dead, and we walked on her still fresh carcass safely across the ravine which had buried her in its treacherous and shifting sands.

The gaits of the Bison are walking, cantering, and galloping, and when at full speed, he can get over the ground nearly as fast as the best horses found in the Indian country. In lying down, this species bends the fore-legs first, and its movements are almost exactly the same as those of the common cow. It also rises with the same kind of action as cattle.

When surprised in a recumbent posture by the sudden approach of a hunter, who has succeeded in nearing it under the cover of a hill, clump of trees or other interposing object, the Bison springs from the ground and is in full race almost as quick as thought, and is so very alert, that one can scarcely perceive his manner of rising on such occasions.

The bulls never grow as fat as the cows, the latter having been occasionally killed with as much as two inches of fat on the boss or hump and along the back to the tail. The fat rarely exceeds half an inch on the sides or ribs, but is thicker on the belly. The males have only one inch of fat, and their flesh is never considered equal to that of the females in delicacy or flavour. In a herd of Buffaloes many are poor, and even at the best season it is not likely that all will be found in good condition; and we have occasionally known a hunting party, when Buffalo was scarce, compelled to feed on a straggling old bull as tough as leather. For ourselves, this was rather uncomfortable, as we had unfortunately lost our molars long ago.

The Bison is sometimes more abundant in particular districts one year than another, and is probably influenced in its wanderings by the mildness or severity of the weather, as well as by the choice it makes of the best pasturage and most quiet portions of the prairies. While we were at Fort Union, the hunters were during the month of June obliged to go out twenty-five or thirty miles to procure Buffalo meat, although at other times, the animal was quite abundant in sight of the fort. The tramping of a large herd, in wet weather, cuts up the soft clayey soil of the river bottoms, (we do not mean the bottom of rivers,) into a complete mush. One day, when on our journey up the Missouri river, we landed on one of the narrow strips of land called bottoms, which formed the margin of the river and was backed by hills of considerable height at a short distance. At this spot the tracks of these animals were literally innumerable, as far as the eye could reach in every direction, the plain was covered with them; and in some places the soil had been so trampled as to resemble mud or clay, when prepared for making bricks. The trees in the vicinity were rubbed by these buffaloes, and their hair and wool were hanging on the rough bark or lying at their roots. We collected some of this wool, we think it might be usefully worked up into coarse cloth, and consider it worth attention. The roads that are made by these animals, so much resemble the tracks left by a large wagon-train, that the inexperienced traveller may occasionally imagine himself following the course of an ordinary wagon-road. These great tracks run for hundreds of miles across the prairies, and are usually found to lead to some salt-spring, or some river or creek, where the animals can allay their thirst.

The captain of the steamboat on which we ascended the Missouri, informed us, that on his last annual voyage up that river, he had caught several Buffaloes, that were swimming the river. The boat was run close upon them, they were lassoed by a Spaniard, who happened to be on board, and then hoisted on the deck, where they were butchered secundum artem.[17] One day we saw several that had taken to the water, and were coming towards our boat. We passed so near them, that we fired at them, but did not procure a single one. On another occasion, one was killed from the shore, and brought on board, when it was immediately divided among the men. We were greatly surprised to see some of the Indians, that were going up with us, ask for certain portions of the entrails, which they devoured with the greatest voracity. This gluttony excited our curiosity, and being always willing to ascertain the quality of any sort of meat, we tasted some of this sort of tripe, and found it very good, although at first its appearance was rather revolting.

The Indians sometimes eat the carcasses of Buffaloes that have been drowned, and some of those on board the Omega one day asked the captain most earnestly to allow them to land and get at the bodies of three Buffaloes which we passed, that had lodged among the drift-logs and were probably half putrid. In this extraordinary request some of the squaws joined. That, when stimulated by the gnawings of hunger, Indians, or even Whites, should feed upon carrion, is not to be wondered at, since we have many instances of cannibalism and other horrors, when men are in a state of starvation, but these Indians were in the midst of plenty of wholesome food and we are inclined to think their hankering after this disgusting flesh must be attributed to a natural taste for it, probably acquired when young, as they are no doubt sometimes obliged in their wanderings over the prairies in winter, to devour carrion and even bones and hides, to preserve their lives. In the height of the rutting-season, the flesh of the Buffalo bull is quite rank, and unfit to be eaten, except from necessity, and at this time the animal can be scented at a considerable distance.

When a herd of Bisons is chased, although the bulls run with great swiftness their speed cannot be compared with that of the cows and yearling calves. These, in a few moments leave the bulls behind them, but as they are greatly preferred by the hunter, he always (if well mounted) pursues them and allows the bulls to escape. During the winter of 1842 and 43, as we were told, Buffaloes were abundant around Fort Union, and during the night picked up straggling handfuls of hay that happened to be scattered about the place. An attempt was made to secure some of them alive, by strewing hay as a bait, from the interior of the old fort, which is about two hundred yards off, to some distance from the gateway, hoping the animals would feed along into the enclosure. They ate the hay to the very gate; but as the hogs and common cattle were regularly placed there, for security, during the night, the Buffaloes would not enter, probably on account of the various odours issuing from the interior. As the Buffaloes generally found some hay scattered around, they soon became accustomed to sleep in the vicinity of the fort, but went off every morning, and disappeared behind the hills, about a mile off.

One night they were fired at, from a four-pounder loaded with musket-balls. Three were killed, and several were wounded, but this disaster did not prevent them from returning frequently to the fort at night, and they were occasionally shot, during the whole winter, quite near the fort.

As various accounts of Buffalo-hunts have been already written, we will pass over our earliest adventures in that way, which occurred many years ago, and give you merely a sketch of the mode in which we killed them during our journey to the West, in 1843.

One morning in July, our party and several persons attached to Fort Union, (for we were then located there,) crossed the river, landed opposite the fort, and passing through the rich alluvial belt of woodland which margins the river, were early on our way to the adjacent prairie, beyond the hills. Our equipment consisted of an old Jersey wagon, to which we had two horses attached, tandem, driven by Mr. CULBERTSON, principal at the fort. This wagon carried Mr. HARRIS, BELL, and ourselves, and we were followed by two carts, which contained the rest of the party, while behind came the running horses or hunters, led carefully along. After crossing the lower prairie, we ascended between the steep banks of the rugged ravines, until we reached the high undulating plains above. On turning to take a retrospective view, we beheld the fort and a considerable expanse of broken and prairie-land behind us, and the course of the river was seen as it wound along, for some distance. Resuming our advance we soon saw a number of antelopes, some of which had young ones with them. After travelling about ten miles farther we approached the Fox river, and at this point one of the party espied a small herd of Bisons at a considerable distance off. Mr. CULBERTSON, after searching for them with the telescope, handed it to us and showed us where they were. They were all lying down and appeared perfectly unconscious of the existence of our party. Our vehicles and horses were now turned towards them and we travelled cautiously to within about a quarter of a mile of the herd, covered by a high ridge of land which concealed us from their view. The wind was favourable, (blowing towards us,) and now the hunters threw aside their coats, tied handkerchiefs around their heads, looked to their guns, mounted their steeds, and moved slowly and cautiously towards the game. The rest of the party crawled carefully to the top of the ridge to see the chase. At the word of command, given by Mr. CULBERTSON, the hunters dashed forward after the bulls, which already began to run off in a line nearly parallel with the ridge we were upon. The swift horses, urged on by their eager riders and their own impetuosity, soon began to overtake the affrighted animals; two of them separated from the others and were pursued by Mr. CULBERTSON and Mr. BELL; presently the former fired, and we could see that he had wounded one of the bulls. It stopped after going a little way and stood with its head hanging down and its nose near the ground. The blood appeared to be pouring from its mouth and nostrils, and its drooping tail showed the agony of the poor beast. Yet it stood firm, and its sturdy legs upheld its ponderous body as if nought had happened. We hastened toward it but ere we approached the spot, the wounded animal fell, rolled on its side, and expired. It was quite dead when we reached it. In the mean time Mr. BELL had continued in hot haste after the other, and Mr. HARRIS and Mr. SQUIRE had each selected, and were following one of the main party.[18] Mr. BELL shot, and his ball took effect in the buttocks of the animal. At this moment Mr. SQUIRE's horse threw him over his head fully ten feet: he fell on his powder-horn and was severely bruised; he called to some one to stop his horse and was soon on his legs, but felt sick for a few moments. Friend HARRIS, who was perfectly cool, neared his bull, shot it through the lungs, and it fell dead on the spot. Mr. BELL was still in pursuit of his wounded animal and Mr. HARRIS and Mr. SQUIRE joined and followed the fourth, which, however, was soon out of sight. We saw Mr. BELL shoot two or three times, and heard guns fired, either by Mr. HARRIS or Mr. SQUIRE, but the weather was so hot that fearful of injuring their horses they were obliged to allow the bull they pursued to escape. The one shot by Mr. BELL, tumbled upon his knees, got up again, and rushed on one of the hunters, who shot it once more, when it paused, and almost immediately fell dead.

The flesh of the Buffaloes thus killed was sent to the fort in the cart, and we continued our route and passed the night on the prairie, at a spot about half way between the Yellow-Stone and the Missouri rivers. Here, just before sundown, seven more bulls were discovered by the hunters, and Mr. HARRIS, Mr. BELL and Mr. CULBERTSON each killed one. In this part of the prairie we observed several burrows made by the swift fox, but could not see any of those animals although we watched for some time in hopes of doing so. They probably scented our party and would not approach. The hunters on the prairies, either from hunger or because they have not a very delicate appetite, sometimes break in the skull of a buffalo and eat the brains raw. At sunrise we were all up, and soon had our coffee, after which a mulatto man called LAFLEUR, an excellent hunter attached to the American Fur-Company, accompanied Mr. HARRIS and Mr. BELL on a hunt for antelopes, as we wanted no more Buffaloes. After waiting the return of the party, who came back unsuccessful, we broke up our camp and turned our steps homeward.

The Buffalo bulls which have been with their fair ones are at this season wretchedly poor, but some of them, which appear not to have much fondness for the latter, or may have been driven off by their rivals, are in pretty good condition. The prairies are in some places whitened with the skulls of the Buffalo, dried and bleached by the summer's sun and the frosts and snows of those severe latitudes in winter. Thousands are killed merely for their tongues, and their large carcasses remain to feed the wolves and other rapacious prowlers on the grassy wastes.

A large Bison bull will generally weigh nearly two thousand pounds, and a fat cow, about twelve hundred. We weighed one of the bulls killed by our party and found it to reach seventeen hundred and twenty-seven pounds, although it had already lost a good deal of blood. This was an old bull and was not fat; it had probably weighed more at some previous period. We were told that at this season a great many half-breed Indians were engaged in killing Buffaloes and curing their flesh for winter use, on Moose river, about 200 miles north of us.

When these animals are shot at a distance of fifty or sixty yards, they rarely, if ever, charge on the hunters. Mr. CULBERTSON told us he had killed as many as nine bulls from the same spot, unseen by these terrible animals. There are times, however, when they have been known to gore both horse and rider, after being severely wounded, and have dropped down dead but a few minutes afterwards. There are indeed instances of bulls receiving many balls without being immediately killed, and we saw one which during one of our hunts was shot no less than twenty-four times before it dropped.

A bull that our party had wounded in the shoulder, and which was thought too badly hurt to do much harm to any one, was found rather dangerous when we approached him, as he would dart forward at the nearest of his foes, and but that his wound prevented him from wheeling and turning rapidly, he would certainly have done some mischief. We fired at him from our six-barrelled revolving pistol, which, however, seemed to have little other effect than to render him more savage and furious. His appearance was well calculated to appal the bravest, had we not felt assured that his strength was fast diminishing. We ourselves were a little too confident, and narrowly escaped being overtaken by him through our imprudence. We placed ourselves directly in his front, and as he advanced, fired at his head and ran back, not supposing that he could overtake us; but he soon got within a few feet of our rear, with head lowered, and every preparation made for giving us a hoist; the next instant, however, we had jumped aside, and the animal was unable to alter his headlong course quick enough to avenge himself on us. Mr. BELL now put a ball directly through his lungs, and with a gush of blood from the mouth and nostrils, he fell upon his knees and gave up the ghost, falling (as usual) on the side, quite dead.

On another occasion, when the same party were hunting near the end of the month of July, Mr. SQUIRE wounded a bull twice, but no blood flowing from the mouth, it was concluded the wounds were only in the flesh, and the animal was shot by Mr. CULBERTSON, OWEN McKENZIE, and Mr. SQUIRE, again. This renewed fire only seemed to enrage him the more, and he made a dash at the hunters so sudden and unexpected, that Mr. SQUIRE, attempting to escape, rode between the beast and a ravine which was near, when the bull turned upon him, his horse became frightened and leaped down the bank, the Buffalo following him so closely that he was nearly unhorsed; he lost his presence of mind and dropped his gun; he, however, fortunately hung on by the mane and recovered his seat. The horse was the fleetest, and saved his life. He told us subsequently that he had never been so terrified before. This bull was fired at several times after SQUIRE's adventure, and was found to have twelve balls lodged in him when he was killed. He was in very bad condition, and being in the rutting season we found the flesh too rank for our dainty palates and only took the tongue with us.

Soon afterwards we killed a cow in company with many bulls and were at first afraid that they would charge upon us, which in similar cases they frequently do, but our party was too large and they did not venture near, although their angry bellowings and their unwillingness to leave the spot showed their rage at parting with her. As the sun was now sinking fast towards the horizon on the extended prairie, we soon began to make our way toward the camping ground and passed within a moderate distance of a large herd of Buffaloes, which we did not stop to molest but increasing our speed reached our quarters for the night, just as the shadows of the western plain indicated that we should not behold the orb of day until the morrow.

Our camp was near three conical hills called the Mamelles, only about thirty miles from Fort Union, although we had travelled nearly fifty by the time we reached the spot.[19] After unloading and unsaddling our tired beasts, all hands assisted in getting wood and bringing water, and we were soon quietly enjoying a cup of coffee. The time of refreshment to the weary hunter is always one of interest: the group of stalwart frames stretched in various attitudes around or near the blazing watch-fires, recalls to our minds the masterpieces of the great delineators of night scenes; and we have often at such times beheld living pictures, far surpassing any of those contained in the galleries of Europe.

There were signs of grizzly bears around us, and during the night we heard a number of wolves howling among the bushes in the vicinity. The service berry was abundant and we ate a good many of them, and after a hasty preparation in the morning, started again after the Buffaloes we had seen the previous evening. Having rode for some time, one of our party who was in advance as a scout, made the customary signal from the top of a high hill, that Buffaloes were in sight; this is done by walking the hunter's horse backward and forward several times. We hurried on and found our scout lying close to his horse's neck, as if asleep on the back of the animal. He pointed out where he had discovered the game, but they had gone out of sight, and (as he said) were travelling fast, the herd being composed of both bulls and cows. The hunters mounted at once, and galloped on in rapid pursuit, while we followed more leisurely over hills and plains and across ravines and broken ground, at the risk of our necks. Now and then we could see the hunters, and occasionally the Buffaloes, which had taken a direction toward the Fort. At last we reached an eminence from which we saw the hunters approaching the Buffaloes in order to begin the chase in earnest. It seems that there is no etiquette among Buffalo hunters, and this not being understood beforehand by our friend HARRIS, he was disappointed in his wish to kill a cow.[20] The country was not as favourable to the hunters as it was to the flying herd. The females separated from the males, and the latter turned in our direction and passed within a few hundred yards of us without our being able to fire at them. Indeed we willingly suffered them to pass unmolested, as they are always very dangerous when they have been parted from the cows. Only one female was killed on this occasion. On our way homeward we made towards the coupee, an opening in the hills, where we expected to find water for our horses and mules, as our supply of Missouri water was only enough for ourselves.[21]

The water found on these prairies is generally unfit to drink, (unless as a matter of necessity,) and we most frequently carried eight or ten gallons from the river, on our journey through the plains. We did not find water where we expected, and were obliged to proceed about two miles to the eastward, where we luckily found a puddle sufficient for the wants of our horses and mules. There was not a bush in sight at this place, and we collected Buffalo dung to make a fire to cook with. In the winter this prairie fuel is often too wet to burn, and the hunters and Indians have to eat their meat raw. It can however hardly be new to our readers to hear that they are often glad to get any thing, either raw or cooked, when in this desolate region.

Young Buffalo bulls are sometimes castrated by the Indians, as we were told, for the purpose of rendering them larger and fatter; and we were informed, that when full grown they have been shot, and found to be far superior to others in the herd, in size as well as flavour. During severe winters the Buffaloes become very poor, and when the snow has covered the ground for several months to the depth of two or three feet, they are wretched objects to behold. They frequently in this emaciated state lose their hair and become covered with scabs; and the magpies alight on their backs and pick the sores. The poor animals in these dreadful seasons die in great numbers.

A singular trait in the Buffalo when caught young, was related to us, as follows: When a calf is taken, if the person who captures it places one of his fingers in its mouth, it will follow him afterwards, whether on foot or on horseback, for several miles.

We now give a few notes from our journal kept at Fort Union, which may interest our readers.

August 7th, 1843, a Buffalo cow was killed and brought into the fort, and to the astonishment of all, was found to be near her time of calving. This was an extraordinary circumstance at that season of the year.

August 8th, The young Buffaloes have commenced shedding their first (or red) coat of hair, which drops off in patches about the size of the palm of a man's hand. The new hair is dark brownish black. We caught one of these calves with a lasso, and had several men to hold him, but on approaching to pull off some of the old hair, he kicked and bounced about in such a furious manner that we could not get near him. Mr. CULBERTSON had it however taken to the press post, and there it was drawn up and held so closely that we could handle it, and we tore off some pieces of its old pelage, which hung to the side with surprising tenacity.[22]

The process of butchering or cutting up the carcass of the Buffalo is generally performed in a slovenly and disgusting manner by the hunters, and the choicest parts only are saved, unless food is scarce. The liver and brains are eagerly sought for, and the hump is excellent when broiled. The pieces of flesh from the sides are called by the French, fillets, or the depouille; the marrow bones are sometimes cut out, and the paunch is stripped of its covering of fat.

Some idea of the immense number of Bisons to be still seen on the wild prairies, may be formed from the following account, given to us by Mr. KIPP, one of the principals of the American Fur Company. "While he was travelling from Travers' Bay to the Mandan nation in the month of August, in a cart heavily laden, he passed through herds of Buffalo for six days in succession. At another time he saw the great prairie near Fort Clark on the Missouri river, almost blackened by these animals, which covered the plain to the hills that bounded the view in all directions, and probably extended farther."

When the Bisons first see a person, whether white or red, they trot or canter off forty or fifty yards, and then stop suddenly, turn their heads and gaze on their foe for a few moments, then take a course and go off at full speed until out of sight, and beyond the scent of man.

Although large, heavy, and comparatively clumsy, the Bison is at times brisk and frolicksome, and these huge animals often play and gambol about, kicking their heels in the air with surprising agility, and throwing their hinder parts to the right and left alternately, or from one side to the other, their heels the while flying about and their tails whisking in the air. They are very impatient in the fly and mosquito season, and are often seen kicking and running against the wind to rid themselves of these tormentors.

The different Indian tribes hunt the Buffalo in various ways: some pursue them on horseback and shoot them with arrows, which they point with old bits of iron, or old knife blades. They are rarely expert in loading or re-loading guns, (even if they have them,) but in the closely contested race between their horse and the animal, they prefer the rifle to the bow and arrow. Other tribes follow them with patient perseverance on foot, until they come within shooting distance, or kill them by stratagem.

The Mandan Indians chase the Buffalo in parties of from twenty to fifty, and each man is provided with two horses, one of which he rides, and the other being trained expressly for the chase, is led to the place where the Buffaloes are started. The hunters are armed with bows and arrows, their quivers containing from thirty to fifty arrows according to the wealth of the owner. When they come in sight of their game, they quit the horses on which they have ridden, mount those led for them, ply the whip, soon gain the flank or even the centre of the herd, and shoot their arrows into the fattest, according to their fancy. When a Buffalo has been shot, if the blood flows from the nose or mouth, he is considered mortally wounded; if not, they shoot a second or a third arrow into the wounded animal.

The Buffalo, when first started by the hunters, carries his tail close down between the legs; but when wounded, he switches his tail about, especially if intending to fight his pursuer, and it behooves the hunter to watch these movements closely, as the horse will often shy, and without due care the rider may be thrown, which when in a herd of Buffalo is almost certain death. An arrow will kill a Buffalo instantly if it takes effect in the heart, but if it does not reach the right spot, a dozen arrows will not even arrest one in his course, and of the wounded, many run out of sight and are lost to the hunter.

At times the wounded Bison turns so quickly and makes such a sudden rush upon the hunter, that if the steed is not a good one and the rider perfectly cool, they are overtaken, the horse gored and knocked down, and the hunter thrown off and either gored or trampled to death. But if the horse is a fleet one, and the hunter expert, the Bison is easily outrun and they escape. At best it may be said that this mode of Buffalo hunting is dangerous sport, and one requires both skill and nerve to come off successfully.

The Gros Ventres, Blackfeet and Assinaboines often take the Buffalo in large pens, usually called parks, constructed in the following manner.

Two converging fences built of sticks, logs and brushwood are made, leading to the mouth of a pen somewhat in the shape of a funnel. The pen itself is either square or round, according to the nature of the ground where it is to be placed, at the narrow end of the funnel, which is always on the verge of a sudden break or precipice in the prairie ten or fifteen feet deep, and is made as strong as possible. When this trap is completed, a young man very swift of foot starts at daylight, provided with a Bison's hide and head, to cover his body and head when he approaches the herd that is to be taken, on nearing which he bleats like a young Buffalo calf, and makes his way slowly towards the mouth of the converging fences leading to the pen. He repeats this cry at intervals, the Buffaloes follow the decoy, and a dozen or more of mounted Indians at some distance behind the herd gallop from one side to the other on both their flanks, urging them by this means to enter the funnel, which having done, a crowd of men, women and children come and assist in frightening them, and as soon as they have fairly entered the road to the pen beneath the precipice, the disguised Indian, still bleating occasionally, runs to the edge of the precipice, quickly descends, and makes his escape, climbing over the barricade or fence of the pen beneath, while the herd follow on till the leader (probably an old bull) is forced to leap down into the pen, and is followed by the whole herd, which is thus ensnared, and easily destroyed even by the women and children, as there is no means of escape for them.

This method of capturing the Bison is especially resorted to in October and November, as the hide is at that season in good condition and saleable, and the meat can be preserved for the winter supply. When the Indians have thus driven a herd of Buffalo into a pen, the warriors all assemble by the side of the enclosure, the pipe is lighted, and the chiefs smoke to the honour of the Great Spirit, to the four points of the compass, and to the herd of Bisons. As soon as this ceremony has ended, the destruction commences, guns are fired and arrows shot from every direction at the devoted animals, and the whole herd is slaughtered before the Indians enter the space where the Buffaloes have become their victims. Even the children shoot tiny arrows at them when thus captured, and try the strength of their young arms upon them.

It sometimes happens, however, that the leader of the herd becomes alarmed and restless while driving to the precipice, and should the fence be weak, breaks through, and the whole drove follow and escape. It also sometimes occurs, that after the Bisons are in the pen, which is often so filled that they touch each other, the terrified crowd swaying to and fro, their weight against the fence breaks it down, and if the smallest gap is made, it is immediately widened, when they dash through and scamper off, leaving the Indians in dismay and disappointment. The side fences for the purpose of leading the Buffaloes to the pens extend at times nearly half a mile, and some of the pens cover two or three hundred yards of ground. It takes much time and labour to construct one of these great traps or snares, as the Indians sometimes have to bring timber from a considerable distance to make the fences and render them strong and efficient.

The Bison has several enemies: the worst is, of course, man; then comes the grizzly bear; and next, the wolf. The bear follows them and succeeds in destroying a good many; the wolf hunts them in packs, and commits great havoc among them, especially among the calves and the cows when calving. Many Buffaloes are killed when they are struggling in the mire on the shores of rivers where they sometimes stick fast, so that the wolves or bears can attack them to advantage; eating out their eyes and devouring the unresisting animals by piecemeal.

When we were ascending the Missouri river, the first Buffaloes were heard of near Fort Leavenworth, some having a short time before been killed within forty miles of that place. We did not, however, see any of these animals until we had passed Fort Croghan, but above this point we met with them almost daily, either floating dead on the river, or gazing at our steamboat from the shore.

Every part of the Bison is useful to the Indians, and their method of making boats, by stretching the raw hide over a sort of bowl-shaped frame work, is well known. These boats are generally made by the women, and we saw some of them at the Mandan village. The horns are made into drinking vessels, ladles, and spoons. The skins form a good bed, or admirable covering from the cold, and the flesh is excellent food, whether fresh or dried or made into pemmican; the fat is reduced and put up in bladders, and in some cases used for frying fish, &c.

The hide of the Buffalo is tanned or dressed altogether by the women, or squaws, and the children; the process is as follows: The skin is first hung on a post, and all the adhering flesh taken off with a bone, toothed somewhat like a saw; this is performed by scraping the skin downwards, and requires considerable labour. The hide is then stretched on the ground and fastened down with pegs; it is then allowed to remain till dry, which is usually the case in a day or two. After it is dry the flesh side is pared down with the blade of a knife fastened in a bone, called a grate, which renders the skin even and takes off about a quarter of its thickness. The hair is taken off with the same instrument and these operations being performed, and the skin reduced to a proper thickness, it is covered over either with brains, liver or grease, and left for a night. The next day the skin is rubbed and scraped either in the sun or by a fire, until the greasy matter has been worked into it, and it is nearly dry; then a cord is fastened to two poles and over this the skin is thrown, and pulled, rubbed and worked until quite dry; after which it is sewed together around the edges excepting at one end; a smoke is made with rotten wood in a hole dug in the earth, and the skin is suspended over it, on sticks set up like a tripod, and thoroughly smoked, which completes the tanning and renders the skin able to bear wet without losing its softness or pliability afterwards.

Buffalo robes are dressed in the same manner, only that the hair is not removed and they are not smoked. They are generally divided into two parts: a strip is taken from each half on the back of the skin where the hump was, and the two halves, or sides, are sewed together after they are dressed, with thread made of the sinews of the animal; which process being finished, the robe is complete and ready for market.

The scrapings of the skins, we were informed, are sometimes boiled with berries, and make a kind of jelly which is considered good food in some cases by the Indians. The strips cut off from the skins are sewed together and make robes for the children, or caps, mittens, shoes, &c. The bones are pounded fine with a large stone and boiled, the grease which rises to the top is skimmed off and put into bladders. This is the favourite and famous marrow grease, which is equal to butter. The sinews are used for stringing their bows, and are a substitute for thread; the intestines are eaten, the shoulder-blades made into hoes, and in fact (as we have already stated) nothing is lost or wasted, but every portion of the animal, by the skill and industry of the Indians, is rendered useful.

Balls are found in the stomach of the Buffalo, as in our common domestic cattle.

Since reading the above letter, we recollect that the Buffalo calves that were kept at Fort Union, though well fed every day, were in the habit of sucking each other's ears for hours together.[23]

There exists a singular variety of the Bison, which is however very scarce, and the skin of which is called by both the hunters and fur traders a "beaver robe."[24] These are valued so highly that some have sold for more than three hundred dollars. Of this variety Mr. Culbertson had the goodness to present us with a superb specimen, which we had lined with cloth, and find a most excellent defence against the cold, whilst driving in our wagon during the severity of our northern winters.

GEOGRAPHICAL DISTRIBUTION.

The range of the Bison is still very extensive; but although it was once met with on the Atlantic coast, it has, like many others, receded and gone west and south, driven onward by the march of civilization and the advance of the axe and plough. His habits, as we have seen, are migratory, and the extreme northern and southern limits of the wandering herds not exactly defined. Authors state, that at the time of the first settlement of Canada it was not known in that country, and Sagard Theodat mentions having heard that bulls existed in the far west, but saw none himself. According to Dr. Richardson, Great Slave Lake, latitude 60°, was at one time the northern boundary of their range; but of late years, according to the testimony of the natives, they have taken possession of the flat limestone district of Slave Point on the north side of that lake, and have wandered to the vicinity of Great Marten Lake, in latitude 63° or 64°. The Bison was not known formerly to the north of the Columbia river on the Pacific coast, and Lewis and Clark found Buffalo robes were an important article of traffic between the inhabitants of the east side and those west of the Rocky mountains.

The Bison is spoken of by HERNANDEZ as being found in New Spain or Mexico, and it probably extended farther south. LAWSON speaks of two Buffaloes that were killed in one season on Cape Fear river, in North Carolina. The Bison formerly existed in South Carolina on the seaboard, and we were informed that from the last herd seen in that State, two were killed in the vicinity of Columbia. It thus appears that at one period this animal ranged over nearly the whole of North America.

At the present time, the Buffalo is found in vast herds in some of the great prairies, and scattered more sparsely nearly over the whole length and breadth of the valleys east and west that adjoin the Rocky Mountain chain.

N°12.

PLATE LIX

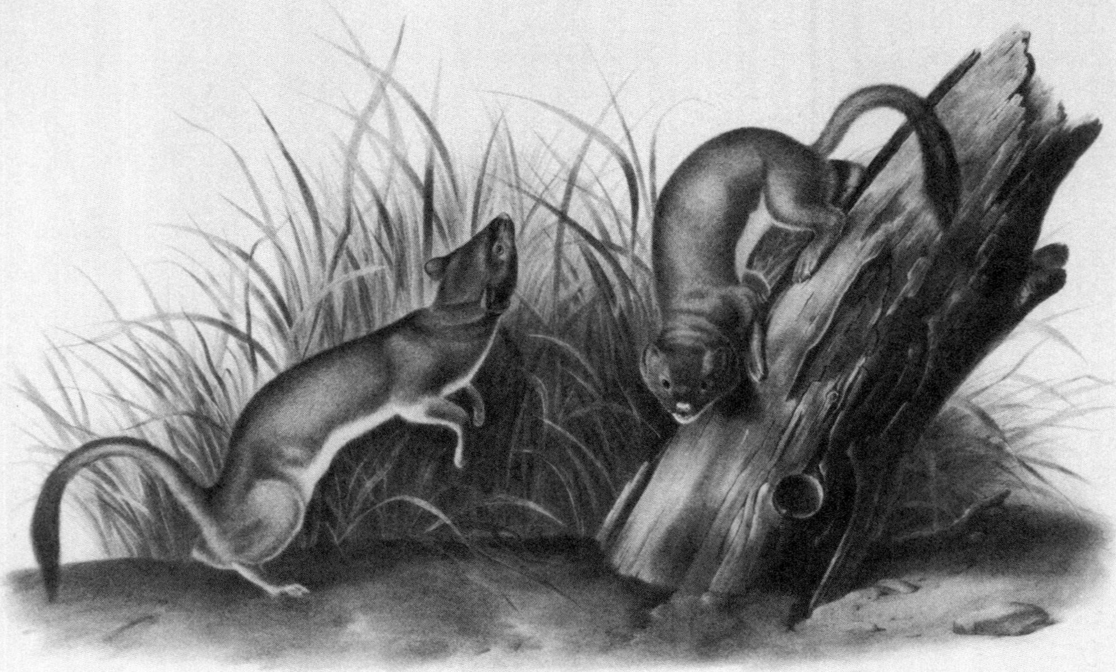

PUTORIUS ERMINEA, LINN.

WHITE WEASEL, STOAT.

White Weasel

ONE OF OUR AUTHORS' PURPOSES IS TO TEACH FACTUAL natural history so that farmers and others can make more informed decisions about pest control. The contrast between the purity and beauty of the ermine's winter color and its alarming bloodthirstiness, however, presents a particular challenge, which the authors meet by arguing that Providence endowed the white weasel with the instinct to kill in order to help farmers whose crops and produce are preyed upon by many species of rodents that the weasel kills. Besides, human condemnation of the apparently unnecessary killing committed by the ermine is quite hypocritical, coming as it does from a species prodigious in sport killing in its own right. Hence this essay's evident delight in some of the bloodiest images in all of the quadruped essays.

---

15. John James Audubon, J. T. Bowen, lithographer, *White Weasel; Stoat*, 1845–46, hand-colored lithograph, plate 59 of *The Viviparous Quadrupeds of North America*. Courtesy of the Jule Collins Smith Museum of Fine Art, Auburn University; Louise Hauss and David Brent Miller Audubon Collection.

PUTORIUS ERMINEA.—LINN.

WHITE WEASEL.—STOAT.

PLATE LIX.—MALE AND FEMALE IN SUMMER PELAGE

HABITS.

The name of Ermine is associated with the pride of state and luxury, its fur having from time immemorial been the favourite ornament of the robes of princes, judges and prelates. From its snowy whiteness it is emblematic of the purity which they ought to possess.

To us the Ermine, in its winter dress, has always appeared strikingly beautiful. On a wintry day, when the earth was covered with a broad sheet of snow, our attention has sometimes been arrested by this little animal peering out from a log heap, or the crevices of a stone fence; its eyes in certain shades of light appearing like sapphires, its colour vieing in whiteness and brilliancy with the snowy mantle of the surrounding landscape.

Graceful in form, rapid in his movements, and of untiring industry, he is withal a brave and fearless little fellow; conscious of security within the windings of his retreat among the logs, or heap of stones, he permits us to approach him to within a few feet, then suddenly withdraws his head; we remain still for a moment, and he once more returns to his post of observation, watching curiously our every motion, seeming willing to claim association so long as we abstain from becoming his persecutor.

Yet with all these external attractions, this little Weasel is fierce and bloodthirsty, possessing an intuitive propensity to destroy every animal and bird within its reach, some of which, such as the American rabbit, the ruffed grouse, and domestic fowl, are ten times its own size. It is a notorious and hated depredator of the poultry house, and we have known forty well grown fowls to have been killed in one night by a single Ermine. Satiated with the blood of probably a single fowl, the rest, like the flock slaughtered by the wolf in the sheepfold, were destroyed in obedience to a law of nature, an instinctive propensity to kill. We have traced the footsteps of this bloodsucking little animal on the snow, pursuing the trail of the American rabbit, and although it could not overtake its prey by superior speed, yet the timid hare soon took refuge in the hollow of a tree, or in a hole dug by the marmot, or skunk. Thither it was pursued by the Ermine, and destroyed, the skin and other remains at the mouth of the burrow bearing evidence of the fact. We observed an Ermine, after having captured a hare of the above species, first behead it and then drag the body some twenty yards over the fresh fallen snow, beneath which it was concealed, and the snow tightly pressed over it; the little prowler displaying thereby a habit of which we became aware for the first time on that occasion. To avoid a dog that was in close pursuit, it mounted a tree and laid itself flat on a limb about twenty feet from the ground, from which it was finally shot. We have ascertained by successful experiments, repeated more than a hundred times, that the Ermine can be employed, in the manner of the ferret of Europe, in driving our American rabbit from the burrow into which it has retreated. In one instance, the Ermine employed had been captured only a few days before, and its canine teeth were filed in order to prevent its destroying the rabbit; a cord was placed around its neck to secure its return. It pursued the hare through all the windings of its burrow and forced it to the mouth, where it could be taken in a net, or by the hand. In winter, after a snow storm, the ruffed grouse has a habit of plunging into the loose snow, where it remains at times for one or two days. In this passive state the Ermine sometimes detects and destroys it. In an unsuccessful attempt at domesticating this grouse by fastening its feet to a board in the mode adopted with the stool pigeon, and placing it high on a shelf, an Ermine which we had kept as a pet, found its way by the curtains of the window and put an end to our experiment by eating off the head of our grouse.

Notwithstanding all these mischievous and destructive habits, it is doubtful whether the Ermine is not rather a benefactor than an enemy to the farmer, ridding his granaries and fields of many depredators on the product of his labour, that would devour ten times the value of the poultry and eggs which, at long and uncertain intervals, it occasionally destroys. A mission appears to have been assigned it by Providence to lessen the rapidly multiplying number of mice of various species and the smaller rodentia.

The white-footed mouse is destructive to the grains in the wheat fields and in the stacks, as well as the nurseries of fruit trees. LE CONTE's pine-mouse is injurious to the Irish and sweet potato crops, causing more to rot by nibbling holes into them than it consumes, and WILSON's meadow-mouse lessens our annual product of hay by feeding on the grasses, and by its long and tortuous galleries among their roots.

Wherever an Ermine has taken up its residence, the mice in its vicinity for half a mile round have been found rapidly to diminish in number. Their active little enemy is able to force its thin vermiform body into the burrows, it follows them to the end of their galleries, and destroys whole families. We have on several occasions, after a light snow, followed the trail of this weasel through fields and meadows, and witnessed the immense destruction which it occasioned in a single night. It enters every hole under stumps, logs, stone heaps and fences, and evidences of its bloody deeds are seen in the mutilated remains of the mice scattered on the snow. The little chipping or ground squirrel, *Tamias Lysteri*, takes up its residence in the vicinity of the grain fields, and is known to carry off in its cheek pouches vast quantities of wheat and buckwheat, to serve as winter stores. The Ermine instinctively discovers these snug retreats, and in the space of a few minutes destroys a whole family of these beautiful little *Tamiæ*; without even resting awhile until it has consumed its now abundant food its appetite craving for more blood, as if impelled by an irresistible destiny it proceeds in search of other objects on which it may glut its insatiable vampire-like thirst. The Norway rat and the common house-mouse take possession of our barns, wheat stacks, and granaries, and destroy vast quantities of grain. In some instances the farmer is reluctantly compelled to pay even more than a tithe in contributions towards the support of these pests. Let however an Ermine find its way into these barns and granaries, and there take up its winter residence, and the havoc which is made among the rats and mice will soon be observable. The Ermine pursues them to their farthest retreats, and in a few weeks the premises are entirely free from their depredations. We once placed a half domesticated Ermine in an outhouse infested with rats, shutting up the holes on the outside to prevent their escape. The little animal soon commenced his work of destruction. The squeaking of the rats was heard throughout the day. In the evening, it came out licking its mouth, and seeming like a hound after a long chase, much fatigued. A board of the floor was raised to enable us to ascertain the result of our experiment, and an immense number of rats were observed, which, although they had been killed on different parts of the building, had been dragged together, forming a compact heap.

The Ermine is then of immense benefit to the farmer. We are of the opinion that it has been over-hated and too indiscriminately persecuted. If detected in the poultry house, there is some excuse for destroying it, as, like the dog that has once been caught in the sheepfold, it may return to commit farther depredations; but when it has taken up its residence under stone heaps and fences, in his fields, or his barns, the farmer would consult his interest by suffering it to remain, as by thus inviting it to a home, it will probably destroy more formidable enemies, relieve him from many petty annoyances, and save him many a bushel of grain.

Let us not too hastily condemn the little Ermine for its bloodthirsty propensities. It possesses well-developed canine teeth, and obeys an instinct of nature. Man, with organs not so decidedly carnivorous, and possessed of the restraining powers of reason and conscience, often commits a wanton havoc on the inferior animals, not so much from want of food, as from a mere love of sport. The buffalo and the elk he has driven across the Mississippi, and their haunts are now restricted to the prairies of the far West. Even now thousands are slaughtered for amusement, and their tongues only are used, whilst their carcasses are left to the wolves. He fills his game bag with more woodcock, partridges and snipe, than he requires; his fishing-rod does not remain idle even after he has provided a full meal for his whole family; and our youngsters are taught to shoot the little warbler and the sparrow as a preparatory training for the destruction of larger game.

The Ermine is far from being shy in its habits. It is not easily alarmed, and becomes tolerably tame when taken young, for we have on several occasions succeeded in our attempts at domesticating it, but it appeared to us that these pets were not quite as gentle as many ferrets that we have seen in Europe. When not kept in confinement, they were apt to stray off into the fields and woods, and finally became wild. The tracks of this species on the snow are peculiar, exhibiting only two footprints, placed near each other, the succeeding tracks being far removed, giving evidences of long leaps. We have frequently observed where it had made long galleries in the deep snow for twenty or thirty yards, and thus in going from one burrow to another, instead of travelling over the surface, it had constructed for itself a kind of tunnel beneath.

The Ermine is easily taken in any kind of trap. We have on several occasions, when observing one peeping at us from its secure hole in the wall, kept it gazing until a servant brought a box trap baited with a bird or piece of meat, which was placed within a few feet of its retreat. The Ermine, after eyeing the trap for a few moments, gradually approached it, then after two or three hasty springs backwards returned stealthily into the trap, seized the bait, and was caught. We find in our note-book the following memorandum: "On the 19th June, 1846, we baited a large wire trap with maize: on visiting the trap on the following day we found it had caught seven young rats and a Weasel; the throats of the former had all been cut by the Weasel, and their blood sucked; but what appeared strange to us, the Weasel itself was also dead. The rats had been attracted by the bait: the Weasel went into the trap and killed them; and whether it met its death by excessive gluttony, or from a wound inflicted by its host of enemies, we are unable to determine."

This species does not appear to be very abundant any where. We have seldom found more than two or three on any farm in the Northern or Eastern States. We have ascertained that the immense number of tracks often seen in the snow in particular localities were made by a single animal, as by capturing one, no signs of other individuals were afterwards seen. We have observed it most abundant in stony regions: in Dutchess and Ontario counties in New-York, on the hills of Connecticut and Vermont, and at the foot of the Alleghanies in Pennsylvania and Virginia. It is solitary in its habits, as we have seldom seen a pair together except in the rutting season. A family of young, however, are apt to remain in the same locality till autumn. In winter they separate, and we are inclined to think that they do not hunt in couples or in packs like the wolf, but that, like the bat and the mink, each individual pursues its prey without copartnership, and hunts for its own benefit.

The only note we have ever heard uttered by the Ermine is a shrill querulous cry: this was heard only when it was suddenly alarmed, or received a hurt, when its sharp scream was always attended with an emission of the offensive odour with which nature has furnished it as a means of defence. Although nocturnal in its habits, the Ermine is frequently met with at all hours of the day, and we have seen it in pursuit of the common rabbit under a bright shining sun at noon-day.

We doubt whether the Ermine ever digs its own burrows, and although when fastened to a chain in a state of confinement we observed it digging shallow holes in the ground, its attempts at burrowing were as awkward as those of the rat; the nests we have seen were placed under roots of trees, in stone heaps, or in the burrows of the ground squirrel, from which the original occupants had been expelled. The rutting season is in winter, from the middle of February to the beginning of March. The young, from four to seven, are born in May, in the latitude of New-York. We were informed by a close observer, that in the upper country of Carolina, the young had been seen as early as the 25th of March. The colour of the young when a week old, is pale yellow on the upper surface.

The Ermine avoids water, and if forcibly thrown into it, swims awkwardly like a cat. It does not, like the fisher and pine marten, pursue its prey on trees, and seems never to ascend them from choice; but from dire necessity, when closely pursued by its implacable enemy, the dog.

One of the most singular characteristics of this species, viz., its change of colour from brown in summer to pure white in winter, and from white in spring to its summer colour, remains to be considered. It is well known that about the middle of October the Ermine gradually loses its brown summer-coat and assumes its white winter-pelage, which about the middle of March is replaced by the usual summer colour.

As far as our observations have enabled us to form an opinion on this subject, we have arrived at the conclusion, that the animal sheds its coat twice a year, i.e., at the periods when these semi-annual changes take place. In autumn, the summer hair gradually and almost imperceptibly drops out, and is succeeded by a fresh coat of hair, which in the course of two or three weeks becomes pure white; while in the spring the animal undergoes its change from white to brown in consequence of shedding its winter coat, the new hairs then coming out brown. We have in our possession a specimen captured in November, in which the change of colour has considerably advanced, but is not completed. The whole of the under surface, the sides, neck and body to within half an inch of the back, together with the legs, are white, as well as the edges of the ears. On the upper surface, the nose, forehead, neck, and an irregular line on the back, together with a spot on the outer surface of the fore-leg, are brown, showing that these parts change colour last.

In reference to the change of pelage and colour as exhibited in spring, we add some notes made by the senior author of this work, in March, 1842, on a specimen sent to him alive by OGDEN HAMMOND, Esq.

The Weasel this evening, the 6th of March, began to show a change of colour; we were surprised to see that all around its nose, the white hair of its winter dress had changed suddenly to a silky black hue, and this extended to nearly between the ears. Here and there also were seen small spots of black about its rump, becoming more apparent toward the shoulders, and forming as it were a ridge along the back of the animal.

March 10th. By noon the change was wonderfully manifested. The whole upper surface of the head had become black to the eye, as well as the ridge of the back, the latter part having become quite clouded, and showing an indescribable motley mixture of closely-blended white, black, and blackish brown.

18th. This day the change of colour reached the root of the tail, where it formed a ring of about one inch, of the same reddish black colour. All other parts remained white, slightly tinged with pale lemon colour. It fed, as we perceived, more voraciously than ever since we have had it in our possession. No less than three or four mice were devoured to-day, and what is very strange, it left no remains of either hair, skull, feet, or any other part of these animals; and on this day, the 18th of March, it ate a very large piece of fresh beef, weighing nearly half a pound.

19th. Last night our Weasel made great progress, for this morning we found the coloured ridge on the back broader and less mottled. The posterior coloured part of the head had joined the ridge of the back. The posterior part of the hind legs had become brown, and we observed a small spot the size of a sixpence on each upper part of the thighs. At this juncture we think the animal is beautiful.

22d. This morning we found all the white hair on the outward ridge of the back had fallen, and portions of the thighs and shoulders had become broader; the coloured parts were of a rich brown to the very nose, and there existed indications of small dark spots coming from the sides of the belly, somewhat like so many beads strung on a thread, separated from the lower edge of the back ridge by a line of white of about half an inch. The weasel continues as lively as ever. When asleep, it curls its body around, and the tail encircles the whole animal, the end covering the nose. The eyes appear to be kept carefully uncovered. The general tints of the coloured parts of this Weasel were very much darker than in any other specimen which we have in our collection. When angry, it emitted a sharp shrill cry, and snapped with all its might at the objects presented to it. It was very cleanly in its habits, never rendering its sleeping apartment disagreeable.

28th. Our Weasel got out of its cage by pushing the wires apart, passing through an aperture not exceeding five-eighths of an inch, as we suppose by putting its head diagonally through the bars. The door and windows of our room were closed, however, and, when we entered, our little fellow looked at us as if well acquainted, but soon ran behind a box. It devoured last night at least half a pound of beef, kept in the room for its day's ration. We placed the cage, with the door open, on the floor, and by walking round the box that concealed it, the animal was induced to run towards the cage, and was again secured in it.

We have often observed this species whilst retreating; if near its place of concealment, it does so backwards, and we observed the same movement when it passed from one section of its cage to the other, dragging its food and concealing it among the straw. While we were sitting at a distance from its retreat, it proceeded by leaps very swiftly to within two or three feet of us, when it suddenly threw itself round and retreated backward, as mentioned before.

The purplish brown was now augmented on the thighs and shoulders to the knee joints, no white hairs remaining mixed with those that were coloured. Beneath the jaws, separate small brown spots appeared at equal distances, leaving an intermediate space of white, as was the case along the flanks. The root of the tail had acquired no farther change. Since last week our animal has diffused a very strong disagreeable odour, musky and fetid, which may be attributable to this being its breeding season; we observed that the smell was more disagreeable in the mornings and evenings, than at mid-day.

April.—On paying our accustomed visit to our Weasel this evening, we found it dead, which put a stop to any further observation of its habits. Its measurements are as follows:

INCHES.

From point of nose to end of tail,—10½
Tail (vertebrae),—5
Tail to end of hair,—6
Height of ear,—¾
Breadth of ear,—⅝
Fore claws and hind claws stretching out to the black hair of the tail,—14¼

N°13

PLATE LXI

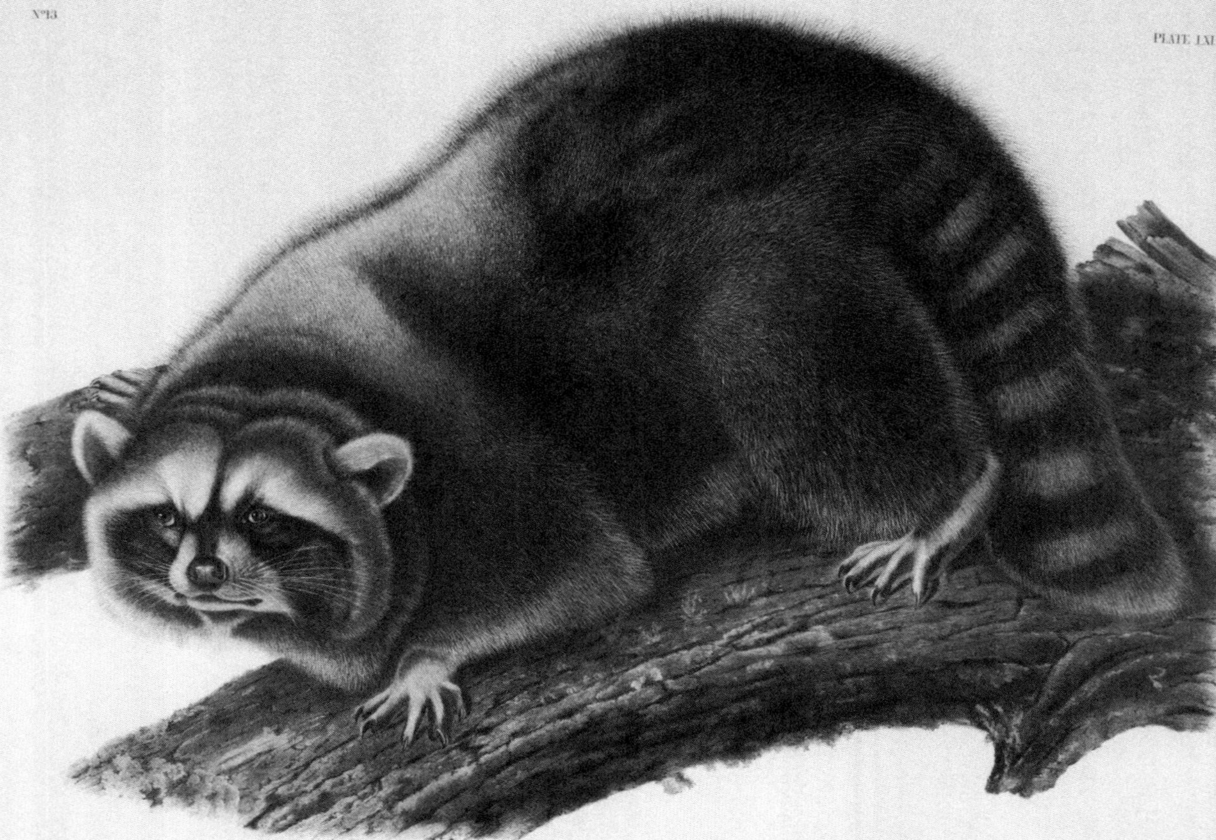

PROCYON LOTOR, CUVIER

RACCOON.

MALE.

Raccoon

THE CENTRAL IMAGE OF "RACCOON" IS ONE SEMITAME SPECImen chained outside among other captive wild and domestic animals. The authors observed this captive raccoon and then developed their description of the species as cunning, curious, and playful, all of which enable the omnivorous raccoon to discover a wide variety of foods in many differing situations. The image of the pet raccoon, even though chained, becomes one of the most charming in all of the mammal essays as the raccoon puzzles over how to make its paws meet inside the fish it is eating and as it handles the young fox to "find out the nature of the animal." The essay shows that the eye of the skilled naturalist can detect behavior adapted for survival in the wild, even in a captive animal. The essay also warns the naturalist against too easily accepting as factual reports of animal behavior from locals whose range of observation and knowledge of natural laws are necessarily limited.

---

16. John James Audubon, J. T. Bowen, lithographer, *Raccoon*, 1845–46, hand-colored lithograph, plate 61 of *The Viviparous Quadrupeds of North America*. Courtesy of the Jule Collins Smith Museum of Fine Art, Auburn University; Louise Hauss and David Brent Miller Audubon Collection.

PROCYON LOTOR.—LINN.

RACCOON.

PLATE LXI.—MALE AND YOUNG.

HABITS.

The Raccoon is a cunning animal, is easily tamed, and makes a pleasant monkey-like pet. It is quite dexterous in the use of its fore-feet, and will amble after its master in the manner of a bear, and even follow him into the streets. It is fond of eggs, and devours them raw or cooked with avidity, but prefers them raw of course, and if it finds a nest will feast on them morning, noon and night without being satiated. It will adroitly pick its keeper's pockets of anything it likes to eat, and is always on the watch for dainties. The habits of the muscles (*unios*) that inhabit our fresh water rivers are better known to the Raccoon than to most conchologists, and their flavour is as highly relished by this animal as is that of the best bowl of clam soup by the epicure in that condiment.

Being an expert climber, the Raccoon ascends trees with facility and frequently invades the nest of the woodpecker, although it may be secure against ordinary thieves, by means of his fore-feet getting hold of the eggs or the young birds. He watches too the soft-shelled turtle when she is about to deposit her eggs, for which purpose she leaves the water and crawling on to the white sand-bar, digs a hole and places them underneath the heated surface. Quickly does the rogue dig up the elastic ova, although ever so carefully covered, and appropriate them to his own use, notwithstanding the efforts of the luckless turtle to conceal them.

Sometimes, by the margin of a pond, shrouded, or crouched among tall reeds and grasses, Grimalkin-like, the Raccoon lies still as death, waiting with patience for some ill-fated duck that may come within his reach. No negro on a plantation knows with more accuracy when the corn (maize) is juicy and ready for the connoisseur in roasting ears, and he does not require the aid of fire to improve its flavour, but attacks it more voraciously than the squirrel or the blackbird, and is the last to quit the cornfield.

The favourite resorts of the Raccoon are retired swampy lands well covered with lofty trees, and through which are small water-courses. In such places its tracks may be seen following the margins of the bayous and creeks, which it occasionally crosses in search of frogs and muscles which are found on their banks. It also follows the margins of rivers for the same purpose, and is dexterous in getting at the shell-fish, notwithstanding the hardness of the siliceous covering with which nature has provided them. In dry seasons, the receding waters sometimes leave the muscles exposed to the heat of the sun, which destroys their life and causes their shells to open, leaving them accessible to the first animal or bird that approaches.

In the dreary months of winter should you be encamped in any of the great Western forests, obliged by the pitiless storm to remain for some days, as we have been, you will not be unthankful if you have a fat Raccoon suspended on a tree above your camp, for when kept awhile, the flesh of this species is both tender and well-flavoured.

The Raccoon when full grown and in good condition we consider quite a handsome animal. We have often watched him with interest, cautiously moving from one trunk to another to escape his view. His bright eye, however, almost invariably detected us ere we could take aim at him, and he adroitly fled into a hollow tree and escaped from us.

We once met with one of these animals whilst we were travelling on horseback from Henderson to Vincennes, on the edge of a large prairie in a copse, and on approaching it ran up a small sapling from which we shook it off with ease; but as soon as it reached the ground it opened its mouth and made directly towards us, and looked so fierce, that drawing a pistol from our holsters, we shot it dead when it was only a few feet from us.

The young are at their birth quite small; (about the size of a half-grown rat;) some that we saw in Texas were not more than two days old and were kept in a barrel. They uttered a plaintive cry not unlike the wail of an infant.

The Raccoon usually produces from four to six young at a time, which are generally brought forth early in May, although the period of their littering varies in different latitudes.

When the Indian corn is ripening, the Raccoons invade the fields to feast on the rich milky grain, as we have just stated, and as the stalks are too weak to bear the weight of these marauders, they generally break them down with their fore-paws, tear off the husks from the ears, and then munch them at their leisure. During this inviting season, the Raccoon is not the only trespasser on the corn fields, but various animals are attracted thither to receive their portion, and even the merry school-boy shares the feast with them, at the risk of paying for his indulgence by incurring the necessity of a physician's prescription the next day. The havoc committed in the Western States by squirrels and other animals is almost incalculable, and no vigilance of the farmer can guard against the depredations of these hungry intruders, which extend from farm to farm, and even penetrate to those embosomed in the forests, where settlements are few and far between.

The Raccoon is not strictly a nocturnal animal; and although it generally visits the corn fields at night, sometimes feeds on the green corn during the day; we have seen it thus employed during the heat of summer, and it will occasionally enter a poultry house at mid-day, and destroy many of the feathered inhabitants, contenting itself with the head and blood of the fowls it kills.

The nest or lair of the Raccoon is usually made in the hollow of some broken branch of a tree. When tamed, these animals are seldom induced to lie or sleep on a layer of straw.

There exists a species of oyster in the Southern States of inferior quality which bears the name of Raccoon Oyster: it lies imbedded in masses in the shallow waters of the rivers. These oysters are covered by high tides, but are exposed at low water. On these the Raccoons are fond of feeding, and we have on several occasions seen them on the oyster banks. We have however never had an opportunity of ascertaining by personal observation the accuracy of a statement which we have frequently heard made with great confidence, viz., that the Raccoon at low tide in endeavouring to extricate these oysters from the shell, is occasionally caught by the foot in consequence of the closing of the valve of the shell fish, when numbers of these being clustered and imbedded together, the Raccoon cannot drag them from their bed, and the returning tide drowns him.

The naturalist has many difficulties to encounter when inquiring into facts connected with his pursuit: every one acquainted with the habits of even our common species must know, that the information gained from most of those who reside near their localities, from their want of particular observation, is generally very limited, and probably the most interesting knowledge gained by such queries, would be the result of a comparison of the accounts given at different places. From the Alleghany mountains, the swamps of Louisiana, and the marshes of Carolina, we have received nearly the same history of the cunning manœuvres and sly tricks of the Raccoon in procuring food.

We add the following notes on a Raccoon kept for a considerable time in a tame state or partially domesticated.

When it first came into our possession it was about one-third grown. By kind treatment it soon became very docile, but from its well known mischievous propensities we always kept it chained.

It was truly omnivorous: never refusing any thing eatable, vegetable or animal, cooked or uncooked, all was devoured with equal avidity. Of some articles however it seemed particularly fond: as sugar, honey, chestnuts, fish and poultry. The animal would become almost frantic when either of the two first was placed near it, but beyond its reach. No means would be left untried to obtain the dainty morsel. It would rush forward as far as the chain permitted, and stretch out a fore-paw toward the object of its wishes to its utmost extent, which failing to reach it, the other was extended; again disappointed, the hind limbs were tried in succession, by which there was a nearer approach to the food, on account of the animal being chained by the neck.

On being offered food when hungry, or roused up suddenly from any cause, or when in active play, the eye was of a lustrous green, changing apparently the whole countenance.

It had a strong propensity to roll food and other things under its paws; segars in particular, especially when lighted. We have observed a similar propensity in young bears.

On placing a pail of water within its reach, it ran to it, and after drinking would examine the contents to the bottom with the fore-paws, seemingly expecting to find some fish or frog. If any thing was found it was speedily brought to the surface and scrutinized. We have seen it throw chips, bits of china and pebbles, &c., into the pail, and then fish them out for amusement, but never saw it put a particle of its food in to soak, except in a few instances when it threw in hard corn, but we do not think it was for this purpose.

After playing for a short time in the water it would commonly urinate in it and then upset the pail.

We gave it a fish weighing two pounds. The Raccoon turned it in all directions in search of a convenient point of attack. The mouth, nose, fins, vent, &c., were tried. At length an opening was made at the vent, into which a paw was deeply inserted; the intestines were withdrawn and eaten with avidity. At the same time an attempt was made to insert the other paw into the mouth of the fish to meet its fellow. This disposition to use the paws in concert, was shown in almost every action, sometimes in a very ludicrous manner. On giving the animal a jug, one paw would be inserted in the aperture, and a hundred twists and turns would be made to join its fellow on the outside.

After devouring as much of the fish as it wished, it placed the paws on the remainder and lay down to doze, until hunger returned, watching the favourite food, and growling at any animal which happened to pass near it. By degrees this propensity to defend its food passed off, and it would allow the dog or fox to partake of it freely. We placed a half-grown fox within its reach: the Raccoon instantly grasped it with its legs and paws and commenced a close examination. It thrust its pointed nose in the ear of the fox to the very bottom, smelling and snuffing as if determined to find out the nature of the animal. During this time it showed no disposition to injure the fox.

The Raccoon can scent an object for some distance with accuracy. We suffered ours to go loose on one occasion, when it made directly for some small marmots confined in a cage in another room.

Our pet Raccoon whose habits we are relating evinced a singular propensity to listen to things at a distance, however many persons were around him, even though he might be at the moment eating a frog, of which food he was very fond. He would apparently hear some distant noise, then raise his head and continue listening, seeming every moment more absorbed; at last he would suddenly run and hide himself in his burrow. This seems to be connected with some instinct of the animal in his wild state, probably whilst sitting on a tree sunning himself, when he is in the habit of listening to hear the approach of an enemy, and then hurrying to his hole in the tree.

Enjoying the hospitality of a friend one night at his plantation, the conversation turned on the habits of animals: and in speaking of the Raccoon he mentioned that it fed on birds and rabbits generally, but in winter robbed the poultry houses. The negroes on his plantation he said kept good dogs, and relied on them for hunting the Raccoon.

Whenever a Raccoon was about to attack the poultry house, the dogs scenting him give a shrill cry, which is the signal for his owner to commence the hunt. He comes out armed with an axe, with a companion or two, resolved on a Raccoon hunt. The dog soon gives chase with such rapidity, that the Raccoon, hard pressed, takes to a tree. The dog, close at his heels, changes his whining cry while running to a shrill short sharp bark. If the tree is small or has limbs near the ground so that it can be easily ascended, the eager hunters climb up after the "coon." He perceives his danger, endeavours to avoid his pursuers by ascending to the farthest topmost branch, or the extremity of a limb; but all his efforts are in vain, his relentless pursuers shake the limb until he is compelled to let go his hold, and he comes toppling heavily to the ground, and is instantly seized by the dogs. It frequently happens however that the trees are tall and destitute of lower branches so that they cannot be climbed without the risk of life or limb. The negroes survey for a few moments in the bright moon-light the tall and formidable tree that shelters the coon, grumble a little at the beast for not having saved them trouble by mounting an easier tree, and then the ringing of their axes resounds through the still woods, awakening echoes of the solitude previously disturbed only by the hooting of the owl, or the impatient barking of the dogs. In half an hour the tree is brought to the ground and with it the Raccoon, stunned by the fall: his foes give him no time to define his position, and after a short and bloody contest with the dogs, he is despatched, and the sable hunters remunerated,—for his skin they will sell to the hatters in the nearest town, and his flesh they will hang up in a tree to freeze and furnish them with many a savoury meal.

The greatest number of Raccoons, however, are killed by log-traps set with a figure of 4 trigger, and baited with a bird or squirrel, an ear of corn, or a fish: either the appetite or curiosity of these animals will entice them into a trap or entangle them in a snare.[25]

Another mode of destroying this species is by fire-hunting, which requires good shooting, as the animal only shows one eye from behind the branch of a tree, which reflecting the light of the fire-hunter's torch, shines like a ball of phosphorus, and is generally knocked out at twenty-five or thirty yards by a good marksman.

The Raccoon, like the bear, hibernates for several months during winter in the latitude of New-York, and only occasionally and in a warm day leaves its retreat, which is found in the hollow of some large tree. We once however tracked in deep snow the footsteps of a pair of this species in the northern parts of New-York, and obtained them by having the tree in which they lay concealed cut down. They had made a circle in company of about a mile, and then returned to their winter domicil.

The specimen from which the representation on our plate was taken was a remarkably fine male, and was sent to us alive by our friend, the late Dr. JOHN WRIGHT of Troy, New-York.

Nº 14.

PLATE LXVI

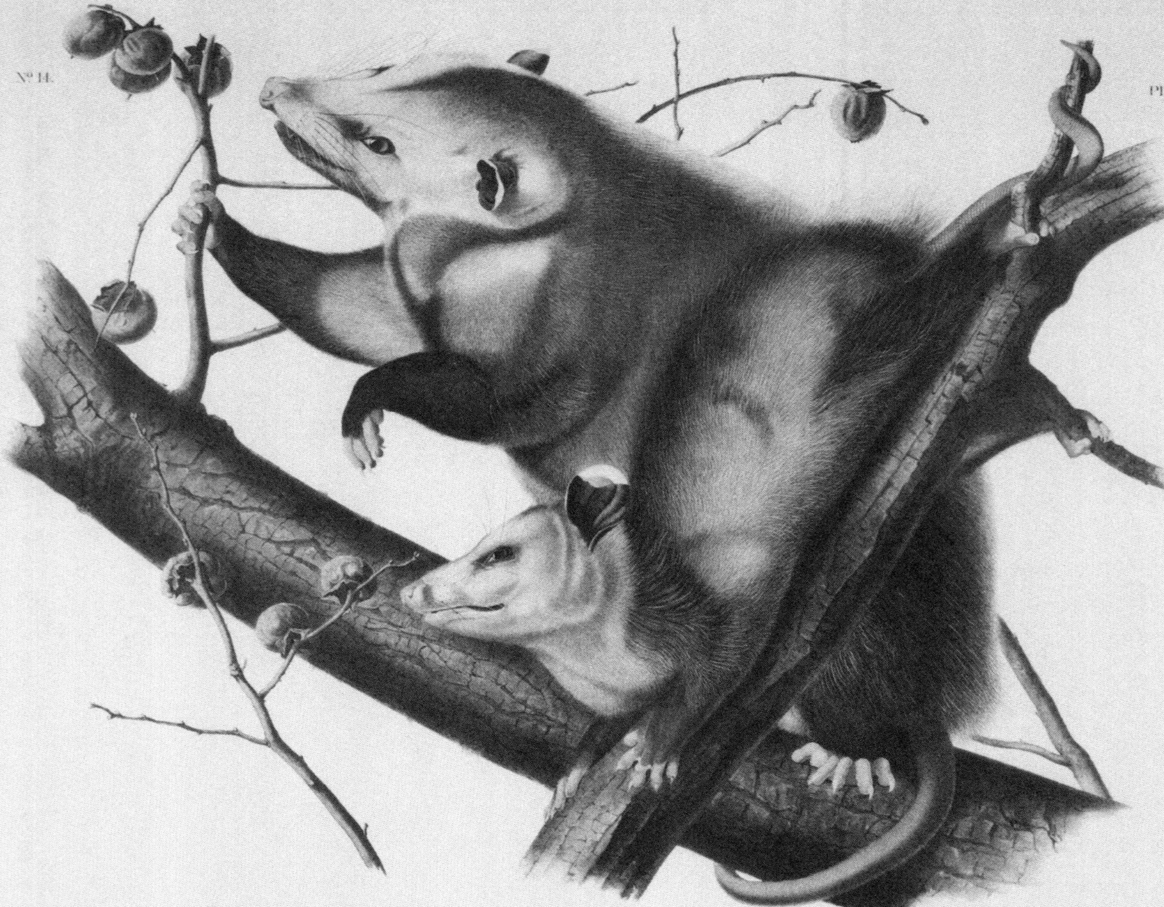

DIDELPHIS VIRGINIANA, PENNANT.

VIRGINIAN OPOSSUM.

Virginian Opossum

THOUGH "VIRGINIAN OPOSSUM" BEGINS WITH AN IMAGINATIVE account of European contact that makes the strangeness of the marsupial new again, the bulk of the essay's content describes the folkways and science of Bachman's contemporary moment. In his description of slaves hunting opossums, Bachman presents what might be called a "slave picturesque," a quaint idealization of slaves' conditions. By contrast, Bachman's experiments concerning opossum "generation" are impressive for their empirical rigor and the cool detachment with which he reports defenseless animals' suffering.

---

17. John James Audubon, J. T. Bowen, lithographer, *Virginian Opossum*, 1845–46, hand-colored lithograph, plate 66 of *The Viviparous Quadrupeds of North America*. Courtesy of the Jule Collins Smith Museum of Fine Art, Auburn University; Louise Hauss and David Brent Miller Audubon Collection.

DIDELPHIS VIRGINIANA.—SHAW.

VIRGINIAN OPOSSUM

PLATE LXVI.—FEMALE, YOUNG MALE SEVEN MONTHS OLD.

HABITS.

In our first volume (pp. 111, 112) we have spoken of the curiosity eagerly indulged, and the sensations excited, in the minds of the discoverers of our country, on seeing the strange animals that they met with. Travellers in unexplored regions are likely to find many unheard-of objects in nature that awaken in their minds feelings of wonder and admiration. We can imagine to ourselves the surprise with which the Opossum was regarded by Europeans when they first saw it. Scarcely any thing was known of the marsupial animals, as New Holland had not as yet opened its unrivalled stores of singularities to astonish the world. Here was a strange animal, with the head and ears of the pig, sometimes hanging on the limb of a tree, and occasionally swinging like the monkey by the tail! Around that prehensile appendage a dozen sharp-nosed, sleek-headed young, had entwined their own tails, and were sitting on the mother's back! The astonished traveller approaches this extraordinary compound of an animal and touches it cautiously with a stick. Instantly it seems to be struck with some mortal disease: its eyes close, it falls to the ground, ceases to move, and appears to be dead! He turns it on its back, and perceives on its stomach a strange apparently artificial opening. He puts his fingers into the extraordinary pocket, and lo! another brood of a dozen or more young, scarcely larger than a pea, are hanging in clusters on the teats. In pulling the creature about, in great amazement, he suddenly receives a gripe on the hand—the twinkling of the half-closed eye and the breathing of the creature, evince that it is not dead, and he adds a new term to the vocabulary of his language, that of "playing 'possum."

Like the great majority of predacious animals, the Opossum is nocturnal in its habits. It suits its nightly wanderings to the particular state of the weather. On a bright starlight or moonlight night, in autumn or winter, when the weather is warm and the air calm, the Opossum may every where be found in the Southern States, prowling around the outskirts of the plantation, in old deserted rice fields, along water courses, and on the edges of low grounds and swamps; but if the night should prove windy or very cold, the best nosed dog can scarcely strike a trail, and in such cases the hunt for that night is soon abandoned.

The gait of the Opossum is slow, rather heavy, and awkward; it is not a trot like that of the fox, but an amble or pace, moving the two legs on one side at a time. Its walk on the ground is plantigrade, resting the whole heel on the earth. When pursued, it by no means stops at once and feigns death, as has often been supposed, but goes forward at a rather slow speed, it is true, but as fast as it is able, never, that we are aware of, increasing it to a leap or canter, but striving to avoid its pursuers by sneaking off to some thicket or briar patch; when, however, it discovers that the dog is in close pursuit, it flies for safety to the nearest tree, usually a sapling, and unless molested does not ascend to the top, but seeks an easy resting place in some crotch not twenty feet from the ground, where it waits silently and immoveably, till the dog, finding that his master will not come to his aid, and becoming weary of barking at the foot of the tree, leaves the Opossum to follow the bent of his inclinations, and conclude his nightly round in search of food. Although a slow traveller, the Opossum, by keeping perseveringly on foot during the greater part of the night, hunts over much ground, and has been known to make a circle of a mile or two in one night. Its ranges, however, appear to be restricted or extended according to its necessities, as when it has taken up its residence near a corn field, or a clump of ripe persimmon trees, (*Diosperos virginiana*,) the wants of nature are soon satisfied, and it early and slowly carries its fat and heavy body to its quiet home, to spend the remainder of the night and the succeeding day in the enjoyment of a quiet rest and sleep.

The whole structure of the Opossum is admirably adapted to the wants of a sluggish animal. It possesses strong powers of smell, which aid it in its search after food; its mouth is capacious, and its jaws possessing a greater number and variety of teeth than any other of our animals, evidencing its omnivorous habits; its fore-paws, although not armed with retractile claws, aid in seizing its prey and conveying it to the mouth. The construction of the hind-foot with its soft yielding tubercles on the palms and its long nailless opposing thumb, enable it to use these feet as hands, and the prehensile tail aids it in holding on to the limbs of trees whilst its body is swinging in the air; in this manner we have observed it gathering persimmons with its mouth and fore-paws, and devouring them whilst its head was downwards and its body suspended in the air, holding on sometimes with its hind-feet and tail, but often by the tail alone.

We have observed in this species a habit which is not uncommon among a few other species of quadrupeds, as we have seen it in the raccoon and occasionally in the common house dog—that of lying on its back for hours in the sun, being apparently dozing, and seeming to enjoy this position as a change. Its usual posture, however, when asleep, is either lying at full length on the side, or sitting doubled up with its head under its fore-legs, and its nose touching the stomach, in the manner of the raccoon.

The Opossum cannot be called a gregarious animal. During summer, a brood composing a large family may be found together, but when the young are well grown, they usually separate, and each individual shifts for himself; we have seldom found two together in the same retreat in autumn or winter.

Although not often seen abroad in very cold weather in winter, this animal is far from falling into that state of torpidity to which the marmots, jumping mice, and several other species of quadrupeds are subject. In the Southern States, there are not many clear nights of starlight or moonshine in which they may not be found roaming about; and although in their farthest northern range they are seldom seen when the ground is covered with snow, yet we recollect having come upon the track of one in snow a foot deep, in the month of March, in Pennsylvania; we pursued it, and captured the Opossum in its retreat—a hollow tree. It may be remarked, that animals like the Opossum, raccoon, skunk, &c., that become very fat in autumn require but little food to support them through the winter, particularly when the weather is cold.

The Opossum, although nocturnal in its general habits, is not unfrequently, particularly in spring and summer, found moving about by day. We have on several occasions met with it in the woods at mid-day, in places where it was seldom molested.

Nature has wisely provided this species with teeth and organs indicating its omnivorous character and its possessing an appetite for nearly all kinds of food; and in this particular it exhibits many of the propensities and tastes of the raccoon. It enters the corn fields (maize), crawls up the stalks, and sometimes breaks them down in the manner of the raccoon, to feed on the young and tender grains; it picks up chestnuts, acorns, chinquapins and beech nuts, and munches them in the manner of the bear. We have, on dissection, ascertained that it had devoured blackberries, whortleberries, and wild cherries, and its resort to the persimmon tree is proverbial. It is also insectivorous, and is seen scratching up the leaves in search of worms, and the larvae of insects, of which it is very fond. In early spring it lays the vegetable kingdom under contribution for its support, and we have observed it digging up the roots of the small atamasco lily, (*Zepherina atamasco*,) and the young and tender shoots of the China brier, (*Smilax rotundifolia*,) as they shoot out of the ground like asparagus. It is moreover decidedly carnivorous, eating young birds that it may detect on the ground, sucking the eggs in all the partridge, towhee-bunting and other nests it can find in its persevering search. It destroys mice and other rodentia, and devours whole broods of young rabbits, scratching about the nest and scattering the hair and other materials of which it was composed. We have observed it squatting in the grass and brier thickets in Carolina, which are the common resort of the very abundant cotton rat, (*Sigmodon hispidum*,) and from patches of skin and other mutilated remains, we satisfied ourselves that the Opossum was one among many other species designed by Providence to keep in check the too rapid increase of these troublesome rats. We must admit that it sometimes makes a sly visit to the poultry house, killing a few of the hens and playing havoc among the eggs. The annoyances of the farmer, however, from this mischievous propensity, are not as great as those sustained from some of the other species, and cannot for a moment be compared with the destruction caused by the weasel, the mink, or the skunk.

The domicile of the Opossum in which it is concealed during the day, and where it brings forth its young, which we have often examined, is found in various localities. This animal is a tolerable digger, although far less expert in this quality than the Maryland marmot, its den is usually under the roots of trees or stumps, when the ground is so elevated as to secure it from rains and inundations. The hollow of a large fallen tree, or an opening at the roots of a standing one, also serve as a convenient place for its nest. The material which we have usually found composing this nest along the seaboard of Carolina is the long moss (*Tillandsia usnoides*); although we have sometimes found it composed of a bushel or more of oak and other leaves.

On firing into a squirrel's nest which was situated in the fork of a tree some forty feet from the ground, we brought down an Opossum, which had evidently expelled its legitimate occupant. The Florida rat is known to collect heaps of sticks and leaves, and construct nests sometimes a yard in diameter and two feet high: these are usually placed on the ground, but very frequently on the entangled vines of the grape, smilax, and supple jack, (*Ziziphus volubilis.*) In these nests an Opossum may occasionally be found, dozing as cozily as if he had a better right than that of mere possession.

Hunting the Opossum is a very favourite amusement among domestics and field labourers on our Southern plantations, of lads broke loose from school in the holidays, and even of gentlemen, who are sometimes more fond of this sport than of the less profitable and more dangerous and fatiguing one of hunting the gray fox by moonlight. Although we have never participated in an Opossum hunt, yet we have observed that it afforded much amusement to the sable group that in the majority of instances make up the hunting party, and we have on two or three occasions been the silent and gratified observers of the preparations that were going on, the anticipations indulged in, and the excitement apparent around us.

On a bright autumnal day, when the abundant rice crop has yielded to the sickle, and the maize has just been gathered in, when one or two slight white frosts have tinged the fields and woods with a yellowish hue, ripened the persimmon, and caused the acorns, chestnuts and chinquapins (*Castanea pumilla*) to rattle down from the trees and strewed them over the ground, we hear arrangements entered into for the hunt. The Opossums have been living on the delicacies of the season, and are now in fine order, and some are found excessively fat; a double enjoyment is anticipated, the fun of catching and the pleasure of eating this excellent substitute for roast pig.

"Come, men," says one, "be lively, let us finish our tasks by four o'clock, and after sundown we will have a 'possum hunt." "Done," says another, "and if an old coon comes in the way of my smart dog, Pincher, I be bound for it, he will shake de life out of him." The labourers work with increased alacrity, their faces are brightened with anticipated enjoyment, and ever and anon the old familiar song of "'Possum up the gum tree" is hummed, whilst the black driver can scarcely restrain the whole gang from breaking out into a loud chorus.

The paraphernalia belonging to this hunt are neither showy nor expensive. There are no horses caparisoned with elegant trappings—no costly guns imported to order—no pack of hounds answering to the echoing horn; two or three curs, half hound or terriers, each having his appropriate name, and each regarded by his owner as the best dog on the plantation, are whistled up. They obey the call with alacrity, and their looks and intelligent actions give evidence that they too are well aware of the pleasure that awaits them. One of these humble rustic sportsmen shoulders an axe and another a torch, and the whole arrangement for the hunt is completed. The glaring torch-light is soon seen dispersing the shadows of the forest, and like a jack o'lantern, gleaming along the skirts of the distant meadows and copses. Here are no old trails on which the cold-nosed hound tries his nose for half an hour to catch the scent. The tongues of the curs are by no means silent—ever and anon there is a sudden start and an uproarious outbreak: "A rabbit in a hollow, wait, boys, till I twist him out with a hickory." The rabbit is secured and tied with a string around the neck: another start, and the pack run off for a quarter of a mile, at a rapid rate, then double around the cotton fields and among the ponds in the pine lands—"Call off your worthless dog, Jim, my Pincher has too much sense to bother after a fox." A loud scream and a whistle brings the pack to a halt, and presently they come panting to the call of the black huntsman. After some scolding and threatening, and resting a quarter of an hour to recover their breath and scent, they are once more hied forwards. Soon a trusty old dog, by an occasional shrill yelp, gives evidence that he has struck some trail in the swamp. The pack gradually make out the scent on the edges of the pond, and marshes of the rice fields, grown up with willows and myrtle bushes (*Myrica cerifera*). At length the mingled notes of shrill and discordant tongues give evidence that the game is up. The race, though rapid, is a long one, through the deep swamp, crossing the muddy branch into the pine lands, where the dogs come to a halt, unite in conclave, and set up an incessant barking at the foot of a pine. "A coon, a coon! din't I tell you," says Monday, "that if Pincher come across a coon, he would do he work?" An additional piece of split light-wood is added to the torch, and the coon is seen doubled up in the form of a hornet's nest in the very top of the long-leaved pine, (*P. palustris*). The tree is without a branch for forty feet or upwards, and it is at once decided that it must be cut down: the axe is soon at work, and the tree felled. The glorious battle that ensues, the prowess of the dogs, and the capture of the coon, follow as a matter of course. See our article on the raccoon, pp. 80, 81, where we have briefly described such a scene.[26]

Another trail is soon struck, and the dogs all open upon it at once: in an instant they rush, pell mell, with a loud burst of mingled tongues, upon some animal along the edge of an old field destitute of trees. It proves to be an Opossum, detected in its nightly prowling expedition. At first, it feigns death, and, rolling itself into a ball, lies still on the ground; but the dogs are up to this "'possum playing," and seize upon it at once. It now feels that they are in earnest, and are not to be deceived. It utters a low growl or two, shows no fight, opens wide its large mouth, and, with few struggles, surrenders itself to its fate. But our hunters are not yet satisfied, either with the sport or the meat: they have large families and a host of friends on the plantation, the game is abundant, and the labour in procuring it not fatiguing, so they once more hie on the dogs. The Opossum, by its slow gait and heavy tread, leaves its foot-prints and scent behind it on the soft mud and damp grass. Another is soon started, and hastens up the first small gum, oak, or persimmon tree, within its reach; it has clambered up to the highest limb, and sits crouching up with eyes closed to avoid the light. "Off jacket, Jim, and shake him down; show that you know more about 'possum than your good-for-nutten fox-dog." As the fellow ascends, the animal continues mounting higher to get beyond his reach; still he continues in pursuit, until the affrighted Opossum has reached the farthest twig on the extreme branches of the tree. The negro now commences shaking the tall pliant tree top; while with its hind hands rendered convenient and flexible by its opposing thumb, and with its prehensile tail, the Opossum holds on with great tenacity. But it cannot long resist the rapidly accumulating jerks and shocks: suddenly the feet slip from the smooth tiny limb, and it hangs suspended for a few moments only by its tail, in the meantime trying to regain its hold with its hind hands; but another sudden jerk breaks the twig, and down comes the poor animal, doubled up like a ball, into the opened jaws of eager and relentless canine foes; the poor creature drops, and yields to fate without a struggle.

In this manner half a dozen or more Opossums are sometimes captured before midnight. The subsequent boasts about the superior noses, speed and courage of the several dogs that composed this small motley pack—the fat feast that succeeded on the following evening, prolonged beyond the hour of midnight, the boisterous laugh and the merry song, we leave to be detailed by others, although we confess we have not been uninterested spectators of such scenes.

"Let not ambition mock their useful toil,
"Their homely joys and destiny obscure,
"Nor grandeur hear with a disdainful smile,
"The simple pleasures of the humble poor."[27]

The habit of feigning death to deceive an enemy is common to several species of quadrupeds, and we on several occasions witnessed it in our common red fox (*V. Fulvus*). But it is more strikingly exhibited in the Opossum than in any other animal with which we are acquainted. When it is shaken from a tree and falls among grass and shubbery, or when detected in such situations, it doubles itself into a heap and feigns death so artfully, that we have known some schoolboys carrying home for a quarter of a mile an individual of this species, stating that when they first saw it, it was running on the ground, and they could not tell what had killed it. We would not, however, advise that the hand should on such occasions be suffered to come too familiarly in contact with the mouth, lest the too curious meddler should on a sudden be startled with an unexpected and unwelcome gripe.

This species has scarcely any note of recognition, and is remarkably silent; when molested, it utters a low growl; at other times its voice resembles the hissing of a cat. The Opossum displays no cunning in avoiding traps set to capture it, entering almost any kind of trap, very commonly being taken in a log trap called a dead fall.

From its very prolific nature it can afford to have many enemies. In addition to the incessant war waged against it by men and dogs, we have ascertained that its chief enemy among rapacious birds is the Virginian owl, (*Strix Virginiana*,) which flying abroad at the same hour in which the Opossum is on foot, pounces on it, and kills it with great ease. We have heard of an instance in which it was seen in the talons of the white-headed eagle, (*Halietus leucocephalus*,) and of two or three in which the great hen-hawk (*F. Borealis*) was observed feeding upon it. We recollect no instance of its having been killed by the wild cat or the fox. The wolf, it is said, seizes on every Opossum it can find, and we have heard of two instances where half-grown animals of this species were found to have been swallowed by the rattlesnake.

Although the dog hunts it so eagerly, yet we have never been able to ascertain that it ever feeds upon its flesh; indeed, we have witnessed the dog passing by the body of a fresh killed Opossum, and going off half a mile farther to feed on some offensive carcass.

The Opossum is easily domesticated when captured young. We have, in endeavouring to investigate one of the very extraordinary characteristics of this species, preserved a considerable number in confinement, and our experiments were continued through a succession of years. Their nocturnal habits were in a considerable degree relinquished, and they followed the servants about the premises, becoming troublesome by their familiarity and their mischievous habits. They associated familiarly with a dog on the premises, which seemed to regard them as necessary appendages of the motley group that constituted the family of brutes in the yard. They devoured all kinds of food: vegetables, boiled rice, hominy, meat both raw and boiled, and the scraps thrown from the kitchen; giving the preference to those that contained any fatty substance.

On one occasion a brood of young with their mother made their escape, concealed themselves under a stable, and became partially wild; they were in the habit of coming out at night, and eating scraps of food, but we never discovered that they committed any depredations on the poultry or pigeons. They appeared however to have effectually driven off the rats, as during the whole time they were occupants of the stable, we did not observe a single rat on the premises. It was ascertained that they were in the habit of clambering over fences and visiting the neighbouring lots and gardens, and we occasionally found that we had repurchased one of our own vagrant animals. They usually, however, returned towards daylight to their snug retreat, and we believe would have continued in the neighbourhood and multiplied the species had they not in their nightly prowlings been detected and destroyed by the neighbouring dogs.

A most interesting part of the history of this animal, which has led to the adoption of many vulgar errors, remains to be considered, viz., the generation of the Opossum.

Our investigations on this subject were commenced in early life, and resumed as time and opportunity were afforded, at irregular, and sometimes after long intervals, and were not satisfactorily concluded until within a month of the period of our writing this article, (June, 1849). The process by which we were enabled to obtain the facts and arrive at our conclusions is detailed in an article published in the Transactions of the Academy of Natural Sciences, April, 1848, p. 40. Subsequent investigations have enabled us to verify some of these facts, to remove some obscurities in which the subject was yet involved, and finally to be prepared to give a correct and detailed history of a peculiarity in the natural history of this quadruped, around which there has hitherto been thrown a cloud of mystery and doubt.

Our early authors—MARCGRAVE, PISON, VALENTINE, BEVERLY, the MARQUIS OF CHASTELLUX, PENNANT, and others, contended that "the pouch was the matrix of the young Opossum, and that the mammae are, with regard to the young, what stalks are to their fruits." DE BLAINVILLE and Dr. BARTON speak of two sorts of gestation, one uterine and the other mammary. BLUMENBACH calls the young when they are first seen on the mammae, abortions; and Dr. BARTON's views (we quote from GRIFFITH) are surprisingly inaccurate: "The Didelphes," he says, "put forth, not fœtuses but gelatinous bodies; they weigh at their first appearance generally about a grain, some a little more, and seven of them together weighed ten grains." In 1819, GEOFFROY ST. HILLAIRE propounded to naturalists the following question: "Are the pouched animals born attached to the teats of the mother?" GODMAN, in his American Natural History, published in 1826, gave to the world a very interesting article on the Opossum, full of information in respect to the habits, &c., comprising all the knowledge that existed at that day in regard to this species. He was obliged, however, to admit, vol. 2, p. 7, "the peculiarities of its sexual intercourse, gestation, and parturition, are to this day involved in profound obscurity. Volumes of facts and conjectures have been written on the subject, in which the proportion of conjecture to fact has been as a thousand to one, and the difficulties still remain to be surmounted." And DEKAY, in the work on the Quadrupeds of the State of N. York, (Nat. Hist. of N. York, 1842, p. 4,) states: "The young are found in the external abdominal sac, firmly attached to the teat in the form of a small gelatinous body, not weighing more than a grain. It was a long time believed that there existed a direct passage from the uterus to the teat, but this has been disproved by dissection. Another opinion is, that the embryo is excluded from the uterus in the usual manner and placed by the mother to the teat; and a third, that the embryo is formed where it is first found. Whether this transfer actually takes place, and if so, the physiological considerations connected with it, still remain involved in great obscurity."

The approaches to truth in these investigations have been very gradual, and the whole unusually slow. COWPER, TYSON, DE BLAINVILLE, HOME and others, by their examinations and descriptions of the organs of the Marsupialiæ, prepared the way for farther developments. A more judicious examination and scientific description by Owen and others, of the corresponding organs in the kangaroo, the largest of all the species composing these genera, and the discovery of the fœtus in utero, enabled naturalists to conclude, that the similar structure in the Opossum would indicate a corresponding result. No one, however, was entitled to speak with positive certainty until the young were actually detected in the uterus, nor could an explanation of the peculiarity in the growth of the foetus be made until it was examined in its original bed.

We have been so fortunate in five instances as to have procured specimens in which the young were observed in this position, and therefore feel prepared to speak with certainty. We are not aware that the young of the Virginian Opossum had been previously detected in the uterus.

All our investigations were made in South Carolina, where this is a very abundant species. For some years we attempted to arrive at the object of our researches by preserving these animals in a state of confinement. But they were subject to many accidents: they frequently made their escape from their cages, and some of them became overburdened with fat and proved sterile, so that we did not succeed in a single instance in obtaining young from females in a state of confinement. From this cause the naturalists of Europe, and especially those of France, who were desirous of making investigations in regard to our Opossum, have been so long unsuccessful. Their usual complaint has been, "Your Opossums do not breed in confinement." In this, Dr. BARTON and our young friend Dr. MICHEL were more fortunate, but in both cases the young were produced before they were enabled to detect them in their previous existing position. We varied our experiments by endeavouring to discern the precise period when young were usually produced. We ascertained, by having a number of females procured with young in their pouches, that about the close of the first week in March, a little earlier or later, according to the age of the individual, or warmth, or coldness, of the previous winter, was the time when in this latitude this event usually occurs. Here, however, another difficulty presented itself, which for several successive seasons, thwarted us in our investigations. In the third week of February 1847, by offering premiums to the servants on several neighbouring plantations we obtained in three nights thirty-five Opossums, but of that number there was not a single female. A week afterwards, however, when the young were contained in the pouch, we received more females than males. From this circumstance we came to the conclusion that during the short period of gestation, the females, like those of some other species of quadrupeds, particularly the American black bear, conceal themselves in their burrows and can seldom be found. We then changed our instructions for capturing them, by recommending that they should be searched for in the day time, in hollow logs and trees and places where they had been previously known to burrow. By this means we were enabled at different times to obtain a small number in the state in which we were desirous of examining them. We feel under great obligations to several gentlemen of Carolina for aiding us in our investigations by procuring specimens, especially our relative Colonel HASKELL, Mr. JOHNSON, and JAMES FISHER, Esq., a close observer and intelligent naturalist. The latter, by his persevering efforts, pursued for some years at Jordan's Mills, on the upper waters of the Edisto, obtained two females in May, 1849, in the particular state in which he knew we were anxious to procure them, and brought them to us without having been previously aware that we had published the facts a year before.

The Opossums we were enabled to examine were dissected on the 11th, 14th and 18th February, 1848, and on the 12th and 22d May, 1849. Somc of these had advanced to near the time of parturition. The young of those brought us by Mr. FISHER each weighed 2½ grains. Those of one, sent us by Col. HASKELL, weighed 3 grains; and the young of another which we obtained by a Cæsarian operation, at a moment when all the rest had been excluded, and this individual alone remained, weighed 4 grains.

We remarked, that this however was a little the largest of six that composed the family, five of which were already in the pouch and attached to the teats. The largest one weighed 3¾, and another 3½ grains. The weight, then, of the young Opossum at the moment of birth, is between 3 and 4 grains, varying a little in different specimens as is the case in the young of all animals.

The degree of life and animation in young Opossums at the moment of birth has been greatly underrated. They are neither abortions, as BLUMENBACH represented them, nor as Dr. BARTON has described them—"not fœtuses, but gelatinous bodies, weighing about a grain more or less, seven of them together weighing 10 grains"—but little creatures that are nearly as well developed at birth as the young of the white-footed mouse and several other species of rodentia. They are covered by an integument, nourished by the mammæ, breathe through nostrils, perform the operations of nature, are capable of a progressive movement at the moment of their birth, and are remarkably tenacious of life. The individual which was dissected from the parent in the manner above detailed, moved several inches on the table by crawling and rolling, and survived two hours; the thermometer in the room was at the time standing at 66° Fahrenheit. The period of gestation is from fifteen to sixteen days. We received a female from a servant who informed us, that he had that morning seen it in intercourse with the male. We first saw the young on the morning of the 17th day. Our friend Dr. MIDDLETON MICHEL, a gentleman of high scientific attainments, and who had long been engaged in investigating the characters and habits of this species, in a communication made to us, (Trans. of the Acad. Nat. Sciences, April, 1848, p. 46,) assured us from his personal observation in which he was careful to note the hour of the day, the exact period is 15 days. As he possessed better opportunities of deciding in regard to the time, the animals being in a state of domestication, we are rather more disposed to yield to his observations than to our own; there is, however, only the difference of a day between us.

The young, when first born, are naked and flesh-coloured; the eyes, together with the ears, are covered by a thin integument through which these organs and the protuberances of the ears are distinctly visible. The mouth is closed, with the exception of a small orifice, sufficiently large to receive the teat, which is so thin and attenuated that it seems no larger than the body of a pin. Length of body, 7-12ths of an inch; of tail, 2-loths. The nails, which can be seen with the naked eye, are very distinct when viewed with a microscope, and are of a dark brown colour, small and much hooked. The nostrils are open; the lungs filled with air, and when placed in water, the young float on the surface.

The number of young usually found in the pouch appear to be less than those that are born. The highest number we have found in the pouch was thirteen, the smallest six; whereas the preserved uterus brought to us by Mr. FISHER, contained fifteen. In all such cases, where a greater number of young are produced than there are teats, the last of the brood must inevitably perish, as those that are attached appear incapable of relinquishing their hold.

The manner in which the young at birth reach the pouch, and become attached to the teats, has been the subject of much speculation and inquiry. We had an opportunity of examining this process in part, without, however, having been aware at the time that it was going on. We intended to dissect a small female Opossum, which had been a few days in our possession, but ascertained in the morning at seven o'clock on the day our examination was to have been made, that she had three young in her pouch; supposing from her small size, that she would produce no additional number, we concluded to spare her life. She was confined in a box in our study; when we occasionally looked at her, we found her lying on one side, her shoulders elevated, her body drawn up in the shape of a ball; the pouch was occasionally distended with her paws—in this position the parts reached the edge of the pouch; she was busily employed with her nose and mouth licking, as we thought, her pouch, but in which we afterwards ascertained, were her young.

At six o'clock in the afternoon we were induced to examine her again, in consequence of having observed that she had for several hours appeared very restless, when we discovered that she had added four more to her previous number, making her young family now to consist of seven. With no inconsiderable labour and the exercise of much patience, we removed three of the young from the teats, one of which perished under the process, we replaced the two living ones in the pouch; at nine o'clock examined her again and found both the young once more attached. We came to the conclusion, that she shoved them into the pouch, and with her nose or tongue moved them to the vicinity of the teats, where by an instinct of nature, the teat was drawn into the small orifice of the mouth by suction. We observed subsequently, that a young one that had been extracted from its parent a few moments before the time when it would have been born, and which had been rolled up in warm cotton, was instinctively engaged in sucking at the fibres of the cotton, and had succeeded in drawing into its mouth a considerable length of thread. A nearly similar process was observed by our friend Dr. MICHEL. He states: "The female stood on her hind legs, and the body being much bent, the young appeared and were licked into the pouch."

There is a great difficulty in deciding the question, whether the mother aids the young in finding the teats, in consequence of the impossibility of the spectators being able to know what she is actually doing, whilst her nose is in the pouch. We believe the majority of naturalists who had an opportunity of witnessing our experiments came to the conclusion, that the mother, after shoving them into the pouch, left them to their own instinct, and they became attached without her assistance. We tried another experiment that suggested itself to us. Believing that the mother would not readily adopt the young of another, or afford them any assistance, we removed six out of ten that composed her brood, returned two of her own to the pouch, together with three others fully double the size, that had been obtained from another female. She was soon observed doubled up with her nose in the pouch, and continued so for an hour, when she was examined and one of her own small ones was found attached to the teat. Seven hours afterwards she was examined again, and both the small ones were attached, but the three larger ones still remained crawling about the pouch. On the following morning, it was ascertained that the mother had adopted the strangers, as the whole family of different sizes were deriving sustenance from her.

On another occasion, a female Opossum had been sent to us caught by a dog and much wounded, in consequence of which she died a few days afterwards, but first producing seven young which to every appearance had been still born. Yet they were in the pouch, and it appeared to us that the mother's uncontrollable attachment to her young, induced her to place her offspring in the pouch, even after they were dead.

An interesting inquiry remains to be answered: Is the Opossum a placental or non-placental animal? Until we were favoured with a recent opportunity of carefully examining a uterus, containing nine young on one side, and six on the other, kindly brought to us by our friend JAMES FISHER, we were unable fully to answer this question. Our dissections and examinations were witnessed by Professors MOULTRIE, HUME, Drs. HORLBECK, MICHEL, PORCHER and others.

The Opossum is, as far as we are able to judge from the specimens examined, a non-placental animal, inasmuch as there could not be detected the slightest adhesion between the exterior membrane of the fœtus and the internal surface of the mother. The membranes consisted of a vitelline sac, filled with ramifications of omphalo-mesenteric vessels, there was a slight appearance of an umbilical cord and umbilical vessels, constituting a true allantois, but no portions of them were attached to the uterus. There was no appearance of a placenta.

The growth of the young Opossum is surprisingly rapid. We weighed the largest young one at a week old and found it had increased from 3¾ grains to 30 grains. Length of head and body exclusive of tail, 1¼ inch; tail, ½ inch. The young at this age were very tenacious of life, as on removing two, they remained alive on the floor without any covering through a cool night, in a room containing no fire, and still exhibited a slight motion at twelve o'clock on the following day. The teats of the mother after the young had been gently drawn off measured an inch in length, having been much distended, and appeared to have been drawn into the stomach of the young. The pouches of the young females were quite apparent; they used their prehensile tails, which could now be frequently seen entwined around the necks of others. At twelve days old the eyes were still closed, a few hairs had made their appearance on the moustache; the orifices of the ears were beginning to be developed, and the nails were quite visible and sharp.

When the young are four weeks old, they begin from time to time to relax their hold on the teats, and may now be seen with their heads occasionally out of the pouch. A week later, and they venture to steal occasionally from their snug retreat in the pouch, and are often seen on the mother's back securing themselves by entwining their tails around hers. In this situation she moves from place to place in search of food, carrying her whole family along with her, to which she is much attached, and in whose defence she exhibits a considerable degree of courage, growling at any intruder, and ready to use her teeth with great severity on man or dog. In travelling, it is amusing to see this large family moving about. Some of the young, nearly the size of rats, have their tails entwined around the legs of the mother, and some around her neck, thus they are dragged along. They have a mild and innocent look, and are sleek, and in fine condition, and this is the only age in which the word pretty can be applied to the Opossum. At this period, the mother, in giving sustenance to so large a family, becomes thin, and is reduced to one half of her previous weight. The whole family of young remain with her about two months, and continue in the vicinity till autumn. In the meantime, a second and often a third brood is produced, and thus two or more broods of different ages may be seen, sometimes with the mother, and at other times not far off.

The Opossum, with the exception of our gray rabbit, is one of the most prolific of our quadrupeds. We consider the early parts of the three months of March, May and July, as the periods in South Carolina when they successively bring forth; it is even probable that they breed still more frequently, as we have observed the young during all the spring and summer months. In the month of May, 1830, whilst searching for a rare species of coleoptera, in removing with our foot some sticks composing the nest of the Florida rat, we were startled on finding our boot unceremoniously and rudely seized by an animal which we soon ascertained was a female Opossum. She had in her pouch five very small young whilst, seven others, about the size of full grown rats were detected peeping from under the rubbish. The females produce young at a year old. The young born in July do not bring forth as early as those born in March, but have their young as soon as the middle of the succeeding May. There is, of course, in this as well as in other species, some degree of irregularity in the time of their producing, as well as in the number of their young. We have reason to believe, also, that this species is more prolific in the southern than in the Middle States.

No. 15 PLATE LXXII

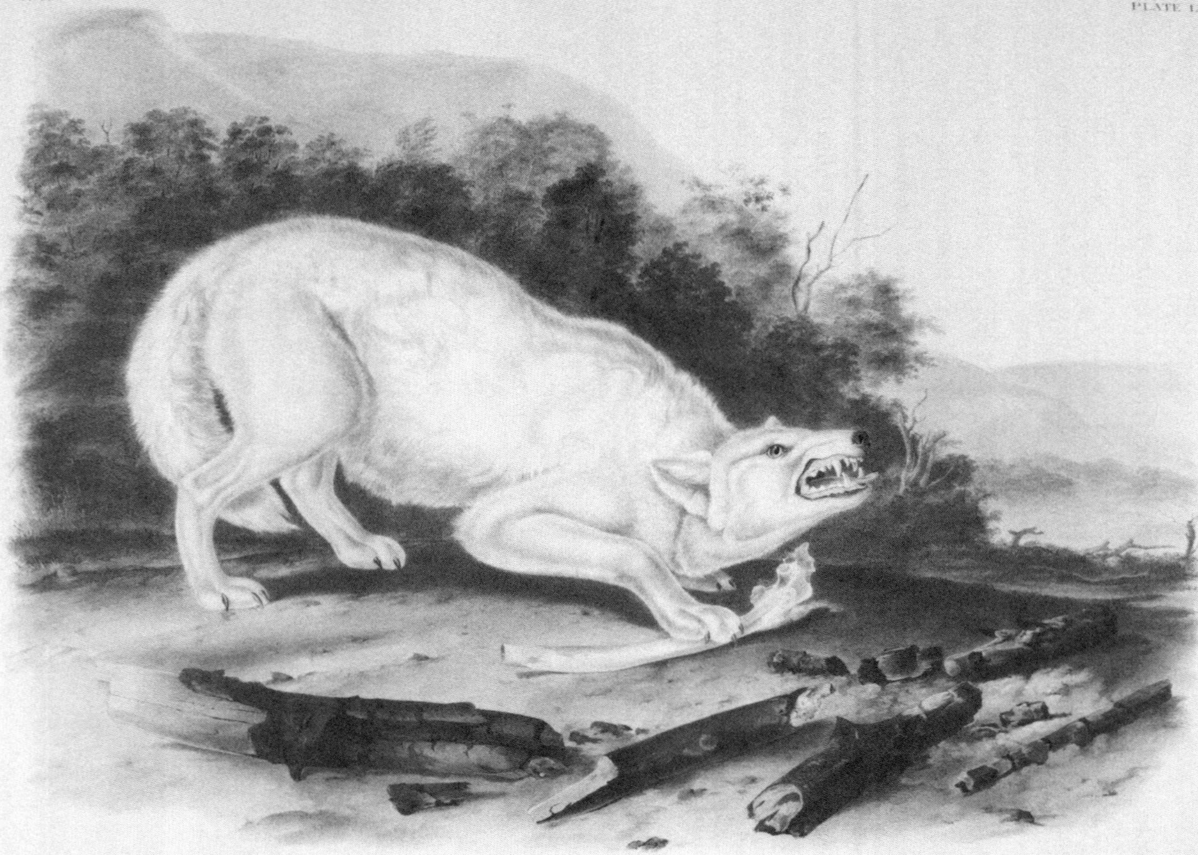

CANIS LUPUS, LINN. VAR. ALBUS.

WHITE AMERICAN WOLF.

White American Wolf

MUCH OF THIS ESSAY IS ADAPTED FROM AUDUBON'S ORIGINAL Missouri River journals, which were later burned. At the heart of the material assembled here is the long-standing human hatred and contempt for the wolf, which is regularly represented as "ravenous" competitor and general menace and in need of being killed. The opening scene captures that traditional attitude toward the wolf as the Fort Union supervisor Alexander Culbertson chases down and kills the first wolf to appear. Culbertson is also displaying for his guests the riding and shooting skills that are necessary and common among successful hunters on the prairies. As do many of the quadruped essays, this one bears witness to the former abundance of its species, in the case of the wolf, "almost beyond credibility." But the essay does not celebrate that abundance, as do the essays about the bison and the pronghorn. Even a report of successful adaptive behavior by wolves, such as is presented as evidence of intelligence in other species, is delivered only inadvertently: wolves on the prairies have learned to approach the sound of gunfire because of the possibility of eating some of what the hunter might leave behind. However, the essay presents this as just another opportunity to kill a wolf rather than as the wolf's successful adaptation.

---

18. John Woodhouse Audubon, J. T. Bowen, lithographer, *White American Wolf*, 1845–46, hand-colored lithograph, plate 72 of *The Viviparous Quadrupeds of North America*. Courtesy of the Jule Collins Smith Museum of Fine Art, Auburn University; Louise Hauss and David Brent Miller Audubon Collection.

CANIS LUPUS.—LINN.—(VAR. ALBUS.)

WHITE AMERICAN WOLF.

PLATE LXXII.—MALE.

HABITS.

The White Wolf is far the most common variety of the Wolf tribe to be met with around Fort Union, on the prairies, and on the plains bordering the Yellow Stone river. When we first reached Fort Union we found Wolves in great abundance, of several different colours, white, grey, and brindled. A good many were shot from the walls during our residence there, by EDWARD HARRIS, Esq., and Mr. J. G. BELL. We arrived at this post on the 12th of June, and although it might be supposed at that season the Wolves could procure food with ease, they seemed to be enticed to the vicinity of the Fort by the cravings of hunger. One day soon after our arrival, Mr. CULBERTSON told us that if a Wolf made its appearance on the prairie, near the Fort, he would give chase to it on horseback, and bring it to us alive or dead. Shortly after, a Wolf coming in view, he had his horse saddled and brought up; but in the meantime the Wolf became frightened and began to make off, and we thought Mr. CULBERTSON would never succeed in capturing him. We waited, however, with our companions on the platform inside the walls, with our heads only projecting above the pickets, to observe the result. In a few moments we saw Mr. CULBERTSON on his prancing steed as he rode out of the gate of the Fort with gun in hand, attired only in his shirt, breeches and boots. He put spurs to his horse and went off with the swiftness of a jockey bent upon winning a race. The Wolf trotted on and every now and then stopped to gaze at the horse and his rider, but soon finding that he could no longer indulge his curiosity with safety, he suddenly gallopped off with all his speed, but he was too late in taking the alarm, and the gallant steed soon began to gain on the poor cur, as we saw the horse rapidly shorten the distance between the Wolf and his enemy. Mr. CULBERTSON fired off his gun as a signal to us that he felt sure of bringing in the beast, and although the hills were gained by the fugitive, he had not time to make for the broken ground and deep ravines, which he would have reached in a few minutes, when we heard the crack of the gun again, and Mr. CULBERTSON galloping along dexterously picked up the slain Wolf without dismounting from his horse, threw him across the pummel of his saddle, wheeled round and rode back to the Fort, as fast as he had gone forth, a hard shower of rain being an additional motive for quickening his pace, and triumphantly placed the trophy of his chase at our disposal. The time occupied, from the start of the hunter, until his return with his prize did not exceed twenty minutes. The jaws of the animal had become fixed, and it was quite dead. Its teeth had scarified one of Mr. CULBERTSON's fingers considerably, but we were assured that this was of no importance, and that such feats as the capture of this wolf were so very common, that no one considered it worthy of being called an exploit.

Immediately after this real wolf hunt, a sham Buffalo chase took place, a prize of a suit of clothes being provided for the rider who should load and shoot the greatest number of times in a given distance. The horses were mounted, and the riders started with their guns empty—loaded in a trice, while at speed, and fired first on one side and then on the other, as if after Buffaloes. Mr. CULBERTSON fired eleven times in less than half a mile's run, the others fired less rapidly, and one of them snapped several times, but as a snap never brings down a Buffalo, these mishaps did not count. We were all well pleased to see these feats performed with much ease and grace. None of the riders were thrown, although they suffered their bridles to drop on their horses' necks, and plied the whip all the time. Mr. CULBERTSON's mare, which was of the full, black foot Indian breed, about five years old, was highly valued by that gentleman, and could not have been purchased of him for less than four hundred dollars.

To return to the wolves.—These animals were in the habit of coming at almost every hour of the night, to feed in the troughs where the offal from the Fort was deposited for the hogs. On one occasion, a wolf killed by our party was devoured during the night, probably by other prowlers of the same species.

The white wolves are generally fond of sitting on the tops of the eminences, or small hills in the prairies, from which points of vantage they can easily discover any passing object on the plain at a considerable distance.

We subjoin a few notes on wolves generally, taken from our journals, made during our voyage up the Missouri in 1843.

These animals are extremely abundant on the Missouri river, and in the adjacent country. On our way up that extraordinary stream, we first heard of wolves being troublesome to the farmers who own sheep, calves, young colts, or any other stock on which these ravenous beasts feed, at Jefferson city, the seat of government of the State of Missouri; but to our great surprise, while there not a black wolf was seen.

Wolves are said to feed at times, when very hard pressed by hunger, on certain roots which they dig out of the earth with their fore-paws, scratching like a common dog in the ground. When they have killed a Buffalo or other large animal, they drag the remains of the carcass to a concealed spot if at hand, then scrape out the loose soil and bury it, and often lie down on the top of the grave they have thus made for their victim, until urged again by hunger, they exume the body and feast upon it. Along the banks of the river, where occasionally many Buffaloes perish, their weight and bulk preventing them from ascending where the shore is precipitous, wolves are to be seen in considerable numbers feeding upon the drowned Bisons.

Although extremely cunning in hiding themselves, at the report of a gun wolves soon come forth from different quarters, and when the alarm is over, you have only to conceal yourself, and you will soon see them advancing towards you, giving you a fair chance of shooting them, sometimes at not more than thirty yards distance. It is said that although they frequently pursue Buffalo, &c., to the river, they seldom if ever follow them after they take to the water. Their gait and movements are precisely the same as those of the common dog, and their mode of copulating, and the number of young brought forth at a litter is about the same. The diversity of their size and colour is quite remarkable, no two being quite alike.

Some days while ascending the river, we saw from twelve to twenty-five wolves; on one occasion we observed one apparently bent on crossing the river, it swam toward our boat and was fired at, upon which it wheeled round and soon made to the shore from which it had started.

At another time we saw a wolf attempting to climb a very steep and high bank of clay, when, after falling back thrice, it at last reached the top and disappeared at once. On the opposite shore another was seen lying down on a sand bar like a dog, and any one might have supposed it to be one of those attendants on man. Mr. BELL shot at it, but too low, and the fellow scampered off to the margin of the woods, there stopped to take a last lingering look, and then vanished.

In hot weather when wolves go to the river, they usually walk in up to their sides, and cool themselves while lapping the water, precisely in the manner of a dog. They do not cry out or howl when wounded or when suddenly surprised, but snarl, and snap their jaws together furiously. It is said when suffering for want of food, the strongest will fall upon the young or weak ones, and kill and eat them. Whilst prowling over the prairies (and we had many opportunities of seeing them at such times) they travel slowly, look around them cautiously, and will not disdain even a chance bone that may fall in their way; they bite so voraciously at the bones thus left by the hunter that in many cases their teeth are broken off short, and we have seen a number of specimens in which the jaws showed several teeth to have been fractured in this way.

After a hearty meal, the wolf always lies down when he supposes himself in a place of safety. We were told that occasionally when they had gorged themselves, they slept so soundly that they could be approached and knocked on the head.

The common wolf is not unfrequently met with in company with the Prairie wolf (*Canis latrans.*) On the afternoon of the 13th of July, as Mr. BELL and ourselves were returning to Fort Union, we counted eighteen wolves in one gang, which had been satiating themselves on the carcass of a Buffalo on the river's bank, and were returning to the hills to spend the night. Some of them had their stomachs distended with food and appeared rather lazy.

We were assured at Fort Union that wolves had not been known to attack men or horses in that vicinity, but they will pursue and kill mules and colts even near a trading post, always selecting the fattest. The number of tracks or rather paths made by the wolves from among and around the hills to that station are almost beyond credibility, and it is curious to observe their sagacity in choosing the shortest course and the most favourable ground in travelling.

We saw hybrids, the offspring of the wolf and the cur dog, and also their mixed broods: some of which resemble the wolf, and others the dog. Many of the Assiniboin Indians who visited Fort Union during our stay there, had both wolves and their crosses with the common dog in their trains, and their dog carts (if they may be so called) were drawn alike by both.

The natural gait of the American wolf resembles that of the Newfoundland dog, as it ambles, moving two of its legs on the same side at a time. When there is any appearance of danger, the wolf trots off, and generally makes for unfrequented hilly grounds, and if pursued, gallops at a quick pace, almost equal to that of a good horse, as the reader will perceive from the following account. On the 16th of July 1843, whilst we were on a Buffalo hunt near the banks of the Yellow Stone river, and all eyes were bent upon the hills and the prairie which is very broad, we saw a wolf about a quarter of a mile from our encampment, and Mr. OWEN McKENZIE was sent after it. The wolf however ran very swiftly and was not overtaken and shot until it had ran several miles. It dodged about in various directions, and at one time got out of sight behind the hills. This wolf was captured, and a piece of its flesh was boiled for supper; but as we had in the mean time caught about eighteen or twenty Cat-fish, we had an abundant meal and did not judge for ourselves whether the wolf was good eating or not, or if its flesh was like that of the Indian dogs, which we have had several opportunities of tasting.

Wolves are frequently deterred from feeding on animals shot by the hunters on the prairies, who, aware of the cautious and timid character of these rapacious beasts, attach to the game they are obliged to leave behind them a part of their clothing, a handkerchief, &c., or scatter gun powder around the carcass, which the cowardly animals dare not approach although they will watch it for hours at a time, and as soon as the hunter returns and takes out the entrails of the game he had left thus protected, and carries off the pieces he wishes, leaving the coarser parts for the benefit of these hungry animals, they come forward and enjoy the feast. The hunters who occasionally assisted us when we were at Fort Union, related numerous stratagems of this kind to which they had resorted to keep off the wolves when on a hunt.

The wolves of the prairies form burrows, wherein they bring forth their young, and which have more than one entrance; they produce from six to eleven at a birth, of which there are very seldom two alike in colour. The wolf lives to a great age and does not change its colour with increase of years.

N°15. PLATE LXXIII

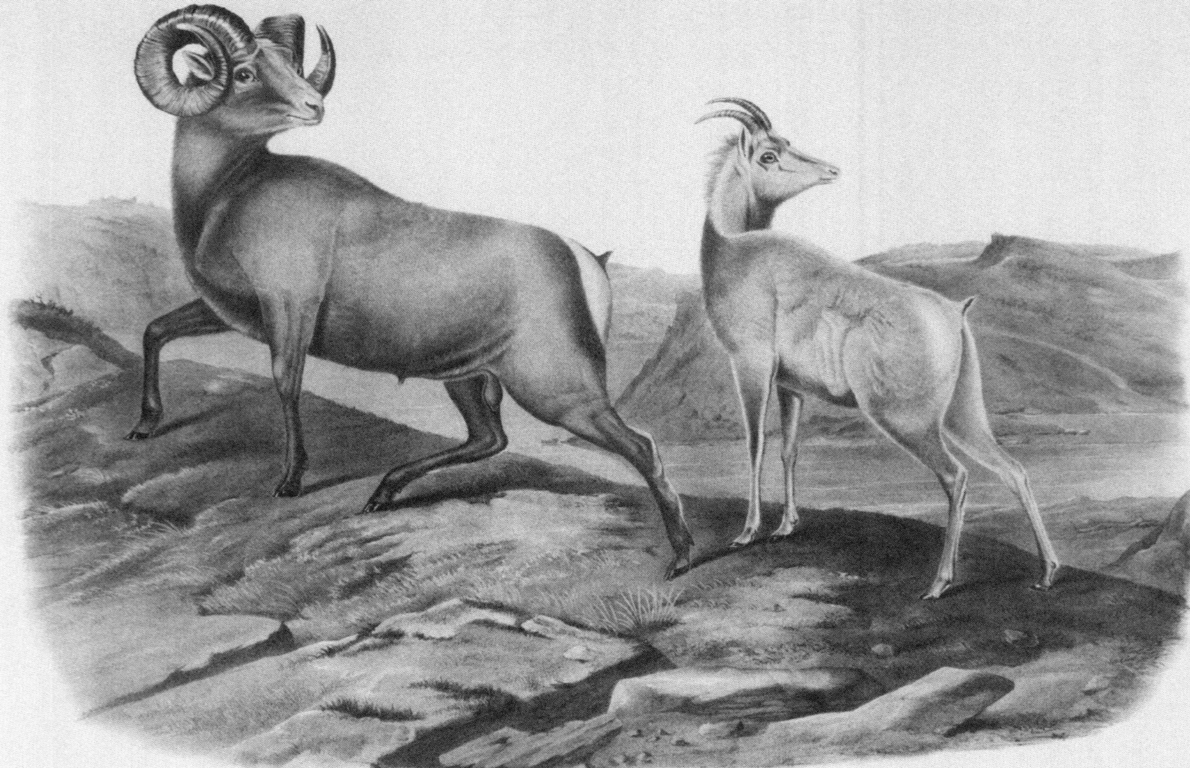

OVIS MONTANA, DESM

ROCKY MOUNTAIN SHEEP

Rocky Mountain Sheep

THE TRUE SUBJECT OF THIS ESSAY IS NOT THE QUADRUPED named in the title; it is instead the landscape on and in which the big horns live out their precarious lives: the Badlands, or "mauvaise terres," of eastern Montana and the western Dakotas. When Audubon visited this region in 1843, he offered top dollar to hunters who brought in specimens. In his nine-week residence there, however, only two managed to do so. Thus, the inaccessibility and elusiveness of the big horn are mirrored by the remote ruggedness and treacherous qualities of the landscape. Bachman developed this extensive description of the badlands from a long entry in Audubon's original journal.[28] In doing so, he introduced readers to one of North America's eeriest and least imaginable landscapes—as well as to the barriers the hunter-naturalist encounters in his attempts to study the big horns.

---

19. John Woodhouse Audubon, J. T. Bowen, lithographer, *Rocky Mountain Sheep*, 1845–46, hand-colored lithograph, plate 73 of *The Viviparous Quadrupeds of North America*. Courtesy of the Jule Collins Smith Museum of Fine Art, Auburn University; Louise Hauss and David Brent Miller Audubon Collection.

OVIS MONTANA.—DESM.

ROCKY MOUNTAIN SHEEP.

PLATE LXXIII.—MALE AND FEMALE.

HABITS.

It was on the 12th of June, 1843, that we first saw this remarkable animal; we were near the confluence of the Yellow Stone river with the Missouri, when a group of them, numbering twenty-two in all, came in sight. This flock was composed of rams and ewes, with only one young one or lamb among them. They scampered up and down the hills much in the manner of common sheep, but notwithstanding all our anxious efforts to get within gun-shot, we were unable to do so, and were obliged to content ourselves with this first sight of the Rocky Mountain Ram.

The parts of the country usually chosen by these animals for their pastures, are the most extraordinary broken and precipitous clay hills or stony eminences that exist in the wild regions belonging to the Rocky Mountain chain. They never resort to the low lands or plains except when about to remove their quarters, or swim across rivers, which they do well and tolerably fast. Perhaps some idea of the country they inhabit (which is called by the French Canadians and hunters, "mauvaise terres") may be formed by imagining some hundreds of loaves of sugar of different sizes, irregularly broken and truncated at top, placed somewhat apart, and magnifying them into hills of considerable size. Over these hills and ravines the Rocky Mountain Sheep bound up and down among the sugar loaf shaped peaks, and you may estimate the difficulty of approaching them, and conceive the great activity and sure-footedness of this species, which, together with their extreme wildness and keen sense of smell, enable them to baffle the most vigorous and agile hunter.

They form paths around these irregular clay cones that are at times from six to eight hundred feet high, and in some situations are even fifteen hundred feet or more above the adjacent prairies, and along these they run at full speed, while to the eye of the spectator below, these tracks do not appear to be more than a few inches wide, although they are generally from a foot to eighteen inches in breadth. In many places columns or piles of clay, or hardened earth, are to be seen eight or ten feet above the adjacent surface, covered or coped with a slaty flat rock, thus resembling gigantic toad stools, and upon these singular places the big horns are frequently seen, gazing at the hunter who is winding about far below, looking like so many statues on their elevated pedestals. One cannot imagine how these animals reach these curious places, especially with their young along with them, which are sometimes brought forth on these inaccessible points, beyond the reach of their greatest enemies, the wolves, which prey upon them whenever they stray into the plains below.

The "mauvaise terres" are mostly formed of greyish white clay, very sparsely covered with small patches of thin grass, on which the Rocky Mountain Sheep feed. In wet weather it is almost impossible for any man to climb up one of these extraordinary conical hills, as they are slippery, greasy and treacherous. Often when a big horn is seen on the top of a hill, the hunter has to ramble round three or four miles before he can reach a position within gun-shot of the game, and if perceived by the animal, it is useless for him to pursue him any further that day.

The tops of some of the hills in the "mauvaise terres" are composed of a conglomerated mass of stones, sand, clay and various coloured earths, frequently of the appearance and colour of bricks. We also observed in these masses a quantity of pumice stone, and these hills, we are inclined to think are the result of volcanic action. Their bases often cover an area of twenty acres; there are regular horizontal strata running across the whole chain of these hills, composed of different coloured clay, coal and earth, more or less impregnated with salt and other minerals, and occasionally intermixed with lava, sulphur, oxide and sulphate of iron; and in the sandy parts at the top of the highest hills, we found shells, but so soft and crumbling as to fall to pieces when we attempted to pick them out. We found in the "mauvaise terres," also, globular shaped masses of heavy stone and pieces of petrified wood, from fragments two or three inches wide, to stumps of three or four feet thick, apparently cotton wood and cedar. On the sides of some of the hills at various heights, are shelf-like ledges or rock projecting from the surface in a level direction, from two to six and even ten feet, generally square or flat. These ledges are much resorted to by the big horns during the heat of the day. Between these hills there is sometimes a growth of stunted cedar trees, underneath which there is a fine sweet grass, and on the summits in some cases a short dry wiry grass is found, and quantities of that pest of the Upper Missouri country, the flat-broad-leaved Cactus, the spines of which often lame the hunter. Occasionally the hills in the "mauvaise terres" are separated by numerous ravines, often not more than ten or fifteen feet wide, but sometimes from ten to fifty feet deep, and now and then the hunter comes to the brink of one so deep and wide as to make his head giddy as he looks down into the abyss below. The edges of the cañons (as these sort of channels are called in Mexico) are overgrown with bushes, wild cherries, &c., and here and there the Bison will manage to cut paths to cross them, descending in an oblique and zig-zag direction; these paths however are rarely found except where the ravine is of great length, and in general the only mode of crossing the ravine is to go along the margin of it until you come to the head, which is generally at the base of some hill, and thus get round.

These ravines exist between nearly every two neighbouring hills, although there are occasionally places where three or more hills form only one. All of them however run to meet each other and connect with the largest, the size of which bears its proportion to that of its tributaries and their number.

Where these ravines have no outlet into a spring or water course they have subterranean drains, and in some of the valleys and even on the tops of the hills, there are cavities called "sink holes;" the earth near these holes is occasionally undermined by the water running round in circles underneath, leaving a crust insufficient to bear the weight of a man, and when an unfortunate hunter treads on the deceitful surface it gives way, and he finds himself in an unpleasant and at times dangerous predicament. These holes sometimes gradually enlarge and run into ravines below them. It is almost impossible to traverse the "mauvaise terres" with a horse, unless with great care, and with a thorough knowledge of the country. The chase or hunt after the big horn, owing to the character of the country, (as we have described it,) is attended with much danger, as the least slip might precipitate one headlong into the ravine below, the sides of the hills being destitute of every thing to hold on by excepting a projecting stone or tuft of worm wood, scattered here and there, without which even the most daring hunter could not ascend them.

In some cases the water has washed out caves of different shapes and sizes, some of which present the most fantastic forms and are naked and barren to a great degree. The water that is found in the springs in these broken lands is mostly impregnated with salts, sulphur, magnesia, &c.; but unpleasant as it tastes, it is frequently the only beverage for the hunter, and luckily is often almost as cold as ice, which renders it less disagreeable. In general this water has the effect very soon of a cathartic and emetic. Venomous snakes of various kinds inhabit the "mauvaise terres," but we saw only one copper-head.

Conceiving that a more particular account of these countries may be interesting, we will here insert a notice of them given to us by Mr. DEWEY, the principal clerk at Fort Union. He begins as follows:

"This curious country is situated, or rather begins half way up White river, and runs from south east to north west for about sixty miles in length, and varying from fifteen to forty miles in width. It touches the head of the Teton river and branches of Chicune, and joins the Black Hills at the south fork of the latter river. The hills are in some places five or six hundred yards high and upwards. They are composed of clay of various colours, arranged in layers or strata running nearly horizontally, each layer being of a different colour, white, red, blue, green, black, yellow, and almost every other colour, appearing at exactly the same height on every hill.

"From the quantity of pumice stone and melted ores found throughout them, one might suppose that they had been reduced to this state by volcanic action. From the head of the Teton river, to cross these hills to White river is about fifteen miles; there is but one place to descend, and the road is not known; the only way to proceed is to go round the end of them on the banks of the White river, and following that stream ascend to the desired point. In four days' march a man will make about fifteen miles in crossing through the 'mauvaise terres.' At first sight these hills look like some ancient city in ruins, and but little imagination is necessary to give them the appearance of castles, walls, towers, steeples, &c. The descent is by a road about five feet broad, winding around and among the hills, made at first probably by the bisons and the big horn sheep, and now rendered practicable by the Indians and others who have occasion to use it. It is however too steep to travel down with a loaded horse or mule, say about one foot in three, for a mile or so, after which the bases of the hills are about level with each other, but the valleys between them are cut up by great ravines in almost every direction from five to twenty and even fifty feet deep.

"In going over this part of the country great precaution is necessary, for a slip of the foot would precipitate either man or horse into the gulf below. When I descended, the interpreter, B. DAUMINE, a half breed, (having his eyes bandaged) was led by the hand of an Indian."

Something like copperas in taste and appearance is found in large quantities, as well as pumice stone, every where. This country is the principal residence of the big horn sheep, the panther and grizzly bear; big horns especially are numerous, being in bands of from twenty to thirty, and are frequently seen at the tops of the highest peaks, completely inaccessible to any other animal. There is but one step from the prairie to the barren clay, and this step marks the difference for nearly its whole length. These "mauvaise terres" have no connexion or affinity to the surrounding country, but are, as it were, set apart for the habitation of the big horns and bears. The sight of this barren country causes one to think that thousands of square miles of earth have been carried off, and nothing left behind but the ruins of what was once a beautiful range of mountains. The principal part of these hills is white clay, which when wet is soft and adhesive, but the coloured strata are quite hard and are never discoloured by the rain, at least not to any extent, for after a hard rain the streams of water are of a pure milk white colour, untinged by any other, and so thick that ten gallons when settled will only yield about two gallons of pure limpid water, which, however, although clear when allowed to stand awhile, is scarcely drinkable, being salt and sulphurous in taste. The sediment has all the appearance of the clay already mentioned, which is nearly as white as chalk. There is only one place where wood and pure sweet water can be found in the whole range, which is at a spring nearly in the centre of the tract, and one day's journey from the White river, towards the Chicune. This appears a little singular, for if it were not for this the voyageur would be obliged to take a circuitous route of from four to five days. This spring is surrounded by a grove of ash trees, about two hundred yards in circumference. It immediately loses itself in the clay at the edge of the timber, and near the spring the road descends about sixty feet and runs through a sort of avenue at least half a mile wide, on each side of which are walls of clay extending horizontally about fifteen miles, and eighty feet high, for nearly the whole distance. Between these walls are small sugar-loaf shaped hills, and deep ravines, such as I have already described. The colours of the strata are preserved throughout. The principal volcano is the "Côte de tonnerre," from the mouth of which smoke and fire are seen to issue nearly at all times. In the neighbourhood and all around, an immense quantity of pumice stone is deposited, and from the noises to be heard, no doubt whatever exists that eruptions may from time to time be expected. There is another smaller hill which I saw giving forth heated vapours and smoke, but in general if the weather is clear the summits of the Black hills are obscured by a mist, from which circumstance many superstitions of the Indians have arisen. The highest of the Black hills are fully as high as the Allegheny mountains, and their remarkable shapes and singular characters deserve the attention of our geologists, especially as it is chiefly among these hills that fossil petrefactions are abundantly met with.

The Rocky Mountain Sheep are gregarious, and the males fight fiercely with each other in the manner of common rams. Their horns are exceedingly heavy and strong, and some that we have seen have a battered appearance, showing that the animal to which they belonged must have butted against rocks or trees, or probably had fallen from some elevation on to the stony surface below. We have heard it said that the Rocky Mountain Sheep descend the steepest hills head foremost, and they may thus come in contact with projecting rocks, or fall from a height on their enormous horns.

As is the case with some animals of the deer tribe, the young rams of this species and the females herd together during the winter and spring, while the old rams form separate flocks, except during the rutting season in December.

In the months of June and July the ewes bring forth, usually one, and occasionally, but rarely, two.

Dr. Richardson, on the authority of Drummond, states that in the retired parts of the mountains where the hunters had seldom penetrated, he (Drummond) found no difficulty in approaching the Rocky Mountain Sheep, which there exhibited the simplicity of character so remarkable in the domestic species; but that where they had been often fired at, they were exceedingly wild, alarmed their companions on the approach of danger by a hissing noise, and scaled the rocks with a speed and agility that baffled pursuit. He lost several that he had mortally wounded, by their retiring to die among the secluded precipices. They are, we are farther informed on the authority of Drummond, in the habit of paying daily visits to certain caves in the mountains that are encrusted with saline efflorescence. The same gentleman mentions that the horns of the old rams attain a size so enormous, and curve so much forwards and downwards, that they effectually prevent the animal from feeding on the level ground.

All our travellers who have tasted the flesh of the Rocky Mountain Sheep, represent it as very delicious when in season, superior to that of any species of deer in the west, and even exceeding in flavour the finest mutton.

We have often been surprised that no living specimen of this very interesting animal has ever been carried to Europe, or any of our Atlantic cities, where it would be an object of great interest.

N°16.

PLATE LXXVII

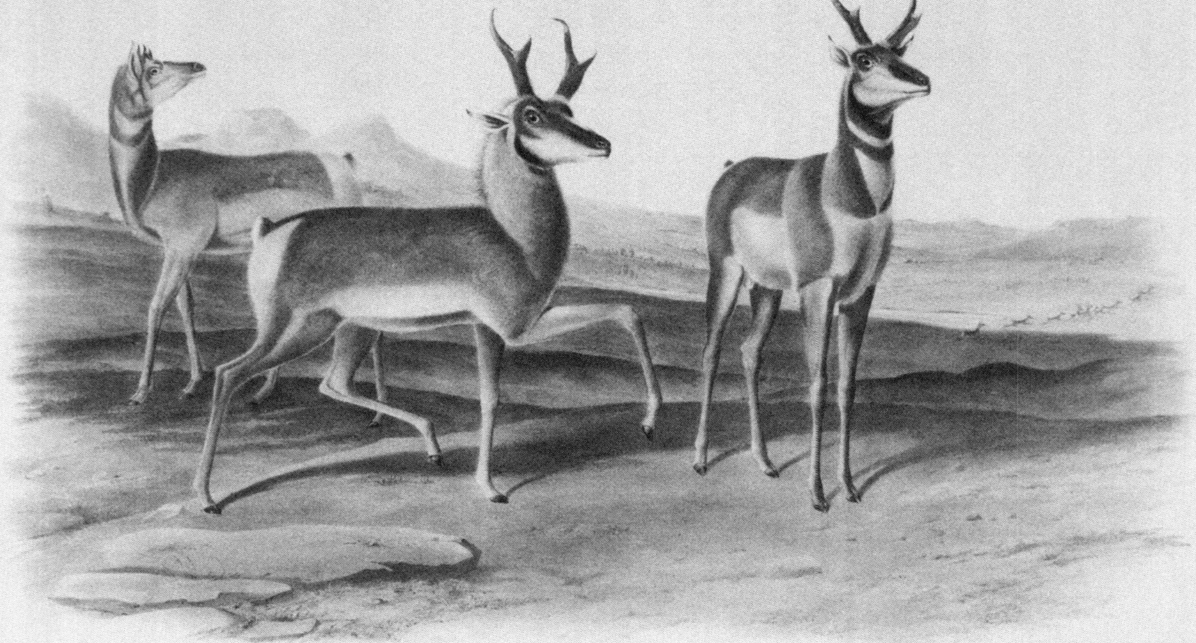

ANTILOPE AMERICANA, ORD.

Prong-horned Antelope.

Males & Female.

Prong-horned Antelope

THE OPENING CELEBRATION, "HURRA FOR THE PRAIRIES AND the swift antelopes," is quickly undercut by the lethal shot of the hunter and by the warning against relying on "the flowery descriptions of travellers" for an accurate understanding of those prairies and of their human inhabitants. George Catlin's written descriptions and graphic depictions of both were greatly disparaged by Audubon and his companions on the 1843 expedition to Fort Union, the main setting for this essay.[29] The pronghorn itself is nonetheless represented as completely admirable in its natural setting: a courageous, protective parent; virile; and incredibly fleet across vast, open, rugged terrain, of which few readers of these essays had any conception. Another value of many of the quadruped essays is in their preservation of material from journals that were later burned by Audubon's granddaughter Maria Rebecca Audubon.[30] In this essay, the accounts of captive young pronghorns and the hunting excursion with Edward Harris of July 21, 1843, were taken from Audubon's original Missouri River journals before those were destroyed. The fact that the long concluding passage from John Woodhouse Audubon's journals does not appear in the edition published in 1906 with Maria's permission and under her supervision suggests that she burned both her father's journals and her grandfather's.

---

20. John Woodhouse Audubon, J. T. Bowen, lithographer, *Prong-horned Antelope*, 1846–48, hand-colored lithograph, plate 77 of *The Viviparous Quadrupeds of North America*. Courtesy of the Jule Collins Smith Museum of Fine Art, Auburn University; Louise Hauss and David Brent Miller Audubon Collection.

ANTILOCAPRA AMERICANA.—ORD.

PRONG-HORNED ANTELOPE.

PLATE LXXVII.—MALE AND FEMALE.

HABITS.

Reader, let us carry you with us to the boundless plains over which the prong-horn speeds. Hurra for the prairies and the swift antelopes, as they fleet by the hunter like flashes or meteors, seen but for an instant, for quickly do they pass out of sight in the undulating ground, covered with tall rank grass. Observe now a flock of these beautiful animals; they are not afraid of man—they pause in their rapid course to gaze on the hunter, and stand with head erect, their ears as well as eyes directed towards him, and make a loud noise by stamping with their forefeet on the hard earth; but suddenly they become aware that he is no friend of theirs, and away they bound like a flock of frightened sheep—but far more swiftly do the graceful antelopes gallop off, even the kids running with extraordinary speed by the side of their parents—and now they turn around a steep hill and disappear, then perhaps again come in view, and once more stand and gaze at the intruder. Sometimes, eager with curiosity and anxious to examine the novel object which astonishes as well as alarms them, the antelopes on seeing a hunter, advance toward him, stopping at intervals, and then again advancing, and should the hunter partly conceal himself, and wave his handkerchief or a white or red rag on the end of his ramrod, he may draw the wondering animals quite close to him and then quickly seizing his rifle send a ball through the fattest of the group, ere the timid creatures have time to fly from the fatal spot.

The Indians, we were told, sometimes bring the antelope to within arrow-shot (bow-shot), by throwing themselves on their backs and kicking up their heels with a bit of a rag fastened to them, on seeing which moving amid the grass the antelope draws near to satisfy his curiosity.

The atmosphere on the western prairies is so pure and clear that an antelope is easily seen when fully one mile off, and you can tell whether it is feeding quietly or is alarmed; but beautiful as the transparent thin air shews all distant objects, we have never found the great western prairies equal the flowery descriptions of travellers. They lack the pure streamlet wherein the hunter may assuage his thirst—the delicious copses of dark, leafy trees; and even the thousands of fragrant flowers, which they are poetically described as possessing, are generally of the smaller varieties; and the Indian who roams over them is far from the ideal being—all grace, strength and nobleness, in his savage freedom—that we from these descriptions conceive him. Reader, do not expect to find any of the vast prairies that border the Upper Missouri, or the Yellow-Stone rivers, and extend to the Salt Lakes amid the Californian range of the Rocky Mountains, verdant pastures ready for flocks and herds, and full of the soft perfume of the violet. No; you will find an immense waste of stony, gravelly, barren soil, stretched before you; you will be tormented with thirst, half eaten up by stinging flies, and lucky will you be if at night you find wood and water enough to supply your fire and make your cup of coffee; and should you meet a band of Indians, you will find them wrapped in old buffalo robes, their bodies filthy and covered with vermin, and by stealing or begging they will obtain from you perhaps more than you can spare from your scanty store of necessaries, and armed with bows and arrows or firearms, they are not unfrequently ready to murder, or at least rob you of all your personal property, including your ammunition, gun and butcher knife!

The Prong-horned Antelope brings forth its young about the same time as the common deer: from early in May to the middle of June; it has generally two fawns at a birth. We have heard of no case in which more than that number has been dropped at a time, and probably in some cases only one is fawned by the dam. The young are not spotted like the fawn of the common deer, but are of a uniform dun colour. The dam remains by her young for some days after they are born, feeding immediately around the spot, and afterwards gradually enlarging her range; when the young are a fortnight old they have gained strength and speed enough to escape with their fleet-footed mother from wolves or other four-footed foes. Sometimes, however, the wolves discover and attack the young when they are too feeble to escape, and the mother then displays the most devoted courage in their defence. She rushes on them, butting and striking with her short horns, and sometimes tosses a wolf heels over head, she also uses her forefeet, with which she deals severe blows, and if the wolves are not in strong force, or desperate with hunger, puts them to flight, and then seeks with her young a safer pasturage, or some almost inaccessible rocky hill side.

The rutting season of this species commences in September, the bucks run for about six weeks, and during this period fight with great courage and even a degree of ferocity. When a male sees another approaching, or accidentally comes upon one of his rivals, both parties run at each other with their heads lowered and their eyes flashing angrily, and while they strike with their horns they wheel and bound with prodigious activity and rapidity, giving and receiving severe wounds,—sometimes like fencers, getting within each other's "points," and each hooking his antagonist with the recurved branches of his horns, which bend considerably inwards and downwards.

The Prong-horned Antelope usually inhabits the low prairies adjoining the covered woody bottoms during spring and autumn, but is also found on the high or upland prairies, or amid broken hills, and is to be seen along the margins of the rivers and streams: it swims very fast and well, and occasionally a herd when startled may be seen crossing a river in straggling files, but without disorder, and apparently with ease.

Sometimes a few of these animals, or even only one or two by themselves may be seen, whilst in other instances several hundreds are congregated in a herd. They are remarkably shy, are possessed of a fine sense of smell, and have large and beautiful eyes, which enable them to scan the surface of the undulating prairie and detect the lurking Indian or wolf, creep he ever so cautiously through the grasses, unless some intervening elevation or copsewood conceal his approach. It is, therefore, necessary for the hunter to keep well to leeward, and to use extraordinary caution in "sneaking" after this species; and he must also exercise a great deal of patience and move very slowly and only at intervals, when the animals with heads to the ground or averted from him, are feeding or attracted by some other object. When they discover a man thus stealthily moving near them, at first sight they fly from him with great speed, and often retire to the broken grounds of the clay hills, from which they are not often tempted to stray a great distance at any time. As we have already mentioned, there are means, however, to excite the timid antelope to draw near the hunter, by arousing his curiosity and decoying him to his ruin. The antelopes of the Upper Missouri country are frequently shot by the Indians whilst crossing the river; and, as we were informed, preferred the northern side of the Missouri; which, no doubt, arises from the prevalence on that bank of the river of certain plants, trees or grasses, that they are most fond of. Males and females are found together at all seasons of the year. We have been told that probably a thousand or more of these animals have been seen in a single herd or flock at one time, in the spring.

It was supposed by the hunters at Fort Union, that the prong-horned antelope dropped its horns; but as no person had ever shot or killed one without these ornamental and useful appendages, we managed to prove the contrary to the men at the fort by knocking off the bony part of the horn, and showing the hard, spongy membrane beneath, well attached to the skull and perfectly immoveable.

The Prong-horned Antelope is never found on the Missouri river below *L'eau qui court*; but above that stream they are found along the great Missouri and its tributaries, in all the country east of the Rocky Mountains, and in many of the great valleys that are to be met with among these extraordinary "big hills." None of these antelopes are found on the shores of the Mississippi, although on the headwaters of the Saint Peter's river they have been tolerably abundant. Their walk is a slow and somewhat pompous gait, their trot elegant and graceful, and their gallop or "run" light and inconceivably swift; they pass along, up or down hills, or along the level plain with the same apparent ease, while so rapidly do their legs perform their graceful movements in propelling their bodies over the ground, that like the spokes of a fast turning wheel we can hardly see them, but instead, observe a gauzy or film-like appearance where they should be visible.

In autumn, this species is fatter than at any other period. Their liver is much prized as a delicacy, and we have heard that many of these animals are killed simply to procure this choice morsel. This antelope feeds on the short grass of the prairies, on mosses, buds, &c.; and suffers greatly during the hard winters experienced in the north-west; especially when the snow is several feet in depth. At such times they can be caught by hunters provided with snow shoes, and they are in this manner killed, even in sight of Fort Union, from time to time.

It is exceedingly difficult to rear the young of this species; and, although many attempts have been made at Fort Union, and even an old one caught and brought within an enclosure to keep the young company, they became furious, and ran and butted alternately against the picket-wall or fence, until they were too much bruised and exhausted to recover. William Sublette, Esq., of St. Louis, Missouri, however, brought with him to that city a female antelope, caught when quite young on the prairies of the far west, which grew to maturity, and was so very gentle, that it would go all over the house, mounting or descending the stairs, and occasionally going on to the roof of the building he lived in. This female was alive when we first reached St. Louis, but not being aware of its existence, we never saw it. It was killed before we left by a buck-elk, belonging to the same gentleman.

Whilst on our journey in the far west, in 1843, on one occasion, we had the gratification of seeing an old female, in a flock of eight or ten antelopes, suckling its young. The little beauty performed this operation precisely in the manner of our common lambs, almost kneeling down, bending its head upwards, its rump elevated, it thumped the bag of its mother, from time to time, and reminded us of far distant scenes, where peaceful flocks feed and repose under the safeguard of our race, and no prowling wolf or hungry Indian defeats the hopes of the good shepherd who nightly folds his stock of the Leicester or Bakewell breed. Our wild antelopes, however, as we approached them, scampered away; and we were delighted to see that first, and in the van of all, was the young one!

On the 21st July, 1843, whilst in company with our friend, Edward Harris, Esq., during one of our hunting excursions, we came in sight of an antelope gazing at us, and determined to stop and try if we could bring him toward us by the trick we have already mentioned, of throwing our legs up in the air and kicking them about, whilst lying on our back in the grass. We kicked away first one foot and then the other, and sure enough, the antelope walked slowly toward us, apparently with great caution and suspicion. In about twenty minutes he had advanced towards us some two or three hundred yards. He was a superb male, and we looked at him for several minutes when about sixty yards off. We could see his fine protruding eyes; and being loaded with buck-shot, we took aim and pulled trigger. Off he went, as if pursued by a whole Black-foot Indian hunting party. Friend HARRIS sent a ball at him, but was as unsuccessful as ourselves, for he only ran the faster for several hundred yards, when he stopped for a few minutes, looked again at us, and then went off, without pausing as long as he was in sight. We have been informed by LAFLEUR, a man employed by the Company, that antelopes will escape with great ease even when they have one limb broken, as they can run fast enough upon three legs to defy any pursuit. Whilst we were encamped at the "Three Mamelles," about sixty miles west of Fort Union, early one morning an antelope was heard *snorting*, and was seen by some of our party for a few minutes only. This snorting, as it is called, resembles a loud whistling, singing sound prolonged, and is very different from the loud and clear snorting of our common deer; but it has always appeared to us to be almost useless to attempt to describe it; and although at this moment we have the sound of the antelope's snort in our ears, we feel quite unable to give its equivalent in words or syllables.

The antelope has no lachrymal pits under the eyes, as have deer and elks, nor has it any gland on the hind leg, so curious a feature in many of those animals of the deer tribe which drop their horns annually, and only wanting (so far as our knowledge extends) in the *Cervus Richardsonii*, which we consider in consequence as approaching the genus *Antilope*, and in a small deer from Yucatan and Mexico, of which we had a living specimen for some time in our possession.

The prong-horned antelope often dies on the open prairies during severe winter weather, and the remains of shockingly poor, starved, miserable individuals of this species, in a state of the utmost emaciation, are now and then found dead in the winter, even near Fort Union and other trading posts.

The present species is caught in pens in the same manner nearly as the bison, (which we have already described at p. [119]) but is generally despatched with clubs, principally by the women. In the winter of 1840, when the snow was deep in the ravines, having drifted, Mr. LAIDLAW, who was then at Fort Union, caught some of them by following them on horse-back and forcing them into these drifts, which in places were as much as ten to twelve feet deep. They were brought to the fort in a sleigh, and let loose about the rooms; they were to appearance so very gentle that the people suffered their children to handle them, although the animals were loose. They were placed in the carpenter's shop, one broke its neck by leaping over a turning-lathe, and the rest all died; for as soon as they had appeased the cravings of hunger, they began to fret for their accustomed liberty, and regained all their original wildness. They leaped, kicked and butted themselves against every obstacle, until too much exhausted to recover.—These individuals were all captured by placing nooses, fixed on the end of long poles, round their necks, whilst they were embedded in the soft and deep snow drifts, to which they had been driven by Mr. LAIDLAW.

There are some peculiarities in the gait of this species that we have not yet noticed. The moment they observe a man or other strange object producing an alarm, they bound off for some thirty or forty yards, raising all their legs at the same time, and *bouncing*, as it were, from two to three feet above the ground; after this they stretch their bodies out and gallop at an extraordinary speed. We have seen some which, when started, would move off and run a space of several miles, in what we thought did not exceed a greater number of minutes!

From what we have already said, it will be inferred that the wolf is one of the most formidable enemies of this species. We have, however, not yet mentioned that in some very cold and backward seasons the young, when first born at such times, are destroyed by these marauders in such numbers that the hunters perceive the deficiency and call them scarce for the next season. Antelopes are remarkably fond of saline water or salt, and know well where the *salt-licks* are found. They return to them daily, if near their grazing grounds, and lay down by them, after licking the salty earth or drinking the salt water. Here they will remain for hours at a time, in fact until hunger drives them to seek in other places the juicy and nourishing grasses of the prairie. This species is fond of taking its stand, when alone, on some knoll, from which it can watch the movements of all wanderers on the plains around, and from which a fair chance to run in any direction is secured, although the object of its fear may be concealed from view occasionally by a ravine, or by another projecting ridge like its own point of sight.

We had in our employ a hunter on the Yellow-Stone River, who killed two female antelopes and broke the leg of a third at one shot from an ordinary western rifle. The ball must have passed entirely through the two first of these animals.

We have represented on our plate two males and a female in the foreground, with a flock of these timid creatures running at full speed in the distance.

We subjoin the following account of the Antelopes seen by J. W. AUDUBON and his party on their overland journey through Northern Mexico and Sonora to California.

"Leaving Altar, Sonora, the country was flat and uninteresting, except that large patches of coarse grass, sometimes miles in length, took the place of the naked clay plains we had been riding through. The tall cactus, described by FREMONT and EMORY, in its eccentric forms was remarkable enough even by daylight, but at night, a very little superstition, with the curved and curiously distorted forms, produced in some cases by disease of the plant, or by the violent gales that periodically sweep those prairies, might make the traveller suppose this was a region in which beings supernatural stalked abroad. The shrill whistle of the Antelope, new to us all, added to the wild and unearthly character of the scene. The Maricapos Indians were said to be friendly, but we *did not know it*, and after our long watchings against Camanche, Apatche, Wako and Paramanii, who among us, as we knew how Indians sometimes personate the animals of the section they live in, but listened with intense interest to the slightest noise foreign to our previous knowledge. The short quick stampings of impatience or nervousness, continually repeated by the animals, were, however, soon distinguished in the stillness of our prairie camp at night, and feeling thus assured that only one of the deer tribe was the cause of our anxiety, blankets and tent soon covered us, and we left the beautiful and innocent creatures, now that we knew them, to their own reflections, if any they made, as to who and what we were, until morning.

"At day light, RHOADES and VAN HORN, two hunters good as ever accompanied a train across the broad prairies ranged over by Buffalo, Elk, or Deer, looked out the trails, and reported Antelopes; but brought none to camp; not expecting to see any more of this herd, we started on our tramp towards the great Sonora Desert.

"STEVENSON had a new horse, and as he had never been mounted without blindfolding him, after the Mexican fashion with young horses, being wild, his owner, by way of making him *more gentle*, commenced beating him with a stick that might have been selected to kill him; before I had time to know what was going on and interfere for the poor horse, he had looked to his own interests, pulled away, and with a bounding gallop went off, like an escaped prisoner, leading four of our best men and horses some ten miles ahead of the train, and when the runaway was at length overtaken, VAN HORN, PENNYPACKER, McCUSKER, and myself were greatly in advance; the curve we had made from the road was slight, and on reaching it again, no trail told that the company had passed, so we had time to look about us, and loitered to rest our tired horses, when simultaneously we saw the back of a deer or Antelope; its head was hidden by the tall grass in which it was grazing on the soft juicy young shoots at the roots of the old tussocks: VAN HORN, with his unerring aim and Mississippi rifle, the eccentric twist of which, no doubt taken from WESSON's patent, renders these guns superior to all we have tried, was told to kill it.[31] For a few seconds he was lost to our sight, though only a hundred yards from us, so low did he squat in the sparse tufts of dead grass and stinking wormwood. How curious it is to stand waiting the result of the skill and caution of the well tried hunter, at such a time; again and again we saw the back of the Antelope, as he passed one bunch of shrubbery after another, but never saw our hunter: at every moment we expected to see the wary animal with sense of smell so keen as nine times out of ten to save him from his enemies, bound away; but how different was his bound when he did leap, not forward, but straight upward. And now we saw VAN HORN, a quarter of a mile off, running to where the last leap was made by his prey, and then came on the sluggish air, the crack of his rifle, almost after we had forgotten to listen for it, as a rifle cracks nowhere except on prairies, where neither woods, rocks or hills send back the sound. When I saw this beautiful creature, a most magnificent male, the first I had ever seen in the flesh, though the drawing for the 'Quadrupeds' had been long made and published, how I wished to redraw it! delicate even to the descriptions of the gazelle, muscular and sinewy as the best bred grey hound that Scotland ever produced.

"I anticipated a treat, as VAN HORN gave me a hind quarter for our men, which I tied doubly secure to my saddle. But when night came, after ten hours' ride, although we enjoyed our steaks, the deer of the Cordilleras was too fresh in our memories to permit us to say that this Antelope was the best meat we had eaten."

* * *[32]

"The eastern spurs of the coast range were just behind us; the black-tailed deer was scarcely past, for a few miles back, high up on one of the conical velvety hills of this range, we had seen three, looking at us from under one of the dwarf oaks that grow at a certain altitude, in forms peculiar to this country; above or below, either a different formation or total absence of shrubbery occurring. We were winding along the base of a moderate line of hills of the *Sierra Nevada*, when what we took for a flock of sheep, the trail of which we had been following for three days on the way to the mines from Los Angeles, was discovered, and we hoped for mutton, to say nothing of the company we anticipated; but our flock of sheep was like the 'Phantom Bark,' for it 'seemed never the nigher,' *au contraire*, turning a hill went out of sight, and we never got another view; we saw another flock some miles on, and at first, supposing it the same, wondered how they could travel so fast.[33] This was probably another portion of the one we had trailed for so many days. We were gratified by the whole flock running near us, from which we argued we were in the chosen country of the Antelope, the broad Tule valley. The flock ran 'shearing' about, as the formation of the land compelled them to turn to the right or left, showing their sides alternately in light and shade. When they are on the mountain sides and discover a foe, or any object that frightens them, the whole flock rush headlong for the plains, whether the enemy is likely to intercept them or not, and they seem to fly with the single idea, that they are in a dangerous place, and must change it for some other, no matter what; at times a whole flock would run to within shot of our company, determined as it were to go through the line, and I believe in one or two instances would have done so, if they had not been shot at by our too impatient party. When on the plains, the same desire possesses them to get to the hills, and back they go a hundred or two in a flock, seldom slackening their speed, except for a few seconds to look again, and be more frightened than ever at what had first startled them. The rolling hills of the western line of the Sierra Nevada were their most favourite locality in this valley, as far as we saw, but LAYTON and myself met an accidental individual or two, nearly up to Sacramento city, as we travelled through the beautiful, park-like scenes of this portion of California to the diggings of the head waters of the 'American Fork.'

"As to the shedding of the horns of this species, I never was able to ascertain it, but a fine buck we killed, late in November, had a soft space between the head and horn, over the bone, that looked as if it had grown that length in one season. A young Antelope is better eating than a deer, but an old one, is *decidedly goaty*."

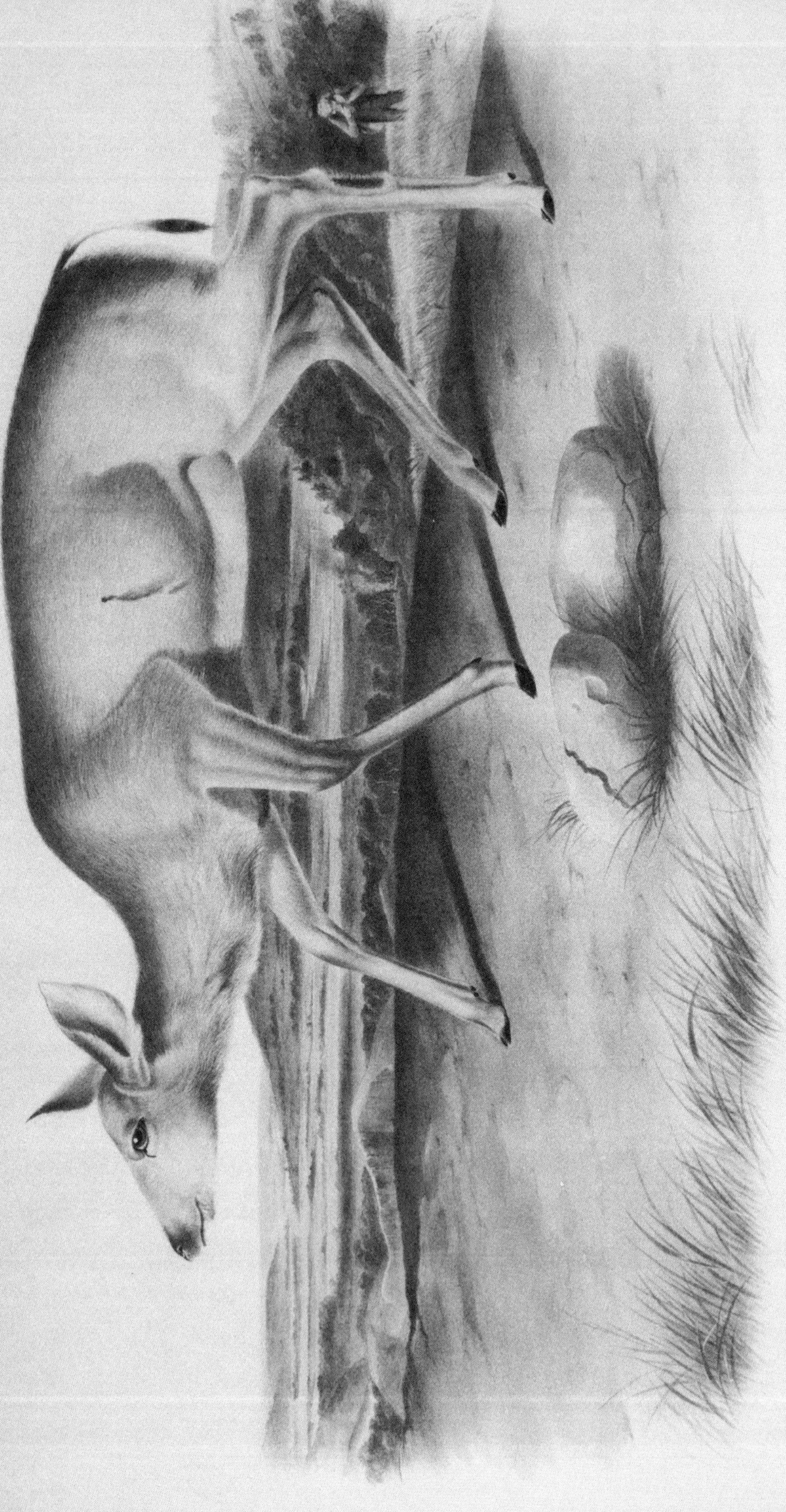

Mule Deer

DESPITE HIS HAVING BROUGHT A YOUNG BLACK-TAILED DEER home from Fort Union in 1843, Audubon's observations of the species on the northern prairies were scant and usually in the form of roasted venison, hence the brevity of the "Habits" section of this essay.[34] John Woodhouse's motive behind depicting his father in the plate as the hunter may have been to prod the American Woodsman jokingly by placing the bullet wound higher on the deer's body than a skilled hunter would have done. We include also the "General Remarks" section not only to illustrate the thoroughness with which our authors applied the principle of priority to the naming of species but also as testimony to the deliberateness to which the quadrupeds essays were intended to advance the state of natural history studies in the United States at the time. The National Institution for the Promotion of Science was born out of a nationalistic desire to do the same; Bachman's expressed and published frustration with its directors is intended to move a young American organization to live up to his ideals of the advancement of science.

---

21. John Woodhouse Audubon, J. T. Bowen, lithographer, *Black-tailed Deer*, 1846–48, hand-colored lithograph, plate 78 of *The Viviparous Quadrupeds of North America*. Courtesy of the Jule Collins Smith Museum of Fine Art, Auburn University; Louise Hauss and David Brent Miller Audubon Collection.

CERVUS MACROTIS.—SAY.

MULE DEER.

PLATE LXXVIII.—FEMALE.—SUMMER PELAGE.

HABITS.

The first opportunity was afforded us of observing this magnificent animal, on the 12th of May as we were ascending the Missouri, about eleven hundred miles above Fort Leavenworth. On winding along the banks, bordering a long and wide prairie, intermingled with willows and other small brush wood, we suddenly came in sight of four Mule or black-tailed Deer, which after standing a moment on the bank and looking at us, trotted leisurely away, without appearing to be much alarmed. After they had retired a few hundred yards, the two largest, apparently males, elevated themselves on their hind legs and pawed each other in the manner of the horse. They occasionally stopped for a moment, then trotted off again, appearing and disappearing from time to time, when becoming suddenly alarmed, they bounded off at a swift pace, until out of sight. They did not trot or run as irregularly as our Virginian Deer, and they appeared at a distance darker in colour, as the common Deer at this season is red. On the 25th of the same month, we met with four others, which in the present instance did not stop to be examined; we saw them at a distance rapidly and gracefully hurrying out of sight. On the evening of the same day, one of our hunters brought to us a young Buck of this species, the horns of which, however, were yet too small to enable us to judge what would be their appearance in the adult animal. When on the Upper Missouri, near Fort Union, we obtained through the aid of our hunters, the female Black-tailed Deer, from which our figure, description and measurements have been made. We regret exceedingly that we were so unfortunate as not to have been able to procure a male, the delineation of which we must leave to our successors.

The habits of this animal approach more nearly those of the Elk, than of either the long-tailed or Virginian Deer. Like the former they remove far from the settlements, fly from the vicinity of the hunter's camp, and when once fairly started, run for a mile or two before they come to a pause.

The female produces one or two young, in the month of June.

We have figured a female in summer pelage, and have represented the animal in an exhausted state, wounded through the body, and about to drop down, whilst the hunter is seen approaching, through the tall grass, anticipating the moment when she will reel and fall in her tracks.

GENERAL REMARKS.

Since the days of LEWIS and CLARK, an impression has existed among naturalists that there were two species of black-tailed Deer; the one existing to the east of the Rocky Mountains, and the other, bordering on the Pacific, and extending through upper California. Although the descriptions of those fearless and enterprising travellers are not scientific, yet their accounts of the various species of animals, existing on the line of their travels, have in nearly every case been found correct, and their description of habits very accurate. They state that "the black-tailed fallow Deer are peculiar to this coast (mouth of the Columbia.) and are a distinct species, partaking equally of the qualities of the Mule and the common Deer (*C. Virginianus.*) The receptacle of the eye more conspicuous, their legs shorter, their bodies thicker and larger. The tail is of the same length with that of the common Deer, the hair on the under side, white; and on its sides and top of a deep jetty black; the horns resembling in form and colour those of the Mule Deer, which it likewise resembles in its gait.[35] The black-tailed Deer never runs at full speed, but bounds with every foot from the ground at the same time, like the Mule Deer. He sometimes inhabits the woodlands, but more often the prairies and open grounds. It may be generally said that he is of a size larger than the common Deer, and less than the Mule Deer. The flesh is seldom fat, and in flavour is far inferior to any other of the species"! It will be seen from the above, that they regarded the Mule Deer of the plains of Western Missouri as a distinct species from the black-tailed Deer, which existed along the Pacific coast near the Columbia river.

SAY gave the first scientific description of the Mule Deer, which he named "*Cervus Macrotis*," which having the priority we have retained.

RICHARDSON, whilst at the Saskatchewan, sought to obtain specimens of this animal for description, but it being a season of scarcity, the appetites of the hunters proved superior to their love of gain, and they devoured the Deer they had shot, even to their skins. When after his return to Europe, in 1829, he published the animals obtained in the expedition, he very properly added such other species as had been collected by the labours of DOUGLASS, DRUMMOND and other naturalists, who had explored the northern and western portions of America. Finding in the Zoological Museum a specimen of black-tailed Deer, procured on the western coast of America, by DOUGLASS, he concluded that it was the species described by SAY, *C. macrotis*; at the close of his article, he refers to the animal mentioned by LEWIS and CLARK, as the black-tailed Deer of the western coast, of which he states, that he had seen no specimen, designating it (F. B. Am. p. 257) *C. macrotis, var. Columbiana*. We have, however, come to the conclusion that the animal described by RICHARDSON was the very western species to which LEWIS and CLARK refer, and that whilst his description of the specimen was correct, he erred in the name, he having described not the Mule Deer of LEWIS and CLARK and SAY, but the Columbian black-tailed Deer, our drawing of which was made from the identical specimen described and figured by RICHARDSON. We have named it, after its first describer, *Cervus Richardsonii*.

The following characters will serve to designate the species.

C. Richardsonii, considerably smaller than *C. macrotis*, the male of the former species being smaller than the female of the latter. The hair of *C. macrotis* is very coarse and spongy, like that of the elk, that of *C. Richardsonii* is much finer and more resembles that of the Virginian Deer. The *C. Richardsonii* has no glandular opening on the outer surface of the hind leg below the knee joint, approaching in this particular the antelopes which are also without such openings, whilst the corresponding portion in *C. macrotis* is longer than that of any known species of Deer, being six inches in length. They differ in the shape of their horns, the *C. macrotis* having the antlers more slender, much less knobbed, and less covered with sharp points than those of the latter. They are also destitute of the basal process, so conspicuous in *C. macrotis*. We regret exceedingly that from circumstances beyond our control, we have been enabled to give a figure of the female only of *C. macrotis*, and of the male only of *C. Richardsonii*. The former was figured from the specimen we obtained at Fort Union, and for the latter we are indebted to the directors of the Zool. Society of London, who very kindly permitted us to make a drawing from the specimen previously described and figured by RICHARDSON.

Note.—In connection with this subject, we are deeply pained to be compelled to notice the obstructions thrown in the way of our pursuits by the directors of the National Institute at Washington, which city we visited shortly after the return of our exploring expedition, when we were kindly invited by Mr. PEALE to an examination of the valuable specimens of Natural History, collected by our adventurous countrymen. We pointed out to him one or two skins of the black-tailed Deer from the Western coast, which we both agreed differed from the *C. Macrotis* of SAY. We proposed to him that he should give a short description of the species, and select the name, which we would afterwards adopt in our work—this is in accordance with the mode usually pursued, and would have only occupied an hour. After the lapse of several years, we made an application by letter to the directors of the Institution for the privilege of making a drawing of the specimen; this we were not only refused, but were even denied the privilege of looking at the specimen, which we were very anxious to see, in order to be enabled to point out in the most satisfactory manner the characteristics by which these two closely allied species of Deer inhabiting our country could be distinguished from each other.

We cannot but contrast the narrow-minded policy pursued towards us in our application at Washington, with the liberality and generosity which was at all times extended to us in Europe under similar circumstances. When we visited England in 1838, the Directors of the Zoological Society opened its museum and assigned to us a private room, of which they gave us the key, and which we occupied for nearly a month—the specimens were taken from the cases by their attendants and brought to us, and when we discovered in the collection undescribed species, we were encouraged and aided in describing them. The same facilities were afforded us in the British museum, and in those of Edinburgh, Paris, Berlin, Dresden, and Zurich.

The British Government, as well as our own, gave us all the assistance which could be rendered by either, consistent with other public services, and we derived material advantages from the aid afforded us by the revenue service and the various military stations we have visited in our researches, in Labrador—in Florida—in the far West, and in Texas.

We know not who were the Directors of the National Institute when our reasonable request was so cavalierly rejected, nor have we inquired whether any changes in policy have since taken place in regard to the collection of animals at Washington, but we feel it our duty publicly to protest against a conduct so narrow, selfish, and inconsistent with the liberality of our free institutions and so little adapted to promote one of the objects sought to be gained by the exploring expedition—viz.: the advancement of natural history.

When the Hudsons Bay Company received an intimation that we would be glad to obtain any specimens they could furnish us from their trading posts in the arctic regions, they immediately gave orders to their agents and we secured from them rare animals and skins, procured at considerable labour and expense, and sent to us without cost, knowing and believing that in benefitting the cause of natural science they would receive a sufficient reward.

No. 16

PLATE LXXIX.

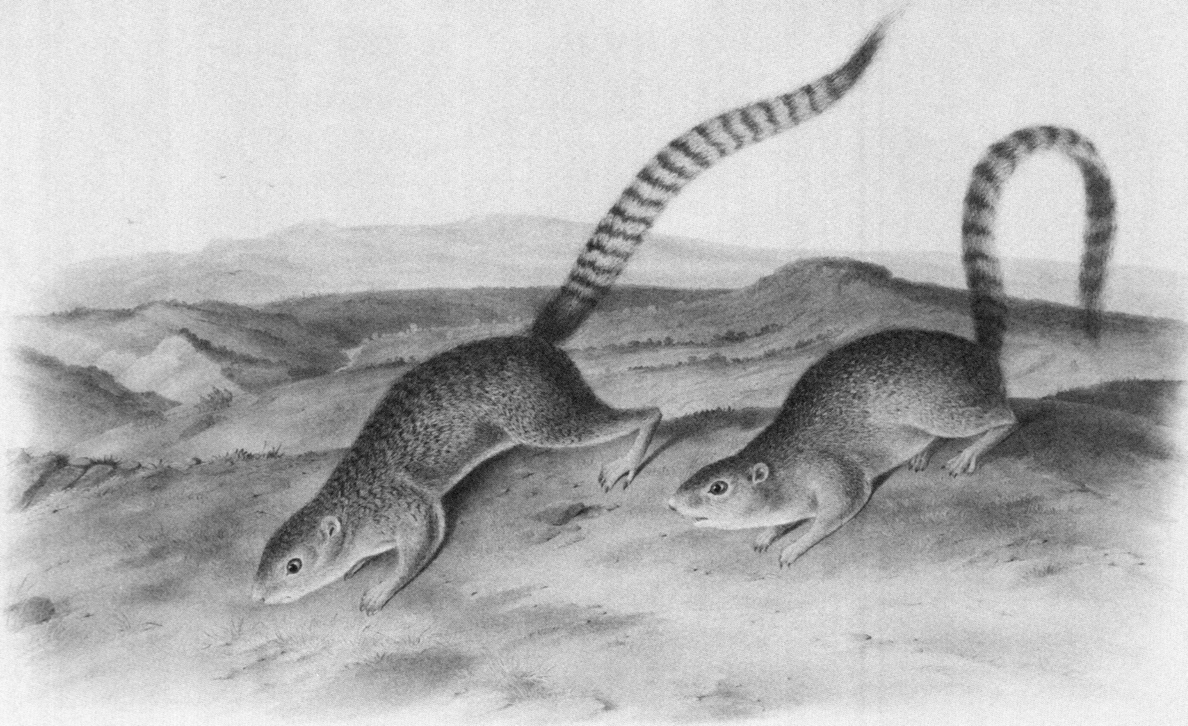

SPERMOPHILUS ANNULATUS, AUD & BACH

ANNULATED MARMOT SQUIRREL.

Annulated Marmot Squirrel

THE SINGLE SPECIMEN OF THIS SPECIES OUR AUTHORS EXAMINED was a gift from Spencer Fullerton Baird, who did not note where it was collected. It is today known as the ring-tailed ground squirrel of central Mexico's Pacific region. We include the "Habits" and "General Remarks" from this species essay to illustrate some of Bachman's thought about the relationships among genera. His conception that a genus can be "intermediate" between two other genera and that a species can be a "connecting link" between two genera reflects his view that all nonmigratory species are associated with specific regions to which their anatomy and behavior are evidently adapted. He was working in the early days of biogeography.

---

22. John James Audubon, J. T. Bowen, lithographer, *Annulated Marmot Squirrel*, 1846–48, hand-colored lithograph, plate 79 of *The Viviparous Quadrupeds of North America*. Courtesy of the Jule Collins Smith Museum of Fine Art, Auburn University; Louise Hauss and David Brent Miller Audubon Collection.

SPERMOPHILUS ANNULATUS.—AUD. AND BACH.

ANNULATED MARMOT SQUIRREL.

PLATE LXXIX.—MALE.—NATURAL SIZE.

HABITS.

We possess no knowledge of the habits of this species, but presume from its form, that it possesses the burrowing propensities of the genus. All the *Spermophili* avoid thickly wooded countries, and are either found in rocky localities, or burrowing in the prairies.

GENERAL REMARKS.

In every department of Natural History, a species is occasionally found which forms the connecting link between two genera, rendering it doubtful under which genus it should properly be arranged. Under such circumstances, the Naturalist is obliged to ascertain, by careful examination, the various predominating characteristics, and finally, place it under the genus to which it bears the closest affinity in all its details.[36] The Spermophili are intermediate in character between the Squirrels and Marmots. They have the lightness of form of the former, and burrow in the ground like the latter. By their cheek pouches, of which the true Squirrels and Marmots are destitute, they are distinguished from both. The second inner toe on the forefoot of the Spermophili is the longest, whilst in the Squirrels the third is longest. But in these closely-allied genera, there are species which approach those of another genus. Thus our Maryland Marmot, (*A. Monax*,) has a rudimentary cheek-pouch, in which a pea might be inserted, yet in every other particular it is a true *Arctomys*. The downy Squirrel, (*Sciurus lanuginosus*, see Journal Acad. Nat. Science, Vol. 8th, part 1st, p. 67,) by its short ears, broad head, and not very distichous tail, approaches the Spermophili, yet by its being destitute of cheek-pouches, by its soft, downy fur, and its hooked, sharp claws, of which the third, as in the Squirrels, is longest, it is more allied to *Sciurus*. On the other hand, the species now under consideration has the long legs, slender form, and sharp, hooked claws of the Squirrel. The two middle toes of the fore-feet being of equal length, prove its affinity to both genera; but in the general shape of its body, its cheek pouches, its short ears, and smooth, rigid hair, it must be regarded as belonging to the genus *Spermophilus*. We consider this species and the downy Squirrel as connecting links between *Sciurus* and *Spermophilus*, as we regard *Sciurus Hudsonius* the connecting link between *Tamias* and *Sciurus*.

N° 16

PLATE LXXX

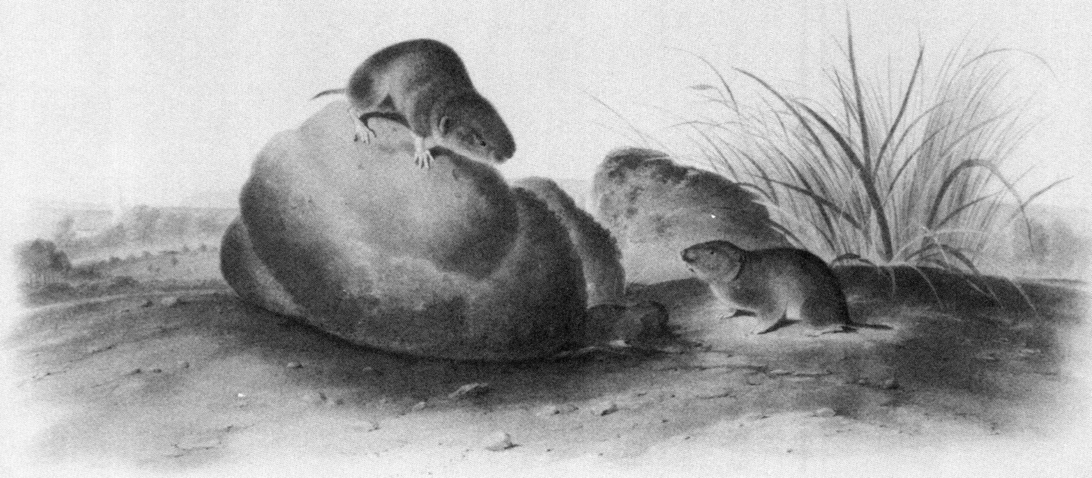

ARVICOLA PINETORUM, LECONTE

LECONTE'S PINE MOUSE.

MALE & FEMALE.

Natural size.

LeConte's Pine Mouse

CONSISTENT WITH ONE OF THE MAIN GOALS OF THEIR MAMMAL project, our authors assume the role of intermediary between farmers and the "poor ground mole" by teaching farmers how to read nature aright. The "mischievous little field mouse" causes damage to crops that farmers erroneously blame on moles because they have not yet learned how to read the evidence before them. Not only do they miss the fact that there is no positive evidence that convicts the moles, but they also miss the importance of the evidence that is there. By not becoming good natural historians, farmers risk economic loss.

---

23. John James Audubon, J. T. Bowen, lithographer, *LeConte's Pine Mouse*, 1846–48, hand-colored lithograph, plate 80 of *The Viviparous Quadrupeds of North America*. Courtesy of the Jule Collins Smith Museum of Fine Art, Auburn University; Louise Hauss and David Brent Miller Audubon Collection.

ARVICOLA PINETORUM.—LECONTE.
LECONTE'S PINE MOUSE.
PLATE LXXX.—MALE AND FEMALE.

HABITS.

The manners of this species do not differ very widely from those exhibited by many other field mice. They however, avoid low grounds, so much the resort of the meadow mice, and prefer higher and drier soils.

This mouse is rather an inhabitant of cultivated fields than of woods, and is seldom found in the forest far removed from the vicinity of plantations, to which it resorts, not only to partake of the gleanings of the fields, but to lay its contributions on the products of the husbandman's labours, claiming a share before the crops are gathered. In the Northern states, it is found in potato fields and in vegetable gardens, gnawing holes into the sides of the potatoes, carrots, rutabaga, and common turnips, following the rows where green peas and corn have been planted, bringing down threats of vengeance from the farmer on the poor ground mole, which, feeding only on worms, is made a kind of cat's-paw by this mischievous little field mouse, which does the injury in most cases, whilst the other is saddled with the blame.[37] In the South it is, next to the Norway rat, the most troublesome visitant of the cellars and banks in which the sweet potato is stored, destroying more than it consumes, by gnawing holes into the tubers, and causing them to rot. Wherever a bed of Guinea corn, Egyptian millet, or Guinea grass is planted, there you will soon observe numerous holes and nests of this species. We have recently seen an instance where a large bed of kohlrabi was nearly destroyed by it; the bulbs appearing above the surface were gnawed into holes, which, in some instances, penetrated to the centre. Our friend, the owner, had, as usual, laid the mischief on the broad shoulders of the hated and persecuted ground mole, of whose galleries not a trace could be seen in the vicinity. A number of small holes at the root of a stump, in the garden, indicated the true author of these depredations, and on digging, about a dozen of LECONTE's field Mice were captured. This species is particularly fond of the pea or ground nut, (*hypogea*.) On examining the beds where this nut is cultivated, we have observed the rows on whole acres perforated in every direction by small holes, giving evidence that this troublesome little pest had been at work. In endeavouring to save and collect the seeds of the Gama grass, (*Tripsacum dactyloides*,) we generally found ourselves forestalled by this active and voracious little rat.

This species has young three or four times during the summer. One which we had in confinement, produced young three times, having three, seven, and four, in the different broods. The young were nearly all raised, but, when full-grown, became pugnacious and persecuted each other so much that we were obliged to separate them. They were almost exclusively fed on ground nuts, corn meal and sweet potatoes, but seemed to relish both boiled rice and bread. We have seen nine young taken from one nest.

The nest of this species is generally found under ground, at the distance of about a foot from the surface; it is small, and composed of light, loose materials, collected in the vicinity.

This prolific field rat possesses many enemies to diminish its numbers. The house cat not only watches for it about the fields and gardens, but is fond of devouring it, whilst the bodies of shrews and ground moles are not eaten. The very common Owl, (*Syrnium nebulosum*,) the Barn Owl (*Strix Americana*,) the Weasel, Ermine, and Mink, all make this species a considerable part of their subsistence.

The only note we have ever heard from this mouse is a low squeak, only uttered when it is either struck suddenly or greatly alarmed. In a state of confinement it was remarkably silent, except when two were engaged in fighting.

No. 17

PLATE LXXXI

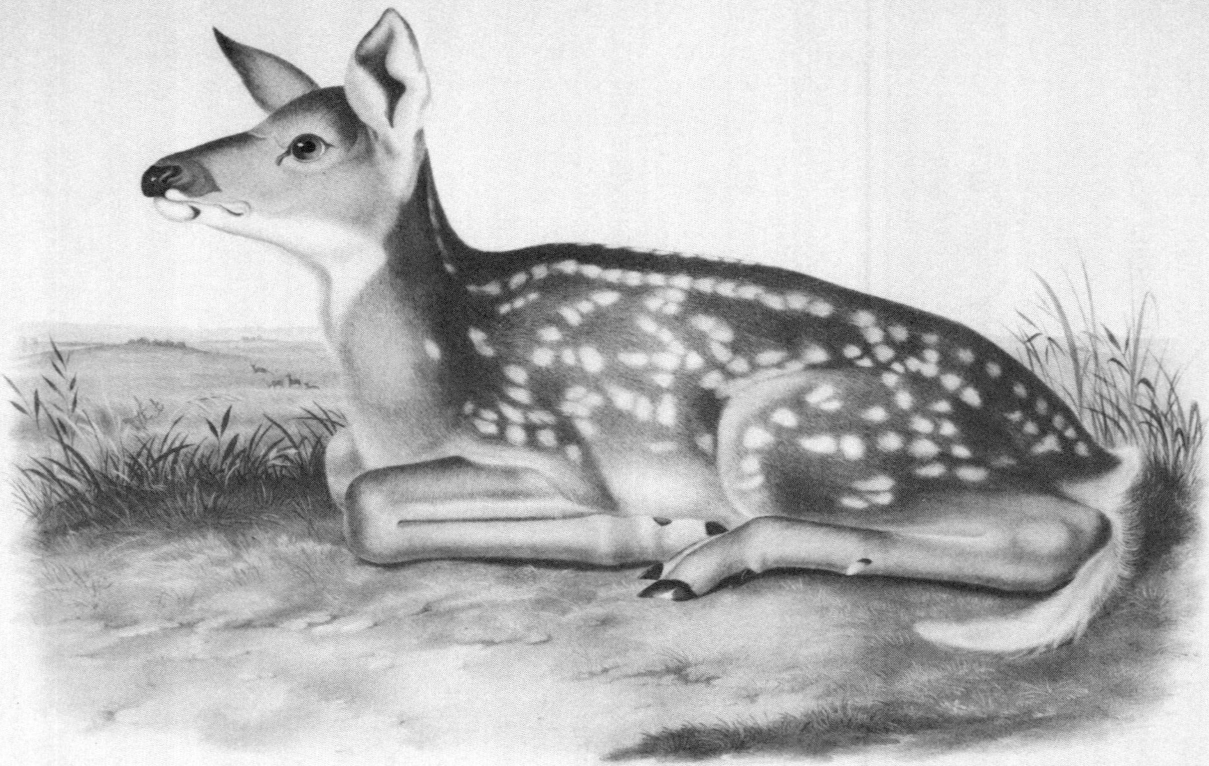

CERVUS VIRGINIANUS, PENNANT

Common American Deer

WHEREAS THE BISON IS THE "MOST IMPORTANT" OF ALL NORTH American mammals, the white-tailed deer is the one of greatest "interest" to humankind. While the species is an important source of meat and leather, this essay makes clear that it is the hunt itself—in all its myriad forms—that defines our relationship to the deer. As with the lengthy "Bison" essay, "Common American Deer" allows adequate space to become a minor treatise on deer hunting, including the history of how Native Americans and then European settlers took the species and how in our authors' time the various races and classes of men developed technologies and strategies peculiar to their particular social conditions, pointing out violations of sportsmanship along the way. How one hunts the deer reflects one's social and cultural standing. Incidents other than hunts are related, of course, and some especially good writing occurs, as in the scene of Bachman's comic nonchalance where he pretends to be reading a book so that the buck submerged in the river will think he has not been detected. There is a fine description of the vibrant and multihued autumnal "charm" of the "Southern States" as well, when "an American forest" is "more striking and beautiful than that of any other country." The spell of this charm, as always, is evoked regardless of the experience of the enslaved "servants" on the plantations.

---

24. John Woodhouse Audubon, J. T. Bowen, lithographer, *Common American Deer*, 1846–48, hand-colored lithograph, plate 81 of *The Viviparous Quadrupeds of North America*. Courtesy of the Jule Collins Smith Museum of Fine Art, Auburn University; Louise Hauss and David Brent Miller Audubon Collection.

PLATE CXXXVI.

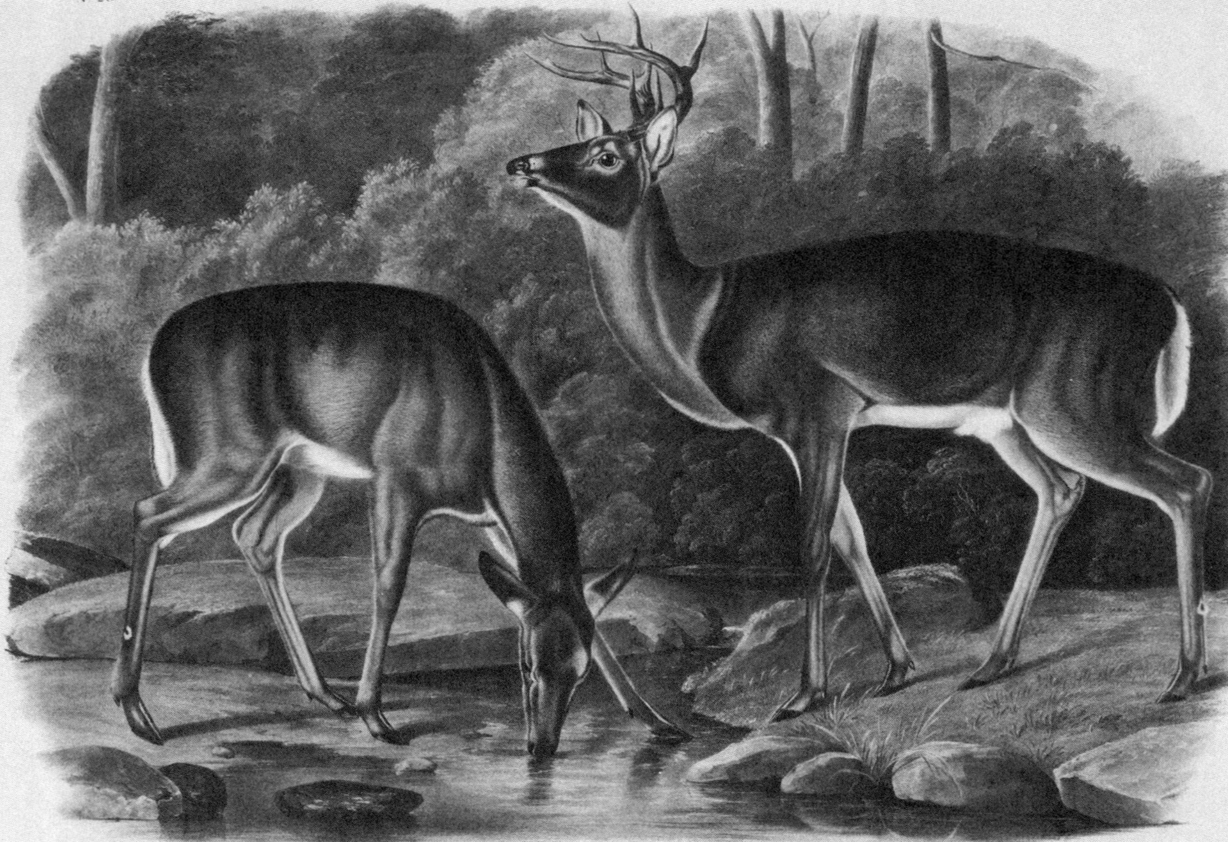

CERVUS VIRGINIANUS, PENNANT.

CERVUS VIRGINIANUS.—PENNANT.

COMMON AMERICAN DEER

PLATE LXXXI.—FAWN.

PLATE CXXXVI.—MALE AND FEMALE.—WINTER PELAGE.

HABITS.

Perhaps no species of wild animal inhabiting North-America, deserves to be regarded with more interest than the subject of our present article, the Common or Virginian Deer; its symmetrical form, graceful curving leap or bound, and its rushing speed, when, flying before its pursuers, it passes like a meteor by the startled traveller in the forest, exciting admiration, though he be ever so dull an observer.

The tender, juicy, savoury, and above all, digestible qualities of its flesh are well known; and venison is held in highest esteem from the camp of the backwoodsman to the luxurious tables of the opulent, and, when not kept too long (a common error in our large cities by the way) a fat haunch with jelly and chafing dishes is almost as much relished, as a "hunter's steak," cooked in the open air on a frosty evening far away in the west. The skin is of the greatest service to the wild man, and also useful to the dweller in towns; dressed and smoked by the squaw, until soft and pliable, it will not shrink with all the wettings to which it is exposed. In the form of mocasins, leggings, and hunting shirts, it is the most material part of the dress of many Indian tribes, and in the civilized world is used for breeches, gloves, gaiters, and various other purposes.

From the horns are made beautiful handles for various kinds of cutlery.

The timidity of the Deer is such, that it hurries away, even from the sight of a child, and it is but seldom that the hunter has any danger to apprehend,

25. John Woodhouse Audubon, J. T. Bowen, lithographer, *Common American Deer*, 1846–48, hand-colored lithograph, plate 146 of *The Viviparous Quadrupeds of North America*. Courtesy of the Jule Collins Smith Museum of Fine Art, Auburn University; Louise Hauss and David Brent Miller Audubon Collection.

even from a wounded buck; it does but little injury to the fields of the planter, and is a universal favourite with old and young of both sexes in our Southern States.

The Virginian, or as we wish to designate it, the Common Deer, is the only large animal, if we except the bear, that is not driven from the vicinity of man by the report of the deer-driver's gun, or the crack of the hunter's rifle; the buffalo and the elk are now rarely seen east of the Mississippi. Hunted by hounds and shot at from day to day, the Deer may retreat from this persecution for a little while, but soon returns again to its original haunts. Although it scarcely ever occupies the same bed on successive nights, yet it is usually found in the same range, or drive as it is called, and often not fifty yards from the place, where it was started before. It is fond of lingering around fences and old fields, that are partially overspread with brush-wood, briar-patches and other cover, to screen it from observation. In the southern States the Deer, especially in summer when they are least disturbed, are fond of leaping the outer fences of plantations, lying through the day in some tangled thicket, overgrown with cane, vines and briars; and in such places you may be so fortunate as to start an old buck in August or September, and many an overgrown denizen of the forest has bowed his huge antlers and fallen a sacrifice to his temerity in seeking a resting-place too near some pea-patch, where his hoofs left traces for many weeks of his nightly depredations.

This habit of resting during the day in the near vicinity of their feeding ground, is however not universal. We during last summer were invited to visit a large cornfield in which a quantity of the Carolina cow-pea had been planted among the corn. This had been the nightly resort of the Deer during the whole summer—their tracks of various sizes covered the ground, as if flocks of sheep had resorted to it, and scarcely a pod or even a leaf was remaining on the vines. The Deer, however, were not in the vicinity, where there were several favourable and extensive covers; they were trailed to some small islands, in a marsh nearly two miles off. We ascertained that the Deer inhabiting the swamps on the east side of the Edisto river, where there are but few cultivated farms, were in the nightly habit of swimming the Edisto and visiting the pea-fields in Barnwell, on the opposite side, returning before day-light to their customary haunts, some four or five miles distant.

The localities selected by Deer as places of rest and concealment during the day are various, such as the season of the year and the nature of the country and climate may suggest to the instincts of the animal. Although we have occasionally in mountainous regions, especially in the higher mountains of Virginia and the Green Mountains of Vermont, detected a Deer lying without concealment on an elevated ledge of bare rock, like the ibex and chamois on the Alps, yet as a general habit, the animal may be said to seek concealment, either among clumps of myrtle or laurel bushes, (*Kalmia*), in large fallen tree-tops, briar-patches, clusters of alder bushes, (*alnus*), or in tall broom-grass, (*Andropogon dissitiflorus*). In cold weather it prefers seeking its repose in some sheltered dry situation, where it is protected from the wind, and warmed by the rays of the sun; and on these occasions it may be found in briar-patches which face the south, or in tufts of broom-grass in old uncultivated fields. In warm weather it retires during the day to shady swamps, and may often be started from a clump of alder or myrtle bushes near some rivulet or cool stream. To avoid the persecution of moschetoes and ticks, it occasionally, like the moose in Maine, resorts to some stream or pond and lies for a time immersed in the water, from which the nose and a part of the head only project. We recollect an occasion, when on sitting down to rest on the margin of the Santee river, we observed a pair of antlers on the surface of the water near an old tree, not ten steps from us. The half-closed eye of the buck was upon us; we were without a gun, and he was, therefore, safe from any injury we could inflict on him. Anxious to observe the cunning he would display, we turned our eyes another way, and commenced a careless whistle, as if for our own amusement, walking gradually towards him in a circuitous route, until we arrived within a few feet of him. He had now sunk so deep in the water that an inch only of his nose, and slight portions of his prongs were seen above the surface. We again sat down on the bank for some minutes, pretending to read a book. At length we suddenly directed our eyes towards him, and raised our hand, when he rushed to the shore, and dashed through the rattling canebrake, in rapid style.

The food of the common Deer varies at different periods of the year. In winter, it feeds on buds of several kinds of shrubs, such as the wild rose, the hawthorn, various species of bramble, (*Rubus*,) the winter green, (*Pyrola*,) the Partridge Berry, (*Mitchella repens*,) the Deer Leaf, (*Hopea tinctoria*,) the bush Honeysuckle, (*Azalea*,) and many others. In spring and summer it subsists on tender grasses, being very select in its choice and dainty in its taste. At these seasons it frequently leaps fences, and visits the fields of the planter, taking an occasional bite at his young wheat and oats, not overlooking the green corn, (*Maize*,) and giving a decided preference to a field planted with cow-peas, which it divests of its young pods and tender leaves; nor does it pass lightly by berries of all kinds, such as the Huckleberry, Blackberry and Sloe, (*Viburnum prunifolium*.) We are informed by a friend that in the vicinity of Nashville, (Tennessee,) there is an extensive park containing about three hundred Deer, the principal food of which is the luxuriant Kentucky blue-grass, (*Poa pratensis*.) In autumn it finds an abundance of very choice food in the chestnuts, chinquapins and beech-nuts strewn over the ground. The localities of the various oaks are resorted to, and we have seen its tracks most abundantly under the Live Oak, (*Quercus virens*,) the acorns of which it appears to prefer to all others. We once observed three deer feeding on these acorns, surrounded by a flock of wild turkeys, all eagerly engaged in claiming their share. The fruit of the Persimmon tree, after having been ripened by the frosts of winter, falls to the ground, and also becomes a favourite food of the Deer.

Possessing such a choice of food, we might suppose this animal would be always fat: this, however, is not the case, and, except at certain seasons of the year, the Deer is rather poor. The bucks are always in fine order from the month of August to November, when we have seen some that were very fat. One which we killed weighed one hundred and seventy-five pounds. We have been informed that some have reached considerably over two hundred pounds. In November, and sometimes a little earlier, the rutting season commences in Carolina, when the neck of the buck begins to dilate to a large size. He is now constantly on foot, and nearly in a full run, in search of the does. On meeting with other males, tremendous battles ensue, when, in some rare instances, the weaker animal is gored to death; generally, however, he flies from the vanquisher, and follows him, crest fallen, at a respectful and convenient distance, ready to turn on his heels and scamper off at the first threat of his victorious rival. In these rencontres, the horns of the combatants sometimes become interlocked in such a manner that they cannot be separated, and the pugnacious bucks are consigned to a lingering and inevitable death by starvation. We have endeavoured to disengage these horns, but found them so completely entwined that no skill or strength of ours was successful. We have several times seen two, and on one occasion, three pairs of horns thus interlocked, and ascertained that the skulls and skeletons of the Deer had always been found attached. These battles only take place during the rutting season, when the horns are too firmly attached to be separated from the skull. Indeed, we have seen a horn shot off in the middle by a ball, whilst the stump still continued firmly seated on the skull. The rutting season continues about two months, the largest and oldest does being earliest sought for, and those of eighteen months at a later period. About the month of January, the bucks drop their horns, when, as if conscious of having been shorn of their strength and honours, they seem humbled, and congregate peaceably with each other, seeking the concealment of the woods, until they can once more present their proud antlers to the admiring herd. Immediately after the rutting season, the bucks begin to grow lean. Their incessant travelling during the period of venery—their fierce battles with their rivals, and the exhaustion consequent on shedding and replacing their horns by a remarkably rapid growth, render them emaciated and feeble for several months. About three weeks after the old antlers have been shed, the elevated knobs of the young horns make their appearance. They are at first soft and tender, containing numerous blood-vessels, and the slightest injury causes them to bleed freely. They possess a considerable degree of heat, grow rapidly, branch off into several ramifications, and gradually harden. They are covered with a soft, downy skin, and are now in what is called "velvet." When the horns are fully grown, which is usually in July or August, the buck shows a restless propensity to rid himself of the velvet covering, which has now lost its heat, and become dry: hence he is constantly engaged in rubbing his horns against bushes and saplings, often destroying the trees by wounding and tearing the bark, and by twisting and breaking off the tops. The system of bony development now ceases altogether, and the horns become smooth, hard, and solid.

The does are fattest from November to January. They gradually get thinner as the season of parturition approaches, and grow lean whilst suckling their young.

The young are, in Carolina, produced in the month of April; young does, however, seldom yean till May or June. In the Northern States, they bring forth a little later, whilst in Florida and Texas the period is earlier. It is a remarkable, but well ascertained fact, that in Alabama and Florida, a majority of the fawns are produced in November. The doe conceals her young under a prostrate tree-top, or in a thick covert of grass, visiting them occasionally during the day, especially in the morning, evening, and at night. The young fawns, when only a few days old, are often found in so sound a sleep that we have, on several occasions, seen them taken up in the arms before they became conscious that they were captives. They are easily domesticated, and attach themselves to their keepers in a few hours. A friend possesses a young deer that, when captured, during the last summer, was placed with a she goat, which reared it, and the parties still live in habits of mutual attachment. We have seen others reared by a cow. A goat, however, becomes the best foster-mother. They breed in confinement, but we have found them troublesome pets. A pair that we had for several years, were in the habit of leaping into our study through the open window, and when the sashes were down they still bounced through, carrying along with them the shattered glasses. They also seemed to have imbibed a vitiated and morbid taste, licked and gnawed the covers of our books, and created confusion among our papers. No shrub in the garden, however valuable to us, was sacred to them; they gnawed our carriage harness, and finally pounced upon our young ducks and chickens, biting off their heads and feet, leaving the body untouched.

The doe does not produce young until she is two years old, when she has one fawn. If in good order, she has two the following year. A very large and healthy doe often produces three, and we were present at Goose Creek when an immense one, killed by J. W. AUDUBON, was ascertained, on being opened, to contain four large and well formed fawns. The average number of fawns in Carolina is two, and the cases where three are produced are nearly as numerous as those in which young does produce only one at a birth.

The wild doe is attached to her young, and its bleat will soon bring her to its side, if she is within hearing. The Indians use a stratagem, by imitating the cry of the fawn, with a pipe made of a reed, to bring up the mother, which is easily killed by their arrows. We have twice observed the doe called up by this imitation of the voice of the young. She is, however, so timid that she makes no effort in defence of her captured offspring, and bounds off at the sight of man.

The common Deer is a gregarious animal, being found on our western prairies in immense scattered herds of several hundred. After the rutting season the males, as we have before stated, herd together and it is only during the season of intercourse that both sexes are found in company. The does, however, although congregating during a considerable portion of the year, are less gregarious than many species of African antelopes, the buffalo, or our domestic sheep; as they are found during the summer separated from the rest of the gang or troop, and are only accompanied by their young.

The Deer is one of the most silent of animals, and scarcely possesses any notes of recognition. The fawn has a gentle bleat that might be heard by the keen ears of its mother at the distance probably of a hundred yards. We have never heard the voice of the female beyond a mere murmur when calling her young, except when shot, when she often bleats loudly like a calf in pain. The buck when suddenly started sometimes utters a snort, and we have at night heard him emitting a shrill whistling sound, not unlike that of the chamois of the Alps, that could be heard at the distance of half a mile. The keen sense of smell the Deer possess enables them to follow each other's tracks. We have observed them smelling on the ground and thus following each other's trail for miles. We were on an autumnal morning seated on a log in the pine lands of Carolina when a doe came running past us. In the course of ten minutes we observed a buck in pursuit, with his nose near the ground, following in all the windings of her course. Half an hour afterwards came a second buck, and during another interval a third small buck pursued the same trail. The sense of sight appears imperfect—as we have often, when standing still, perceived the Deer passing within a few yards without observing us, but we have often noticed the affrighted start when we moved our position or when they scented us by the wind.

On one occasion we had tied our horse for some time at a stand;—on his becoming restless we removed him to a distance—a Deer pursued by dogs ran near the spot where the horse had originally stood, caught the scent, started suddenly back, and passed within a few feet of the spot where we were standing, without having observed us. Their sense of hearing is as keen as that of smell. In crawling towards them in an open wood, against the wind, you may approach within gun shot, but if you unfortunately break a stick, or create a rustling among the leaves, they start away in an instant.

This animal cannot exist without water, being obliged nightly to visit some stream or spring for the purpose of drinking. During the present year (1850) a general drought prevailed throughout our southern country. On the Hunting Islands between Beaufort and Savannah, the Deer, we were informed, nearly all perished in consequence of the streams on these Islands having dried up. Deer are fond of salt, and like many other wild animals resort instinctively to salt-licks or saline springs. The hunters, aware of this habit, watch at these "licks," as they are called, and destroy vast numbers of them. We have visited some of these pools, and seen the Deer resorting to them in the mornings and evenings and by moon light. They did not appear to visit them for the mere purpose of drinking, but after walking around the sides, commenced licking the stones and the earth on the edges, preferring in this manner to obtain this agreeable condiment, to taking a sudden draught and then retiring. On the contrary they lingered for half an hour around the spring, and after having strayed away for some distance, they often returned a second and even a third time to scrape the sides of it, and renew the licking process. Our common Deer may be said to be nocturnal in its habits, yet on the prairies, or in situations where seldom disturbed, herds of Deer may be seen feeding late in the morning and early in the afternoon. Their time for rest, in such situations, is generally the middle of the day. In the Atlantic States, where constantly molested by the hunters, they are seldom seen after sunrise, and do not rise from their bed until the dusk of the evening. The Deer is more frequently seen feeding in the day time during spring and summer, than in winter; a rainy day, and snowy wintery weather, also invite it to leave its uncomfortable hiding place and indulge in its roaming habits. We have no doubt, that in localities where

Deer have been constantly hunted, they, from a sense of fear, allow you to approach much nearer to their place of concealment than in situations where they are seldom disturbed. They continue lying still, not because they are asleep or unaware of your approach, but because they are afraid to expose themselves to view, and hope by close concealment to be passed without being observed. We have seen them lying with their hind legs drawn under them ready for a spring—their ears pressed flat on the sides of the neck, and their eyes keenly watching every movement of the intruder. Under these circumstances your only chance of success is to ride slowly around the animal as if he was not observed, and suddenly fire before he leaps from his bed. This effect of fear, on your near approach, is not confined to our Deer; it may be seen in the common partridge, the snipe, and other game birds. Before being hunted, they are restless—are unwilling to assume the crouching posture called setting, and rise at a distance from their pursuers; but after having been a few times disturbed and shot at, they, in the language of sportsmen, become tame, and permit themselves to be nearly trodden on before they can be induced to rise; this apparent tameness is in reality wildness, and their squatting and hiding the effect of terror to which they are prompted by an instinct of self-preservation.

The gait of this Deer is various. In walking it carries its head very low, and pursues its course cautiously and silently, occasionally moving its ears and whisking its tail; the largest animal is usually the leader of the herd, which travel in what is called Indian file, there seldom being two abreast. Walking is the ordinary pace of the Deer unless frightened, or in some state of excitement. When first started, without being much alarmed, it gives two or three springs, alighting with apparent awkwardness on three feet—and immediately afterwards resting on the opposite side, erecting its white tail and throwing it from side to side. A few high bounds succeed, whilst the head is turned in every direction to enable it to detect the cause of alarm. The leaps and high boundings of the Deer are so graceful, that we have never witnessed them without excitement and admiration. When, however, the Deer observes you before it is routed from its bed, it bolts off with a rush, running low to the ground, with its head and tail on a line with the body, and for a few hundred yards rivalling the speed of a race horse. But this rattling pace cannot be kept up for any length of time—after the first burst its speed slackens, it foams at the mouth, and exhibits other evidences of fatigue. We have sometimes seen it overtaken and turned by an active rider in the open wood, and under other favourable circumstances, and on one occasion a fat buck was headed by a fearless driver, lashed with his whip, brought to bay, and finally knocked in the head and taken without having been shot. We have witnessed a few instances where a pack of hounds, after a four hours' chase, succeeded in running down a Deer. These cases are, however, rare, nor would we give any encouragement to this furious Sylvan race, in which the horse and his mad rider are momentarily exposed to the danger of a broken neck from the many holes in the pine lands. The Deer, after an attempt at bringing it to bay, frequently succeeds in escaping from the hunter and the hounds, by dashing into a swamp or crossing a river, and even should it be captured, after a long chase the venison is found to be insipid and of no value.

In riding through the woods at night in the vicinity of Deer, we have often heard them stamp their feet, the bucks on such occasions giving a loud snort, then bounding off for a few yards and again repeating the stamping and snorting, which appear to be nocturnal habits.

Deer take the water freely, and swim with considerable rapidity; their bodies are on such occasions submerged, their heads only being visible above the surface. We have witnessed them crossing broad rivers and swimming the distance of two miles. When thus under way, they cleave the water with such celerity that a boat can scarcely overtake them.

Along our southern sea-board the Deer, when fatigued by the hounds, plunge into the surf and swim off for a mile or two, floating or swimming back with the returning tide, when they ascend the beach near the same place where they entered the water.

As already remarked, the flesh of our common Deer is the best flavoured and most easy of digestion of all the species with which we are acquainted, except the black-tailed Deer; it is superior to the Elk or Moose of our country, or the red Deer or Roebuck of Europe. It is, however, only a delicacy when it is fat, which is generally the case from the beginning of August to the month of December. In Carolina, the haunch and loin only are served up on the tables of the planters, the shoulders and skin are the perquisites of the driver, or negro huntsman. The Indians eat every part of the Deer, not omitting the entrails and the contents of the stomach—the latter many of the tribes devour raw, without subjecting them to any cooking or roasting process. It is stated, even by white men, that the stomach, with all its half-digested ingredients, is very palatable. Hunger and hardships seldom fail to give a zest to the appetite. Vegetable food is scarce in the wilderness or on the prairies. The traveller who has long been obliged to sleep in a tent and make his toilet in the woods, soon becomes indifferent to the etiquette of civilized life, and does not inquire whether his dish has been prepared according to the recipe of the cookery-books. A Deer paunch contains a mixture of many ingredients, picked up from various shrubs, seeds, and grasses, and may become a substitute for vegetables where the kitchen-garden has not yet been introduced. According to a northern traveller (Lyon's Narrative, p. 242), who referred, however, to another animal, the reindeer of our continent, it is "acid and rather pungent, resembling a mixture of sorrel and radish leaves," its smell like "fresh brewer's grains." As we have never been subjected to the necessity of testing the virtues of this primitive chowder, we are unable to pronounce it a delicacy, and must leave the decision to those who may be disposed to make the experiment.

The capture of the common Deer exercised the ingenuity and patience of the Indian, ages before the pale faces intruded on his hunting-grounds, with their rifles, their horses, and hounds. He combatted with the wolf and the cougar for their share of the prey, leaving on our minds a melancholy impression of the near approach of the condition of savage life to that of the brute creation. Different modes of hunting were suggested by the peculiar face of the localities of the country, and the degrees of intelligence or native cunning of the several tribes. The bow and arrow evidently must have been in common use throughout the whole length and breadth of our land, as the numerous arrow-heads still every where turned up by the plough abundantly attest.

The Rein Deer, inhabiting the extensive, cold, and inhospitable regions of the British possessions to the north of Quebec, were caught in snares manufactured from the hide, and sometimes of the sinews, of the animal.

During the season of their annual migrations, rude fences of brush-wood were constructed, which were a mile or two apart at the entrance, narrowing down to nearly a point at the other end, in which the snares were placed, and at the termination of this "cul de sac" was erected a high fence or pound, secured by stakes, stones, and other strong materials, in which the Deer that escaped from the snares were finally enclosed and shot with arrows. The common Deer, however, is more suspicious and timid, and will seldom suffer itself to be circumvented in this manner.

The American Rein Deer is also brought near to the hunter lying in wait behind the concealment of a clump of bushes, or heap of stones, by the waving of a small flag of cloth, or a deer's tail, which, exciting its attention, it falls a sacrifice to its curiosity. This stratagem is also successfully practised on our western prong-horned Antelope.

The Common Deer is frequently brought within bow-shot by the Indians, who call up the does, as we have already mentioned, by imitating, with a pipe made of a reed, the bleating of the fawn, and also the bucks, by an imitation of the shrill, whistling sound which they emit during the rutting season. The wily savage often clothes himself in the hide of a Deer, with the horns and ears attached—imitating the walk and other actions of the animal, by which means he is enabled to approach and almost mingle with the herd, and kill several with his arrows before they take the alarm. Since the introduction of fire-arms, however, many tribes of Indians have laid aside the bow and arrow, and adopted the gun. The traders who visit them, usually supply them with an inferior article, and we have never seen any considerable number of Indians expert in the use of the rifle. The late Dr. LEITNER informed us that the Florida Indians seldom shot at a Deer beyond twenty-five or thirty yards, exercising great patience and caution before they ventured on firing; the result, however, under these favourable circumstances, was usually successful. We believe the Indians of North America never used poisoned arrows in the destruction of game, like the natives of Caffraria and other portions of Africa, or the aborigines of Brazil and the neighbouring regions of South America.

The white man conducts his hunting excursions in various modes suited to his tastes and adapted to the nature of the country in which he resides. In mountainous, rocky regions, where horses cannot be used with advantage, he goes on foot, armed with a rifle, carries no dog, and seeks for the Deer in such situations as his sagacity and experience suggest. He either espies him in his bed, or silently steals upon him behind the covert of the stem of a large tree whilst he is feeding, and leisurely takes a steady and fatal aim. On the contrary, in situations adapted to riding, where the woods are thickly clothed with underbrush, where here and there wide openings exist between briar-patches, and clumps of myrtle-bushes, as in the Southern States, the Deer are almost universally chased with hounds, and instead of the rifle, double-barrelled deer-guns, of different sizes, carrying from twelve to twenty buck-shot, are alone made use of by the hunters.

It may not be uninteresting to our readers if we point out the different modes in which Deer hunts are conducted.

In the early settlement of our country, when men hunted for food, and before they accustomed themselves to study their ease and comfort even in the chase, "still hunting," as it is termed, was universally practised. The wolves and other depredating animals, by which the colonists were surrounded, as well as the proximity of hostile Indians, almost precluded them for many years from raising a sufficient supply of sheep, hogs, and poultry. The cultivation of a small field furnished them with bread, while for meat they were chiefly dependent on the gun. Hence a portion of their time was from a kind of necessity devoted to the chase. The passion for hunting seems however to be innate with many persons, and we have observed that it often runs in families and is transmitted to their posterity, as is known to be the case with the descendants of the hunters in the Alps. There are even now many persons in our country, who devote weeks and months to the precarious employment of Deer hunting, when half the industry and fatigue in regular labour would afford their families every necessary and comfort. Hunting is a pleasant recreation, but a very unprofitable trade; it often leads to idleness, intemperance, and poverty.

For success in still-hunting it is essential that the individual who engages in it, should be acquainted with the almost impenetrable depths of the forest, as well as the habits of the Deer. He must be expert in the use of the rifle, possess a large stock of patience, and be constitutionally adapted to endure great fatigue. Before the dawn of day, he treads the paths along which the animal strays in returning from its nightly rambles to the covert usually its resting-place for the day. He ascends an elevation, to ascertain whether he may not observe the object of his search feeding in the vallies. If the patience and perseverance of the morning are not attended with success, he seeks for the Deer in its bed—if it should be startled by his stealthy tread and spring up, it stops for a moment before bounding away, and thus affords him the chance of a shot; even if the animal should keep on its course without a pause, he frequently takes a running, or what is called a chance shot, and is often successful.

There is another mode of deer hunting we saw practised many years ago in the Western parts of the State of New-York, which we regard as still more fatiguing to the hunter, and as an unfair advantage taken of the unfortunate animals. The parties sally out on a deep snow, covered by a crust, which sometimes succeeds a rain during winter. They use light snow-shoes and seek the Deer in situations where in the manner of the moose of Nova Scotia, they have trampled paths through the snow in the vicinity of the shrubs on which they feed. When started from these retreats they are forced to plunge into the deep snow; and breaking through the crust leave at every leap traces of blood from their wounded legs; they are soon overtaken, sometimes by dogs, at other times by the hunters, who advance faster on their snow-shoes than the exhausted Deer, which fall an easy prey either to the hunter's knife or his gun. In this manner thousands of Deer were formerly massacred in the Northern States.

We have ascertained that our common Deer may be easily taken by the grey-hound. A pair of the latter, introduced into Carolina by Col. CATTEL, frequently caught them after a run of a few hundred yards. The Deer were trailed and started by beagles—the grey-hounds generally kept in advance of them, making high leaps in order to get a glimpse of the Deer which were soon overtaken, seized by the throat, and thrown down. The nature of the country, however, from its swamps and thick covers often prevented the huntsmen from coming up to the captured animal before it was torn and mutilated by the hounds, and many Deer could not be found, as the pack becomes silent as soon as the Deer is taken. We predict, however, that this will become the favourite mode of taking Deer on the open western prairies, where there are no trees or other obstructions, and the whole scene may be enacted within view of the hunters.

Some hunters, who are engaged in supplying the salt and red Sulphur Springs of Virginia with venison during summer, practise a novel and an equally objectionable mode in capturing the Deer. A certain number of very large steel-traps made by a blacksmith in the vicinity, are set at night in the waters of different streams at the crossing-places of the Deer. The animal when thus captured instead of tearing off its leg by violent struggles is said to remain standing still, as passive as a wolf when similarly entrapped. Another and still more cruel mode is sometimes practiced in the South: The Deer have particular places where they leap the fences to visit the pea-fields; a sharpened stake is placed on the inside of the fence—the Deer in leaping over is perforated through the body by this treacherous spike, and is found either dead or dying on the following morning. It is also a frequent practice in the South for the hunter during clear nights to watch a pea-field frequented by Deer. To make sure of this game he mounts some tree, seats himself on a crotch or limb which is above the current that would convey the scent to the keen olfactories of the Deer, and from this elevation leisurely waits for an opportunity to make a sure shot.

In some parts of the Northern and Middle States the Deer are captured by the aid of boats. We observed this mode of hunting pursued at Saratoga and other lakes, and ascertained that it was frequently attended with success. The hounds are carried to the hills to trail, and start the Deer before day light. Some of the hunters are stationed at their favourite crossing places to shoot them should they approach within gun shot. After being chased for an hour or two the Deer pushes for the lake. Here on some point of land a party lie in wait with a light and swift boat: after the Deer has swam to a certain distance from the shore he is headed and approached by the rowers, a noose is thrown over the head, and the unfortunate animal drawn to the side of the boat, when the captors proceed to cut its throat in violation of all the rules of legitimate sporting.

Fire hunting is another destructive mode of obtaining Deer. In this case two persons are essential to success. A torch of resinous wood is carried by one of the party, the other keeps immediately in front with his gun. The astonished Deer instead of darting off seems dazzled by the light, and stands gazing at this newly kindled flame in the forest. The hunter sees his eyes shining like two tapers before him; he fires and is usually successful; sometimes there are several Deer in the gang, who start off for a few rods at the report of the gun, and again turn their eyes to the light. In this manner two or three are frequently killed within fifty yards of each other. This kind of hunting by firelight is often attended with danger to the cattle that may be feeding in the vicinity, and is prohibited by a law of Carolina, which is however frequently violated. The eyes of a cow are easily mistaken for those of a deer. We conversed with a gentleman who informed us that he had never indulged in more than one fire-hunt, and was then taught a lesson which cured him of his passion for this kind of amusement. He believed that he saw the eyes of a Deer and fired, the animal bounded off, as he was convinced, mortally wounded. In the immediate vicinity he detected another pair of eyes and fired again. On returning the next morning to look for his game, he found that he had slaughtered two favourite colts. Another related an anecdote of a shot fired at what was supposed to be the shining eyes of a Deer, and ascertained to his horror that it was a dog standing between the legs of a negro, who had endeavoured to keep him quiet. The dog was killed and the negro slightly wounded.

There is still another mode of Deer hunting which remains to be described. It is called "driving," and is the one in general practice, and the favourite pastime among the hospitable planters of the Southern States. We have at long intervals, occasionally joined in these hunts, and must admit that in the manner in which they were conducted, this method of Deer hunting proved an exciting and very agreeable recreation. Although we regret to state that it is pursued by some persons at all seasons of the year, even when the animals are lean and the venison of no value, yet the more thoughtful and judicious huntsmen are satisfied to permit the Deer to rest and multiply for a season, and practice a little self-denial, during summer when the oppressive heats which usually prevail—the danger of being caught in heavy showers—and the annoyance of gauzeflies, mosquetoes, and ticks, present serious drawbacks to its enjoyment. The most favourable season for this kind of amusement is from the beginning of October to January. The Deer are then in fine order; the heats of summer are over; the crops of rice gathered, and the value of the planter's crop can be calculated. The autumn of the Southern States possesses a peculiar charm; high winds seldom prevail, and the air is soft and mellow; although many of the summer Warblers have migrated farther to the south, yet they have been replaced by others: The blue-bird, cat-bird, and mocking-bird have not yet lost their song, and the swallows and night-hawks are skimming through the air in irregular and scattered groups on their way to the tropics. Vegetation has been checked, but not sufficiently destroyed to give a wintry aspect to the landscape. The *Gentians Gerardias* and other autumnal flowers are still disclosing a few lingering blossoms and emitting their fragrance. The forest trees present a peculiar and most striking appearance. A chemical process has been going on among the leaves, since the first cool nights have suspended the circulation, giving to those of the maple and sweet gum, a bright scarlet hue, which contrasted with the yellow of the hickory, and the glossy green of the magnolia grandiflora, besides every shade of colour that can be imagined, render an American forest, more striking and beautiful than that of any other country. It is the season of the year that invites to recreation and enjoyment. The planters have been separated during the summer; some have travelled from home—others have resided at their summer retreats;—they are now returning to their plantations, and the intercourse of the neighbourhood, that has been suspended for a season, is renewed. We recall with satisfaction some past scenes of pleasureable associations of this kind. The space already taken up by this article will preclude us from entering into minute detail, and restrict us to a few incidents which will present the general features of a Carolina Deer hunt. We comply with the oft-repeated invitation to make our annual visit to our early and long-tried friend Dr. DESEL at his hospitable residence some twenty miles from the city, which his friends have named Liberty Hall. The mind requires an occasional relaxation as well as the body. We have resolved to fly for a day or two, from the noise and turmoil of the city—to leave books and cares behind us—to break off the train of serious thought—to breathe the fresh country air, and mingle in the innocent sports of the field and the forest. Reader, you will go with us and enter into our feelings and enjoyments. As we approach the long avenue a mile from the residence of the companion of thirty-five years, we are espied by his domestics who welcome us with a shout, and inform us that their "Boss" is looking out for us. Our friend soon perceives us, and hurries to the gate. How pleasant are the greetings of friendship—the smiling look of welcome, the open hand, and the warm heart of hospitality.

The usual invitation is sent to a neighbour, to lunch, dine, and meet a friend. The evening is spent in social converse and closed with the family bible, and offerings of gratitude and praise to the Giver of all good. The sleep of him, who has escaped from the din of the city to the quiet of the country, is always refreshing. The dawn of day invites us to a substantial breakfast. The parties now load their double-barrelled guns, whilst the horses are being saddled. The horn is sounded, and the driver, full of glee, collects his impatient hounds. The party is unexpectedly augmented by several welcome guests. Our intelligent friend HARRIS, from New-Jersey, has come to Carolina, to be initiated into the mysteries of Deer hunting, as a preparation to farther exploits on the Western prairies, among the elk and the buffalo; with him comes AUDUBON, the Nestor of American ornithology, and his son, together with Dr. WILSON. After the first greetings are over, we hasten to saddle additional horses for those of our guests, who are disposed to join us. The old ornithologist, having no relish for such boyish sports, sallies to the swamps in search of some rare species of woodpecker. We proceed to the drives, as they are called, viz., certain woods, separated by old fields and various openings, in some parts of which the Deer have their usual run, where the parties take their stands. These drives are designated by particular names, and we are familiar with Crane pond, Gum thicket, the Pasture, the Oak swamp, and a number of bays, one of which we would be willing to forget, for there we missed a Deer, and the bay was named after us, to our mortification. The driver is mounted on a hardy, active, and sure-footed horse, that he may be enabled to turn the course of the Deer, if he attempts to run back, or to stop the dogs. We were carried round to our stands by our host, when a Deer bounced up before us; in an instant a loud report is heard waking the echoes of the forest—the animal leaps high into the air, and tumbles to the ground. Thus, our venison is secured, and we carry on our farther operations from the mere love of sport. Anxious to give our friend HARRIS an opportunity of killing his first Deer, we place him at the best stand. Our mutual wishes are soon gratified. He is stationed at the edge of a bay—a valley overgrown with bay-trees (*Magnolia glauca*)—which from that day received the cognomen of Harris' bay. The hounds after considerable trailing rouse two noble bucks, one of them bounds out near our friend. He is obliged to be ready in a moment, before the Deer comes in the line with another hunter. At the report of his gun we perceive that the buck is wounded. "Mind," cries out friend WILSON, "your shot have whistled past me." Friend H. grows pale at the thought of having endangered the life of another, but we comfort him by stating, that his shot had not reached within fifty yards of the nervous hunter, and moreover, that the old buck was wounded and would soon be his. We observed where he had laid down in the grass, and was started up again by the dogs. Now for a chase of a wounded buck. He takes through an old field once planted with cotton, now full of ruts and ditches, and grown up with tall broom-grass. We agree to let the boys have the pleasure of the chase whilst we are the silent spectators. They bound over ditches and old corn-fields, firing as they run. Suddenly the hounds become silent, and then the loud sounding of the horn is heard mingled with the whoops of the hunters, which inform us, that the game is secured; it proves to be a majestic buck. The successful hunter is now obliged to submit to the ordeal of all who have fleshed their maiden sword, and killed their first Deer. "I submit," he said good naturedly, "but spare my spectacles and whiskers." So his forehead and cheeks were crossed with the red blood of the buck, and the tail was stuck in his cap. The hunt proceeded merrily and successfully. Young AUDUBON, however, had not yet obtained a shot. At length a Deer was started near our host. He would not shoot it, but strove to drive it to his neighbour. He ran after it, and shouted, stumbled over a root, and in the fall threw off his spectacles: but as he was groping for them among the leaves, he ascertained that his generous efforts had been successful; the Deer had been turned to Mr. AUDUBON. One barrel snapped—then came a sharp report from the other—a loud whoop succeeded, and we soon ascertained that another Deer had fallen. We now conceived that we had our wishes for a successful hunt fully gratified; the dinner hour had arrived. Five noble Deer were strung upon the old pecan-nut tree in sight of our festive hall. The evening passed off in pleasant conversation—some of those present displayed their wit and poetical talents by giving the details of the hunt in an amusing ballad, which however has not yet found its way into print. Thus ended a Carolina Deer hunt.

We regret to be obliged to state, that the Deer are rapidly disappearing from causes that ought not to exist. There are at present not one-fifth of the number of Deer in Carolina that existed twenty years ago. In the Northern and Middle States, where the farms have been sub-divided, and the forests necessarily cleared, the Deer have disappeared because there was no cover to shelter them. In the Southern States, however, where there are immense swamps subject to constant inundations and pine barrens too poor for cultivation, they would remain undiminished in numbers were it not for the idle and cruel practice of destroying them by firelight, and hunting them in the spring and summer seasons by overseers and idlers. There is a law of the State forbidding the killing of Deer during certain months in the year. It is, however, never enforced, and Deer are exposed for sale in the markets of Charleston and Savannah at all seasons. In some neighbourhoods, where they were formerly abundant, now none exist, and the planters have given up their hounds. In New-Jersey and Long Island, where the game laws are strictly enforced, Deer are said to be on the increase. In some parts of Carolina, where the woods are enclosed with fences, not sufficiently high to prevent the Deer from straying out, but sufficient to prevent the hunters from persecuting them in summer, they have greatly multiplied and stocked the surrounding neighbourhoods. If judicious laws were framed and strictly enforced the Deer could be preserved for ages in all our Southern States, and we cannot refrain from submitting this subject to the consideration of our southern legislators.

Nº 17 PLATE LXXXII

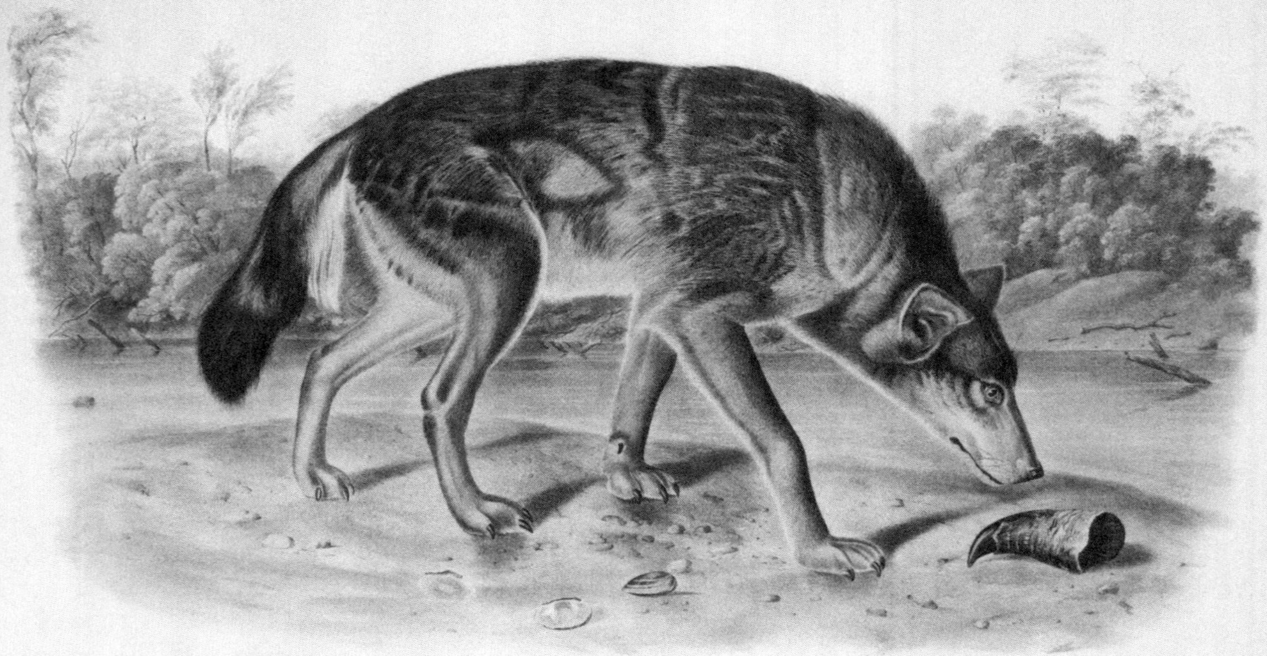

CANIS LUPUS, LINN, VAR RUFUS

RED TEXAN WOLF.

MALE.

Red Texan Wolf

THIS ESSAY BRINGS THE MEDIEVAL EUROPEAN TRADITION OF the "beasts of battle" to an American setting. While no raven or eagle appears, the red wolf is shown preying insidiously upon the corpses of fallen soldiers in the 1846–48 war against Mexico, the aftermath of which John Woodhouse witnessed and heard about on his expedition across Texas and northern Mexico in 1849. Having already described general wolf behavior in "White American Wolf," our authors feel they have room here for entertaining anecdotes from John's original journal from that expedition. The war was controversial because unprovoked, but the essay shows approval for the aggression against Mexico in its characterization of the Americans killed as "the brave, the young, and the patriotic, who have fallen for their country's honour!" Some of the Anglo bias against Mexicans is also reflected in the willingness to repeat the admittedly unfounded claim that wolves won't eat Mexican corpses because of "the quantity of pepper used by" them. We have seen before that the quality of Bachman's science did not always steer him clear of racism.

---

26. John Woodhouse Audubon, J. T. Bowen, lithographer, *Red Texan Wolf*, 1846–48, hand-colored lithograph, plate 82 of *The Viviparous Quadrupeds of North America*. Courtesy of the Jule Collins Smith Museum of Fine Art, Auburn University; Louise Hauss and David Brent Miller Audubon Collection.

CANIS LUPUS.—LINN: VAR. RUFUS.

RED TEXAN WOLF

PLATE LXXXII.—MALE.

HABITS.

This variety is by no means the only one found in Texas, where Wolves, black, white and gray, are to be met with from time to time. We do not think, however, that this Red Wolf is an inhabitant of the more northerly prairies, or even of the lower Mississippi bottoms, and have, therefore, called him the Red Texan Wolf.

The habits of this variety are nearly similar to those of the black and the white Wolf, which we have already described, differing somewhat, owing to local causes, but showing the same sneaking, cowardly, yet ferocious disposition.

It is said that when visiting battle-fields in Mexico, the Wolves preferred the slain Texans or Americans, to the Mexicans, and only ate the bodies of the latter from necessity, as owing to the quantity of pepper used by the Mexicans in their food, their flesh is impregnated with that powerful stimulant. Not vouching for this story, however, the fact is well known that these animals follow the movements of armies, or at least are always at hand to prey upon the slain before their comrades can give them a soldier's burial, or even after that mournful rite; and if anything could increase the horrors displayed by the gory ensanguined field, where man has slain his fellows by thousands, it would be the presence of packs of these ravenous beasts disputing for the carcasses of the brave, the young, and the patriotic, who have fallen for their country's honour!

No corpse of wounded straggler from his troop, or of unfortunate traveller, butchered by Camanches, is ever "neglected" by the prowling Wolf, and he quarrels in his fierce hunger in his turn over the victim of similar violent passions exhibited by man!

The Wolf is met on the prairies from time to time as the traveller slowly winds his way. We will here give an extract from the journal kept by J. W.

AUDUBON while in Texas, which shows the audacity of this animal, and gives us a little bit of an adventure with a hungry one, related by POWELL, one of the gallant Texan Rangers.

"Like all travellers, the ranger rides over the wide prairie in long silences of either deep thought or listless musings, I have never been able to decide which; but when, riding by the side of WALKER or HAYS, who would like to say that a vacant mind was ever in the broad brow or behind the sparkling eye either of him with the gray, or of him with the brown? but at times when watching closely I have thought I could trace in the varying expression, castle after castle mounting higher and higher, till a creek 'to water at,' or a deer which had been sound asleep and to windward of us, started some 30 or 40 yards off our path to wake up the dreamers of our party. No one is certain that his queries will be welcome to the backwoodsman on a march through a strange country, any more than would be those of a passenger, put to the captain of a vessel as he leans over the weather-rail looking what the wind will be, or thinking of the disagreeable bustle he will have, when he gets into port, compared to his lazy luxury on shipboard: but as I rode by the side of POWELL we started no deer, nor came to a 'water hole,' but a Red Wolf jumped up some two or three hundred yards from us, and took to the lazy gallop so common to this species; 'Run you ——,' cried POWELL, and he sent a yell after him that would have done credit to red or white man for its shrill and startling effect, the Wolf's tail dropped lower than usual, and now it would have taken a racer to have overtaken him in a mile; a laugh from POWELL, and another yell, which as the sound reached the Wolf made him jump again, and POWELL turned to me with a chuckle, and said, 'I had the nicest trick played me by one of those rascals you ever heard of.' The simple, how was it, or let's have it, was all that he wanted, and he began at the beginning. 'I was out on a survey about 15 miles west of Austin, in a range that we didn't care about shooting in any more than we could help, for the Camanches were all over the country; and having killed a deer in the morning, I took the ribs off one side and wrapping them in a piece of the skin, tied it to my saddle and carried it all day, so as to have a supper at night without hunting for it; it was a dark, dismal day, and I was cold and hungry when I got to where I was to camp to wait for the rest of the party to come up next day; I made my fire, untied my precious parcel, for it was now dark, with two sticks put up my ribs to roast, and walked off to rub down and secure my horse, while they were cooking; but in the midst of my arrangements I heard a stick crack, and as that in an Indian country means something, I turned and saw, to my amazement, for I thought no animal would go near the fire, a large Red Wolf actually stealing "my ribs" as they roasted; instinct made me draw a pistol and "let drive" at him; the smoke came in my face and I saw nothing but that my whole supper was gone. So not in the most philosophical manner I lay down, supperless, on my blanket; at daylight I was up to look out for breakfast, and to my surprise, my half-cooked ribs lay within twenty feet of the fire, and the Wolf about twenty yards off, dead; my ball having been as well aimed as if in broad daylight.'"

We have represented a fine specimen of this Wolf, on a sand-bar, snuffing at the bone of a buffalo, which, alas! is the only fragment of "animal matter" he has in prospect for breakfast.

N° 18

PLATE LXXXVII

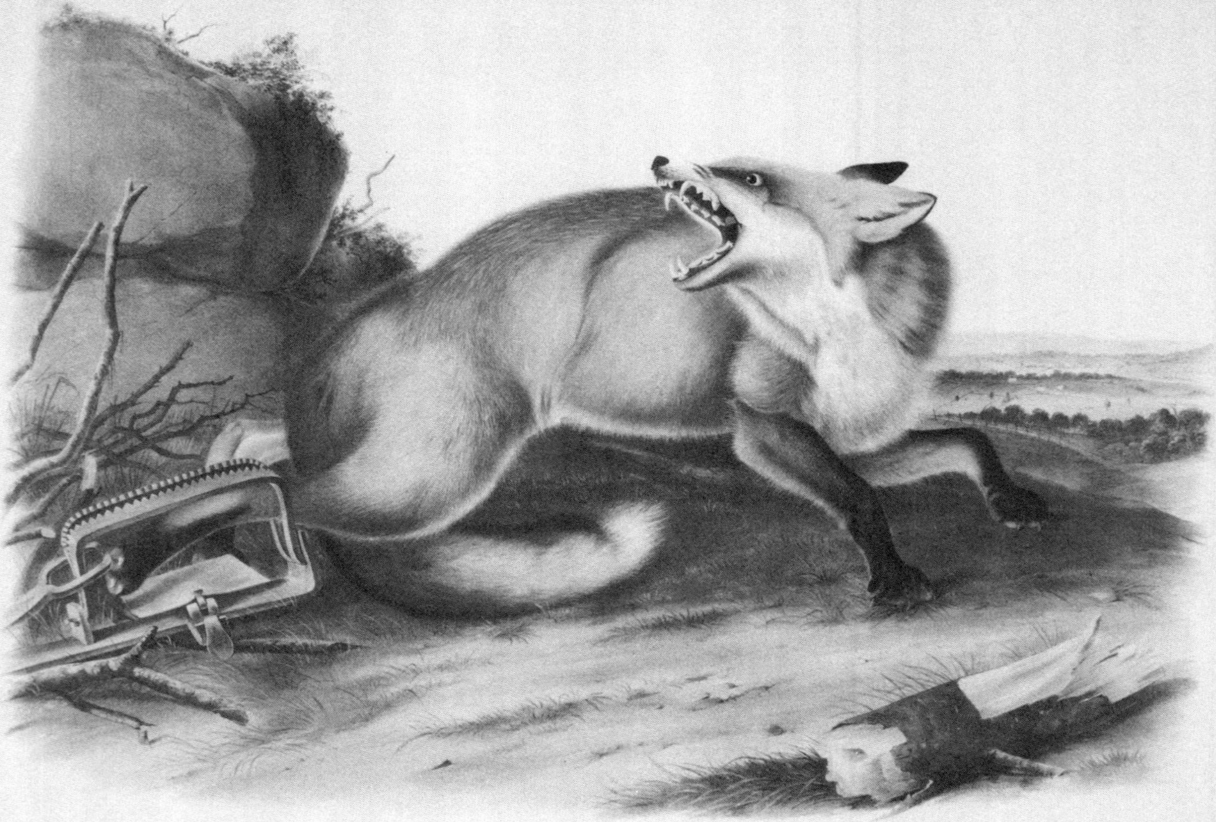

VULPES FULVUS, (DESM.)

AMERICAN RED-FOX.

American Red Fox

AFTER INTRODUCING THE ARRIVAL OF THE RED FOX INTO A new area as an indicator of the advance of agriculture, the essay turns to the fox's chief interest for humans: as an object of the hunt. Bachman, who complained to Victor Audubon about having to rewrite material he sent for the essays, was no doubt pleased to receive a competently written letter from friend Edward Harris narrating how the fox "is chased and killed with dogs, in fine style" on the beaches of New Jersey. Harris's letter does Bachman a literary favor by giving the tale of how foxes have learned to find ingested birds inside the bodies of dead fish on shore the tone and structure of a fable by Aesop: "Moral—never swallow what you cannot digest." The essay advises, however, that the tale ought not be casually dismissed, since the fox is known to have the ability to learn from experience and alter its behavior to its benefit. To refute the claim of the Scottish naturalist John Richardson that the endurance of the American red fox is inferior to that of the English fox, the essay offers an account of hunters and hounds pursuing one for thirteen hours before abandoning the hunt and thereby scores a nationalistic point.

---

27. John James Audubon, J. T. Bowen, lithographer, *American Red Fox*, 1846–48, hand-colored lithograph, plate 87 of *The Viviparous Quadrupeds of North America*. Courtesy of the Jule Collins Smith Museum of Fine Art, Auburn University; Louise Hauss and David Brent Miller Audubon Collection.

VULPES FULVUS.—DESM.

AMERICAN RED FOX

PLATE LXXXVII.—MALE.

HABITS.

This Fox, in times gone by, was comparatively rare in Virginia, and farther south was unknown. It is now seldom or never to be met with beyond Kentucky and Tennessee. Its early history is not ascertained, it was probably for a long time confounded with the Gray Fox, (which is in many parts of the country the most abundant species of the two,) and afterwards was supposed to have been imported from England, by some Fox-hunting governor of one of the "colonies." It was first distinguished from the Gray Fox and hunted, in Virginia; but now is known to exist in all the Northern States, and we are somewhat surprised that it should so long have been overlooked by our forefathers. No doubt, however, the cultivation and improvement of the whole country, is the chief reason why the Red Fox has become more numerous than it was before the Revolution, and it will probably be found going farther south and west, as the woods and forests give place to farms, with hens, chickens, tame turkeys, ducks, &c., in the barn-yards.

The Red Fox is far more active and enduring than the Gray, and generally runs in a more direct line, so that it always gives both dogs and hunters a good long chase, and where the hounds are not accustomed to follow, it will frequently beat-out the whole pack, and the horses and huntsmen to boot.

In some parts of the country, however, it is chased and killed with dogs, in fine style. The following account of the mode of taking the Red Fox, at the sea side in New-Jersey, near Cape May, is from an interesting letter written to us in December, 1845, by our friend EDWARD HARRIS, Esq., of Moorestown, in the neighbourhood of Philadelphia; it is quite different from the ordinary mode of hunting the Red Fox. He begins thus:

"On Saturday, a week ago, I went to Cape May Court-house, where I spent Monday and Tuesday among the quails, (*Perdrix virginianus*), which I found exceedingly abundant, but the ground so bad for shooting, that in both days two of us shot but thirty-three birds. On Wednesday my friend Mr. HOLMES took me to BEASLEY's Point at the northern extremity of the county; here I was sorry to learn that young BEASLEY, who was to have returned from Philadelphia on the Saturday previous, had not yet made his appearance; his father, however, showed a great desire to forward my views in regard to 'Monsieur Reynard.' The next day it rained cats and dogs, and TOM BEASLEY did not arrive in the stage. In the afternoon it cleared off sufficiently to make 'a drive' in the point, where we started a noble specimen in beautiful pelage, but alas! he would not come near the standers.

"The next morning, we drove the same ground, being the only place on the main land where there was any prospect of driving a Fox to standers without dogs, (of which there are none in the vicinity). This time we saw none. After dinner I took my pointer, and bagged eight brace and a half of quails, having this time found them on good ground. The next day, Saturday, with three drivers, and three standers, we drove the beach for five and a-half miles, without seeing a fox, and so ended this unsuccessful expedition. I had great hopes of this beach, (PECK's), as it had not been hunted since the winter before the last, although some of the gunners told me they had seen but few 'signs' since that time.

"The mode of driving, which requires no dogs, is for the drivers to be furnished with two boards, or shingles, which they strike together, or with what is better, a rattle, similar to a watchman's. The standers are sent ahead to a narrow part of the beach, where the creeks of the salt-marshes approach nearest to the sand-hills: when they are supposed to have reached their stands, the drivers enter, and walk abreast among the bushes, between the sand-hills and the marshes, making all the noise they can, with their lungs, as well as their boards or rattles; and these unusual noises are almost sure to drive the Foxes to the standers, where if they pass harmless, they have again to run the gauntlet to the end of the beach, at the inlet, where, Mr. BEASLEY assures me, he has known seven Red Foxes cornered, out of which four were killed, and three escaped from bad shooting. We made four drives in the five and a-half miles.

"The facts in regard to the history of the Red Fox on the Jersey coast that I have been able to collect, are few; such as they are I will give them to you.

"Certain it is that they frequent the beaches in great numbers, and so far as I can learn, the Gray Fox is not found in the same places, nor is the raccoon, which we know to be so abundant on the sea islands and beaches of our southern coast. They pass to the beaches on the ice, in the winter season, when the 'sounds' are frozen, and have frequently been seen in the day time, making their passage, though doubtless it is more frequently performed in the night. Their means of subsistence there are ample, consisting of wild fowl of various kinds, upon which they spring while they are asleep upon the ponds and creeks, but more particularly upon the wounded fowl which escape from the numerous gunners, also crabs and fish, which are thrown up dead by the surf, and rabbits and wading birds, in the summer. A marvellous story is told of their sagacity in selecting the food they like best, which is vouched for by Mr. BEASLEY, and all the gunners along shore, but which I think requires confirmation, at least so far as to have the fish in question, seen by some naturalist in the state, described by the narrators, in order to ascertain its name, or describe it, if new, before its publication is ventured on. The story is, that a certain fish, called the cramp-fish, from its supposed power of paralizing the hand which touches it while living, is thrown ashore dead, by the surf in the winter season, that every one of these fishes contains a bird, such as the coot, (either *fusca* or *perspicillata*), or a gull, which appears to have destroyed the fish, by its proving rather hard to digest, without having been plucked. Mr. Fox finds the fish that has come to this deplorable end, and either in the vain hope of restoring animation to the unfortunate defunct, or for the gratification of a less noble impulse, he makes a longitudinal incision into the peritonæum of the subject, and extracts the bird, of which he makes a meal; but; mind you, Mr. Fox has profited by the awful example before him—he picks the bird before he eats it. Moral—never swallow what you cannot digest. But, to be serious, I do not mean to ridicule the fact, which I cannot but believe with the testimony which accompanies it, but if it be new, which I cannot answer for, it might in its plain, unvarnished form, without being announced in pedantic Latin, afford too tempting a morceau for the snarling critic. The fish are said to reach sometimes the length of four-feet, with a mouth twenty-two inches wide, they are scaled, and are said to resemble, somewhat, the sea cat-fish, with which I am not acquainted. The Fox on the beach when hunted by hounds, resorts to his usual trick of taking the water, to throw the dogs off the scent, by following the retreating surf, so that its return may efface his trail, then lying down among the sand hills to rest, while the dogs are at fault. In the woods on the main land both Red and Gray Foxes are abundant, the latter rather predominating. The Foxes are abundant on some of the beaches, and generally may be procured. Mr. SPENCER, of Mount Holly, has been on a party when five were killed, but I do not know where, nor whether it was this season or before."

We have not been able to procure the fish which is alluded to in the foregoing, but have no doubt of the correctness of the account. The Red Fox will eat fish as well as birds, and when hard pressed does not refuse even carrion. It is, therefore, probable that the discovery of the bird within the dead fish, may be the result of accident rather than of instinct, reason, or keenness of smell on the part of the Fox; for when he begins to devour a fish he must soon find the more savoury bird in its stomach, and being fonder of fowl than of fish, he would of course eat the bird and leave the latter. A Fox after having in this way discovered coots, gulls, or any other bird, would undoubtedly examine any dead fish that he came across, in hopes of similar good luck. Hence the foxes on the beaches have, we suppose, acquired the habit of extracting birds from the stomachs of such fish as have swallowed them, and are cast ashore dead by the storms on the coast; and they also at times get a plentiful meal from the dead birds that float ashore. We received a beautiful specimen of the Red Fox, in the flesh, from our friend Mr. HARRIS, not long after the foregoing letter, and our figure was drawn from it. We represented the animal just caught in a steel-trap.

The Red Fox brings forth from four to six young at a litter, although not unfrequently as many as seven. The young are covered, for some time after they are born, with a soft woolly fur, quite unlike the coat of the grown animal, and generally of a pale rufous colour. Frequently, however, the cubs in a litter are mixed in colour, there being some red and some black-cross Foxes together: when this is the case it is difficult to tell which are the red and which the cross Foxes until they are somewhat grown. In these cases the parents were probably different in colour.

This animal feeds upon rats, rabbits, and other small quadrupeds, and catches birds, both by lying in wait for them, and by trailing them up in the manner of a pointer dog, until watching an opportunity he can pounce or spring upon them. In our article on the Gray Fox, (vol. 1., p. 164) we have described the manner in which this is done by that species, and the Red Fox hunts in the same way.

The Red Fox also eats eggs, and we have watched it catching crickets in an open field near an old stone wall. It is diverting to witness this—the animal leaps about and whirls round so quickly as to be able to put his foot on the insect, and then gets hold of it with his mouth; we did not see him snap at them; his movements reminded us of a kitten playing with a mouse.

We once knew a Red Fox that had been chased frequently, and always escaped at the same spot, by the hounds losing the track: the secret was at last found out, and proved to be a trick somewhat similar to the stratagem of the Gray Fox related in our first volume, p. 171; the Red Fox always took the same course, and being ahead of the dogs so far that they could not see him, leaped from a fallen log on to a very sloping tree, which he ascended until concealed by the branches, and as soon as the dogs passed he ran down and leaping on to his old track ran back in his former path. So dexterously was this "tour" performed that he was not suspected by the hunters, who once or twice actually whipped their dogs off the trail, thinking they were only following the "back track."

The Red Fox is in the habit of following the same path, which enables the fox hunters to shoot this species from "stands," even in a country where the animal has room enough to take any course he may choose to run. The "hunters" who go out from the city of New-York, are a mixed set, probably including Germans, Frenchmen, Englishmen, and Irishmen, and each one generally takes his own dog along, (on the speed and prowess of which he is ready to bet largely,) and the hunt is organized on the height beyond Weehawken in "the Jerseys," where a good many Red Foxes are to be found, as well as more Gray ones.

The men are all on foot, and station themselves along ridges, or in gaps in the rocky hilly country, now running to a point, to try and get a shot, now yelling to their dogs, and all excitement and hubbub. If the Fox doubles much, he is very apt to get shot by some one before he passes all the "standers," and the hunters then try to start another; but the Fox often gets away, as the underbrush is thick and a good deal of the ground swampy, and in that case he makes for a large rocky hill which stands in the Newark marshes, familiarly known as Rattle-snake hill. When running across the low level to this stronghold the Fox is frequently seen by the whole company of hunters, and the chase is lengthened out to a run of many miles, as Reynard will turn again toward the high ridges nearer the Hudson River.

We will give an account of one of these hunts as related by some young friends, who having two fine harriers (to contribute their share of dogs to the pack,) were gladly hailed by the other gentlemen in the field.

"After some beating about among the thickets and ravines, we found the dogs had strayed away down the side of the hills nearly to the level of the marshes, and raising our horn to call them up, observed that they were running toward a cur-dog that appeared to have come from somewhere in them; we immediately gave a loud halloo, and urged all the hounds to the chase. The cur turned tail at once, the whole pack 'opened' after him in full cry, and all the hunters came running forth from the woods to the brow of the hill, whence we had a view of the whole scene. The cur looked a good deal like a Fox, at a distance, and most of the hunters thought he was one 'certain,' he shewed good bottom, took several leaps over the stone walls and fences, and dodged about and round patches of briars and rocks with extraordinary agility, until he got fairly off towards his home, when he positively 'streaked it,' until, to the utter amazement of the hunters, he jumped on to a wall enclosing a small farm yard, and disappeared within, immediately setting up a loud bark of defiance, while some of the hunters who had expressed most confidence, were loudly laughed at by their comrades, who banteringly asked what they would take for their dogs, &c., and broke out in fresh roars of merriment."

The Red Fox is taken in traps, but is so very wary that it is necessary to set them with great nicety.

Dr. RICHARDSON tells us that the best fox hunters in the fur countries use *assfœtida*, *castoreum*, and other strong smelling substances, with which they rub their traps and the small twigs set up in the neighbourhood, alleging that Foxes are fond of such perfumes.

The same author informs us that their flesh is ill tasted, and is eaten only through necessity.

Red Foxes have gradually migrated from the Northern to the Southern States. This change of habitation may possibly be owing to the more extensive cultivation to which we have alluded, as a reason for this species having become more numerous than it was before the Revolution. This idea, however, would seem to be overthrown by the continued abundance of Gray Foxes in the Eastern States. In the early history of our country the Red Fox was unknown south of Pennsylvania, that State being its Southern limit. In process of time it was found in the mountains of Virginia, where it has now become more abundant than the Gray Fox. A few years afterwards it appeared in the more elevated portions of North Carolina, then in the mountains of South Carolina, and finally in Georgia; where we have recently observed it.

This species was first seen in Lincoln County, Georgia, in the year 1840, since then it has spread over the less elevated parts of the country, and is not rare in the neighbourhood of Augusta. We are informed by Mr. BEILE, an intelligent observer of the habits of animals, that on one occasion near Augusta, as he was using a call for wild turkeys, a little before sunrise, in the vicinity of Augusta, two Red Foxes came to the call, supposing it to be that of a wild turkey, and were both killed by one discharge of his gun.

In order to ascertain whether the speed of the Red Fox was as great in the south as in the colder regions of the north, several gentlemen near Augusta, in the winter of 1844, resolved to test the question by a regular Fox chase. They congregated to the number of thirty, with one hundred hounds, many of them imported dogs, and all in fine running order. They started a Fox at two o'clock on a moonlight morning. He took to a pretty open country on the west bank of the Savannah river. A number of gentlemen were mounted on fleet horses. Mr. BEILE rode in succession three horses during the chase, two of which were good hunters. The pursuit of the flying beast was kept up till three o'clock in the afternoon, having continued thirteen hours, when the horses and the whole pack of hounds were broken down, and the hunt was abandoned. This account does not accord with that given by RICHARDSON, who states (Fauna Boreali. Am. p. 93,) "The Red Fox does not possess the wind of its English congener. It runs for about a hundred yards with great swiftness, but its strength is exhausted in the first burst, and it is soon overtaken by a wolf or a mounted huntsman." It is quite evident that our estimable friend never had an opportunity of participating in the chase of the American Red Fox.

Whilst the Gray Fox seldom is known to dig a burrow, concealing its young usually beneath the ledges of rocks, under roots, or in the hollow of some fallen tree, the Red Fox on the contrary, digs an extensive burrow with two or three openings. To this retreat the Fox only flies after a hard chase and as a last resort. If, as often happens, the burrow is on level ground it is not very difficult by ascertaining the direction of the galleries and sinking a hole at intervals of seven or eight feet, to dig out and capture the animal. When thus taken he displays but little courage—sometimes, like the Opossum, closing his eyes and feigning death.

The young, from four to six at a birth, are born in February and March, they are blind when born, and are not seen at the mouth of the den for about six weeks.

It is at this period, when the snows in the Northern States are still on the ground, that the Fox, urged by hunger and instinct, goes out in search of prey. At a later period, both the parents hunt to provide food for their young. They are particularly fond of young lambs, which they carry off for miles to their burrows. They also kill geese, turkeys, ducks, and other poultry, and have a bad reputation with the farmer. They likewise feed on grouse and partridges, as well as on hares, squirrels, and field-rats of various species, as we have previously mentioned.

Nº 18

PLATE XC.

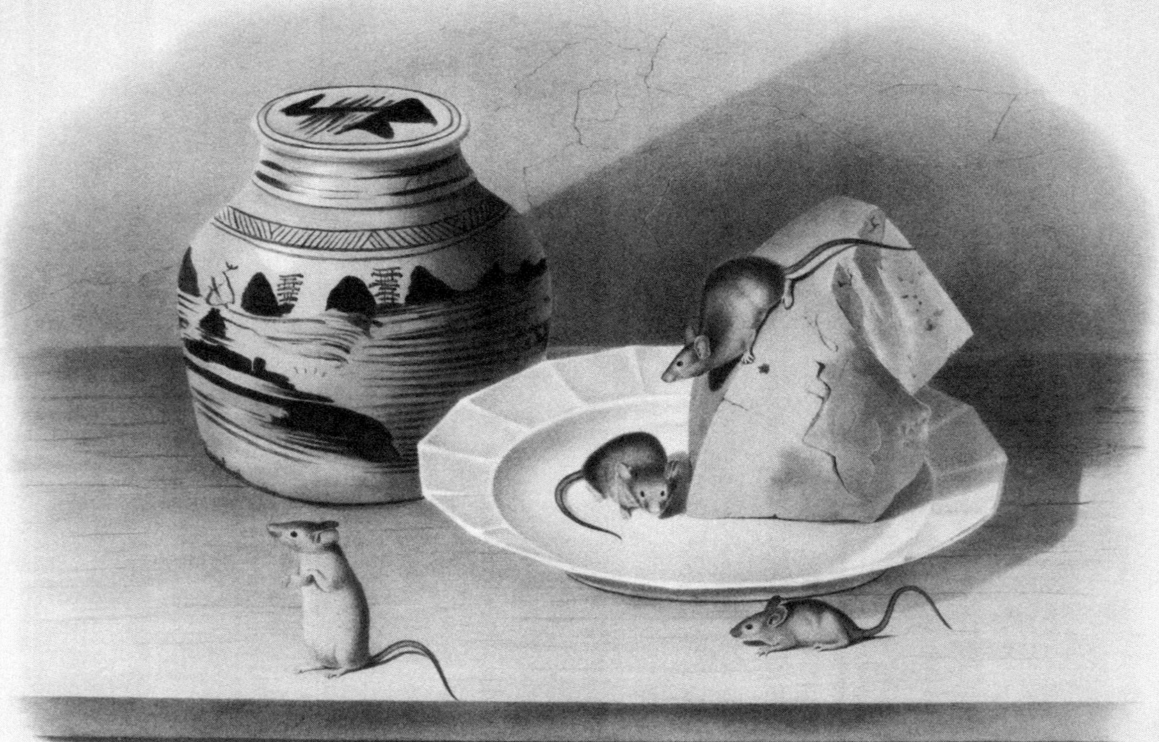

MUS MUSCULUS, LINN

COMMON MOUSE.

Common Mouse

AMONG ALL THE QUADRUPED ESSAYS, THIS ONE IS UNIQUE IN its close relationship with the image on the plate and the extent to which our authors use the devices of prose fiction to develop scenes of life with mice. To understand the moment depicted in the plate, one must read the essay, just as in the case of program music one must know the story it is designed to tell. This plate and its essay might also be the most painterly of them all with the reference to a painted landscape on the "china jar" (a painting within the painting) that evokes no natural landscape but rather the sweetmeats contained within and not visible. And is the plate depicted in plate XC a pun? The authors chose a playful tone for this small essay about the most familiar of all North American wild mammals, conjuring the shared experience of many who have been annoyed or frightened by an unexpected mouse noise more than once. And being the most familiar, the species' life history is already most known.

---

28. John Woodhouse Audubon, J. T. Bowen, lithographer, *Common Mouse*, 1846–48, hand-colored lithograph, plate 90 of *The Viviparous Quadrupeds of North America*. Courtesy of the Jule Collins Smith Museum of Fine Art, Auburn University; Louise Hauss and David Brent Miller Audubon Collection.

MUS MUSCULUS.—LINN.

COMMON MOUSE

PLATE XC.—MALE, FEMALE, AND YOUNG.

HABITS.

We have attempted to shew a portion of a shelf in a pantry, on which stands a china jar, with its indigo-blue peaked mountains, its fantastic trees and its (take them altogether) rather remarkable landscapes, reminding us more of the sweetmeats it contains than of aught in the way of nature; and we have also portrayed a plate, with a piece of hard old cheese in it, on which a Mouse is standing in the act of listening, while another in the plate, and two more on the shelf, likewise appear a little startled, and are expecting to be disturbed ere they can make their intended meal; the little rascals have reason to fear, for the careful housekeeper has heard them of late, squealing in their squabblings with each other, has found the marks of their teeth on the bread and butter, and is determined to get rid of them instanter, if possible; she is calling now to her faithful pussy cat, and inquiring for the trap.

But although the thievish Mouse is often frightened, and may be said to eat his dinner with "a cat" over his head,[38] although he is assailed with pokers, broomsticks, &c., whenever he unluckily runs across the floor, and in fact is killed as often as his death can be compassed by the ingenuity of man, or the cunning and quickness of his ally the cat, the Mouse will not retire from the house, and even where the supply of food for him is small, or in rooms that have long been shut up, he may be found; and would he let our drawings and books alone, we should willingly allow him the crumbs from our table; but he will sometimes gnaw into shreds valuable papers, to make a bed behind some bureau or old chest. He in his turn frightens man at times, and should the hard-hearted hoarding wretch who has made gold his God, while with aged, trembling hands, locked in his inmost chamber, he counts his money-bags, but hear a little Mouse; what a feeling of terror shoots through his frame; despair seems for an instant to be written on his face, and he clutches convulsively the metal to which he is a slave; another moment, and he recovers, but he is still agitated, and hastily secures with locks and bolts the treasure which is to him more precious than the endearments of a wife, the love of children, the delights of friendship and society, the blessings and prayers of the poor, or the common wants of humanity in his own person.

Many a young lady will scream at sight of a poor little Mouse, and many a brave young man might be startled in the stillness of the night by the noise made by this diminutive creature, especially if given to the reading of the "Mysteries of Udolpho" or the "Castle of Otranto," late in the hours of darkness, alone in a large old lumbering house.[39]

The Common Mouse is a graceful, lively little animal—it is almost omnivorous, and is a great feeder, although able to live on but little food if the supply is scanty. This species has from four to ten young at a litter, and the female suckles her young with tender care. When first born, they are very small, almost naked, and of a pinkish colour. The Mouse has several litters every year. We kept a pair in confinement, which produced four times, having from four to nine in each litter. Dr. GODMAN quotes Aristotle, who says that "a pregnant female being shut up in a chest of grain; in a short time a hundred and twenty individuals were counted."

On examining our corn-crib in the spring, and cleaning it out; although it was constructed with a special view to keep off rats and vermin, being on posts, and the floor raised from the ground some three feet, with boards outside inclining downwards all round, we found and killed nearly fifty Mice. A basket in the crib, hanging by a rope from a cross-beam, in which we had put some choice corn for seed, had been entered by them, and every grain of corn in it devoured. We found in the basket nothing but husks, and the remains of a Mouse's nest. The animal must therefore have climbed up to the roof of the crib, and then descended the cord by which the basket of corn was suspended.

The activity, agility, and grace of the Mouse, have made it a favourite pet with the prisoner in his solitary cell, and it has been known to answer his call, and come out of its hiding places to play with the unfortunate captive, showing the greatest fondness for him, and eating out of his hand without fear.

Of late years, white Mice have been in request in London, where they are taught various tricks, and are exhibited by boys in the streets. It is stated that in order to increase the number of this variety, persons exclude them from the light, this they pretend causes a great many of them to be born albinos. We are however satisfied from personal experience that a pair of albinos, accidentally produced, would continue to propagate varieties of the same colour without the aid of darkness; as is the case in the albino variety of the English rabbit.

N.º 20. PLATE XCVI.

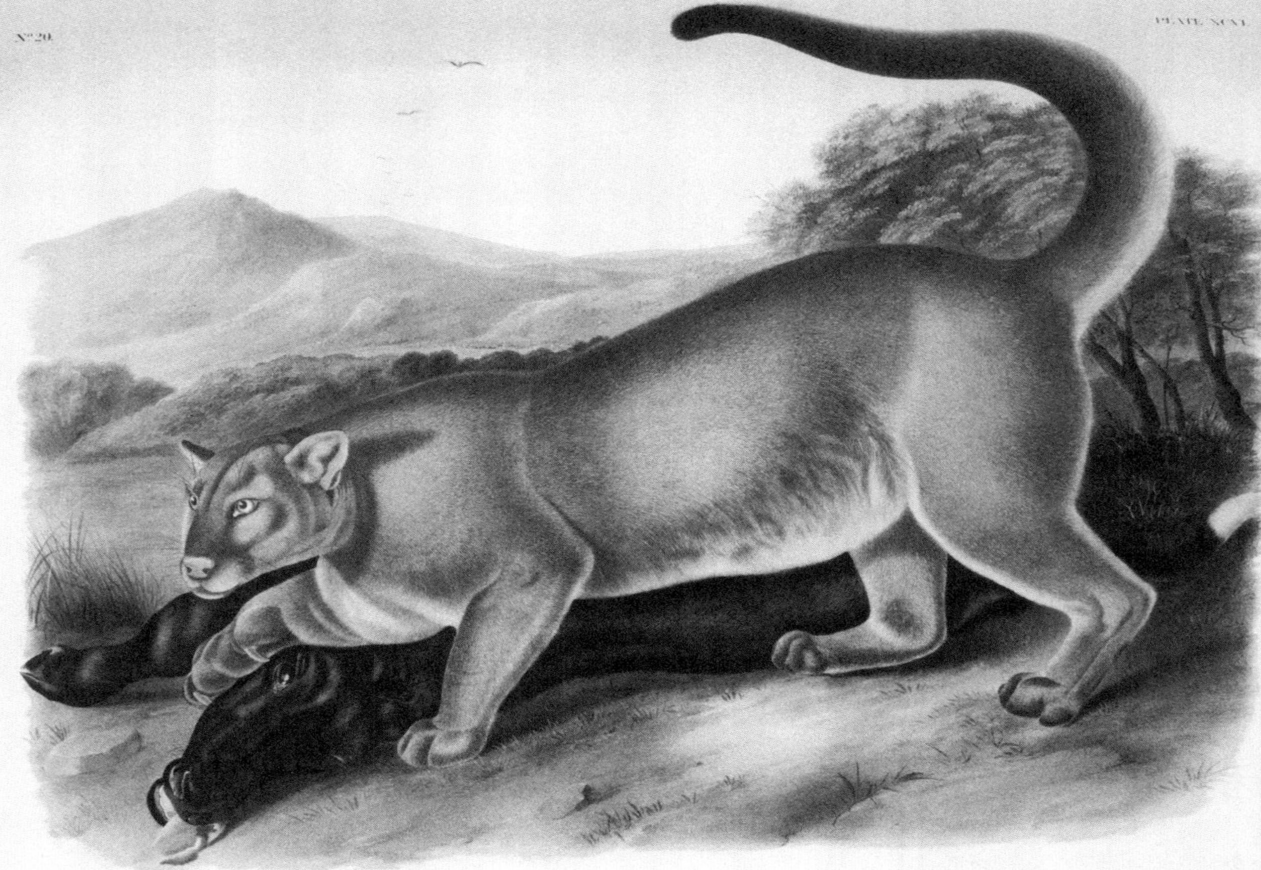

FELIS CONCOLOR, LINN.

THE COUGAR.

Cougar

THE PERVASIVELY ANTHROPOCENTRIC ETHIC OF *QUADRUPEDS* is on full display in this essay. Despite the acknowledged fact that cougars rarely attack humans and that a female cougar with young is an admirably dedicated and capable parent, like the wolf, a cougar is an animal to be killed. It is also an animal to fuel lurid, sensationalized narratives of danger and horror. Even while noting that only "the minds of the ignorant and timid" are susceptible to the sensation of terror from this species, our authors are willing to exploit that very sensation, at one point describing the eyes of a cougar on the point of attacking a man as "glaring upon him like two burning coals." They (probably Audubon or Victor more so than Bachman) seem willing to allow moments of poor natural history for the sake of appealing to a broader readership. The anthropocentrism is also apparent in the use of "cowardly" to describe the cougar. Whereas Audubon strongly objected in his *Ornithological Biography* to the characterization of bird behavior as stupid, reflecting only the ignorance of those who did so, the ethic of *Quadrupeds* allows "cowardly" to stand in the description of the cougar, even though the behavior described shows simply wariness or intelligence enough to know to run from its most dangerous enemy.[40] The popular culture representation of this species has not matured much since this essay was published.

---

29. John Woodhouse Audubon, J. T. Bowen, lithographer, *The Cougar*, 1846–48, hand-colored lithograph, plate 96 of *The Viviparous Quadrupeds of North America*. Courtesy of the Jule Collins Smith Museum of Fine Art, Auburn University; Louise Hauss and David Brent Miller Audubon Collection.

FELIS CONCOLOR.—LINN.

THE COUGAR.—PANTHER.

PLATE XCVI.—MALE

PLATE XCVII.—FEMALE AND YOUNG.

HABITS.

The Cougar is known all over the United States by the name of the panther or painter, and is another example of that ignorance or want of imagination, which was manifested by the "Colonists," who named nearly every quadruped, bird, and fish, which they found on our continent, after species belonging to the Old World, without regard to more than a most slight resemblance, and generally with a total disregard of propriety. This character of the "Colonists," is, we are sorry to say, kept up to a great extent by their descendants, to the present day, who in designating towns and villages throughout the land, have seized upon the names of Rome, Carthage, Palmyra, Cairo, Athens, Sparta, Troy, Babylon, Jericho, and many other ancient cities, as well as those of Boston, Portsmouth, Plymouth, Bristol, Paris, Manchester, Berlin, Geneva, Portland, &c., &c., from which probably some of the founders of our country towns may have emigrated. We sincerely hope this system of nomenclature will henceforth be discarded; and now let us go back to the Cougar, which is but little more like the true *panther* than an opossum is like the kangaroo! Before, however, entirely quitting this subject, we may mention that for a long time the Cougar was thought to be the lion; the supposition was that all the skins of the animal that were brought into the settlements by the Indians were skins of females; and the lioness, having something the same colour and but little mane, it occurred to the colonists that the skins they saw could belong to no other animal!

The Cougar is found sparsely distributed over the whole of North America up to about latitude 45°. In former times this animal was more abundant than at present, and one was even seen a few miles from the city of New-York within the recollection of Dr. DEKAY, who speaks of the consternation occasioned by its appearance in Westchester County, when he was a boy.

The Cougar is generally found in the very wildest parts of the country, in deep wooded swamps, or among the mountain cliffs and chasms of the Alleghany range. In Florida he inhabits the miry swamps and the watery everglades; in Texas, he is sometimes found on the open prairies, and his tracks may be seen at almost every cattle-crossing place on the sluggish bayous and creeks with their quick-sands and treacherous banks. At such places the Cougar sometimes finds an unfortunate calf, or perhaps a cow or bullock, that has become fast in the oozy, boggy earth, and from exhaustion has given up its strugglings, and been drowned or suffocated in the mire.

This species at times attacks young cattle, and the male from which our drawing was made, was shot in the act of feeding upon a black heifer which he had seized, killed, and dragged into the edge of a thicket close adjoining the spot. The Cougar is, however, generally compelled to subsist on small animals, young deer, skunks, raccoons, &c., or birds, and will even eat carrion when hard pressed by hunger. His courage is not great, and unless very hungry, or when wounded and at bay, he seldom attacks man.

J. W. AUDUBON was informed, when in Texas, that the Cougar would remain in the vicinity of the carcass of a dead horse or cow, retiring after gorging himself, to a patch of tall grasses, or brambles, close by, so as to keep off intruders, and from which lair he could return when his appetite again called him to his dainty food. In other cases he returns, after catching a pig or calf, or finding a dead animal large enough to satisfy his hungry stomach, to his accustomed haunts, frequently to the very place where he was whelped and suckled.

Dr. DEKAY mentions, that he was told of a Cougar in Warren County, in the State of New-York, that resorted to a barn, from whence he was repeatedly dislodged, and finally killed. "He shewed no fight whatever. His mouth was found to be filled with the spines of the Canada porcupine, which was probably the cause of his diminished wariness and ferocity, and would in all probability have finally caused his death."

The panther, or "painter," as the Cougar is called, is a nocturnal animal more by choice than necessity, as it can see well during the day time. It steals upon its intended prey in the darkness of night, with a silent, cautious step, and with great patience makes its noiseless way through the tangled thickets of the deepest forest. When the benighted traveller, or the wearied hunter may be slumbering in his rudely and hastily constructed bivouac at the foot of a huge tree, amid the lonely forest, his fire nearly out, and all around most dismal, dreary, and obscure, he may perchance be roused to a state of terror by the stealthy tread of the prowling Cougar; or his frightened horse, by its snortings and struggles to get loose, will awaken him in time to see the glistening eyes of the dangerous beast glaring upon him like two burning coals. Lucky is he then, if his coolness does not desert him, if his trusty rifle does not miss, through his agitation, or snap for want of better flint; or well off is he, if he can frighten away the savage beast by hurling at him a blazing brand from his nearly extinguished camp-fire. For, be sure the animal has not approached him without the gnawing hunger—the desire for blood; engendered by long fasting and gaunt famine. Some very rare but not well authenticated instances have been recorded in our public prints, where the Cougar at such times has sprang upon the sleeper. At other times the horses are thrown into such a fright, that they break all fastenings and fly in every direction. The late Mr. Robert Best of Cincinnati, wrote to Dr. Godman, that one of these animals had surprised a party of travellers, sprung upon the horses, and so lacerated with its claws and teeth their flanks and buttocks, that they with the greatest difficulty succeeded in driving the poor creatures before them next morning, to a public house some miles off. This party, however, had no fire, and were unarmed.

A planter on the Yazoo river, some years ago, related the following anecdote of the Cougar to us. As he was riding home alone one night, through the woods, along what is called a "bridle-path" (i.e. a horse-track), one of these animals sprang at him from a fallen log, but owing to his horse making a sudden plunge forward, only struck the rump of the gallant steed with one paw, and could not maintain his hold. The gentleman was for a moment unable to account for the furious start his horse had made, but presently turning his head saw the Cougar behind, and putting spurs to his horse, galloped away. On examining the horse, wounds were observed on his rump corresponding with the claws of the Cougar's paw, and from their distance apart, the foot must have been spread widely when he struck the animal.

Another respectable gentleman of the State of Mississippi gave us the following account. A friend of his, a cotton planter, one evening, while at tea, was startled by a tremendous outcry among his dogs, and ran out to quiet them, thinking some person, perhaps a neighbour, had called to see him. The dogs could not be driven back, but rushed into the house; he seized his horsewhip, which hung inside the hall door, and whipped them all out, as he thought, except one, which ran under the table. He then took a candle and looking down, to his surprise and alarm discovered the supposed refractory dog to be a Cougar. He retreated instanter, the females and children of his family fled frightened half out of their senses. The Cougar sprang at him, he parried the blow with the candle-stick, but the animal flew at him again, leaping forward perpendicularly, striking at his face with the fore-feet, and at his body with the hind-feet. These attacks he repelled by dealing the Cougar straight-forward blows on its belly with his fist, lightly turning aside and evading its claws, as he best could. The Cougar had nearly overpowered him, when luckily he backed toward the fire-place, and as the animal sprang again at him, dodged him, and the panther almost fell into the fire; at which he was so terrified that he endeavoured to escape, and darting out of the door was immediately attacked again by the dogs, and with their help and a club was killed.

Two raftsmen on the Yazoo river, one night encamped on the bank, under a small tent they carried with them, just large enough to cover two. They had a merry supper, and having made a large fire, retired, "turned in" and were soon fast asleep. The night waned, and by degrees a drizzling rain succeeded by a heavy shower pattering on the leaves and on their canvas roof, which sheltered them from its fury, half awakened one of them, when on a sudden the savage growl of a Cougar was heard, and in an instant the animal pounced upon the tent and overthrew it. Our raftsmen did not feel the full force of the blow, as the slight poles of the tent gave way, and the impetus of the spring carried the Panther over them; they started up and scuffled out of the tent without further notice "to quit," and by the dim light of their fire, which the rain had nearly extinguished, saw the animal facing them and ready for another leap; they hastily seized two of the burning sticks, and whirling them around their heads with loud whoops, scared away the midnight prowler. After this adventure they did not, however, try to sleep under their tent any more that night!

We have given these relations of others to show that at long intervals, and under peculiar circumstances, when perhaps pinched with hunger, or in defence of its young, the Cougar sometimes attacks men. These instances, however, are very rare, and the relations of an affrightened traveller must be received with some caution, making a due allowance for a natural disposition in man to indulge in the marvellous.

Our own experience in regard to the habits of this species is somewhat limited, but we are obliged to state that in the only three instances in which we observed it in its native forests, an impression was left on our minds that it was the most cowardly of any species of its size belonging to this genus. In our boyhood, whilst residing in the northern part of New-York, forty-eight years ago, on our way to school through a wood, a Cougar crossed the path not ten yards in front of us. We had never before seen this species, and it was, even at that early period, exceedingly rare in that vicinity. When the Cougar observed us he commenced a hurried retreat; a small terrier that accompanied us gave chase to the animal, which, after running about a hundred yards, mounted an oak and rested on one of its limbs about twenty feet from the ground. We approached and raised a loud whoop, when he sprang to the earth and soon made his escape. He was, a few days afterwards, hunted by the neighbours and shot. Another was treed at night, by a party on a raccoon hunt; supposing it to be a raccoon, one of the men climbed the tree, when the Cougar leaped to the ground, overturning one of the young hunters that happened to be in his way, and made his escape. A third was chased by cur-dogs in a valley in the vicinity of the Catskill mountains, and after half an hour's chase ascended a beech-tree. He placed himself in a crotch, and was fired at with duck-shot about a dozen times, when he was finally killed, and fell heavily to the ground. A Mr. RANDOLPH, of Virginia, related to us an amusing anecdote of a rencontre which he and a Kentuckian had in a valley of one of the Virginia mountains with a Cougar. This occurrence took place about thirty years ago. They had no guns, but meeting him near the road, they gave chase with their horses, and after a run of a few hundred yards he ascended a tree. RANDOLPH climbed the tree, and the Cougar sprang down, avoiding the Kentuckian, who stood ready to attack him with his club. The latter again followed, on his horse, when he treed him a second time. RANDOLPH again climbed after him, but found the animal was coming down, and disposed to fight his way to the ground. He stunned him with a blow, when the Cougar let go his hold, fell to the earth, and was killed by his comrade, who was waiting with his club below.

From all the conversations we have had with hunters who were in the habit of killing the Cougar, we have been brought to the conviction that a man of moderate courage, with a good rifle and a steady arm, accompanied by three or four active dogs, a mixture of either the fox-hound or grey-hound, might hunt the Cougar with great safety to himself, and with a tolerable prospect of success.

This animal, which has excited so much terror in the minds of the ignorant and timid, has been nearly exterminated in all our Atlantic States, and we do not recollect a single well authenticated instance where any hunter's life fell a sacrifice in a Cougar hunt.

Among the mountains of the head-waters of the Juniatta river, as we were informed, the Cougar is so abundant, that one man has killed for some years, from two to five, and one very hard winter, he killed seven. In this part of the country the Cougar is hunted with half-bred hounds, the full-blooded dogs lacking courage to attack so large and fierce looking an animal when they overtake it. The hunt is conducted much in the manner of a chase after the common wild-cat. The Cougar is "treed" after running about fifteen or twenty minutes, and generally shot, but sometimes it shews fight before it takes to a tree, and the hunters consider it great sport: we heard of an instance of one of these fights, in which the Cougar got hold of a dog, and was killing it, when the hunter in his anxiety to save his dog, rushed upon the Cougar, seized him by the tail and broke his back with a single blow of an axe.

According to the relations of old hunters, the Cougar has three or four young at a litter. We have heard of an instance of one being found, a very old female, in whose den there were five young, about as large as cats, we believe, however, that the usual number of young, is two.

The dens of this species are generally near the mouth of some cave in the rocks, where the animal's lair is just far enough inside to be out of the rain;

and not in this respect like the dens of the bear, which are sometimes ten or twelve yards from the opening of a large crack or fissure in the rocks. In the Southern States, where there are no caves or rocks, the lair of the Cougar is generally in a very dense thicket, or in a cane-brake. It is a rude sort of bed of sticks, weeds, leaves, and grasses or mosses, and where the canes arch over it; as they are evergreen, their long pointed leaves turn the rain at all seasons of the year. We have never observed any bones or fragments of animals they had fed upon, at the lairs of the Cougar, and suppose they always feed on what they catch near the spot where they capture the prey.

The tales related of the cry of the Cougar in the forest in imitation of the call of a lost traveller, or the cry of a child, must be received with much caution, and may in many of their exaggerations be set down as vulgar errors. In a state of captivity, we have never heard the male uttering any other note than a low growl; the female, however, we have frequently heard uttering a kind of mewing like that of a cat, but a more prolonged and louder note, that could be heard at the distance of about two hundred yards. All the males, however, of the cat kind, at the season when the sexes seek each other, emit remarkable and startling cries, as is evidenced by the common cat, in what is denominated caterwauling. We have observed the same habit in the leopard, the ocelot, and in our two species of lynx. It is not impossible, therefore, that the male Cougar, may at the rutting season have some peculiar and startling notes. The cries, however, to which persons have from time to time directed our attention, as belonging to the Cougar, we were well convinced were uttered by other animals. In one instance, we ascertained them to proceed from a red fox which was killed in the hunt, got up for the purpose of killing the Cougar. In other cases the screams of the great horned, the barred, or the screech owl are mistaken for the cries of this animal.

The female Cougar is a most affectionate mother, and will not leave her young cubs, unless occasionally to procure food to support her own strength; she therefore often becomes very lean and poor. The female we have figured, was in this condition; we procured one of her cubs and figured it, presenting its beautiful spots, seldom before noticed. The other made its escape.

The whelps are suckled by the dam until about half grown, and then hunt with the old ones (which generally go in pairs) until the mother is with young again, or the young ones find mates for themselves, and begin to breed.

The period of gestation of the Cougar is ninety-seven days, as has been ascertained at the Zoological Society of London, (Proceedings, 1832, p. 62.) In the Northern and Middle States, the young are produced in the spring. In the Southern States, however, where the animal is supplied with an abundance of food, and not much incommoded by the cold, the young have in some instances been discovered in autumn. J. W. AUDUBON found, in Texas, young Cougars nearly half grown in February.

Nº 20

PLATE XCVIII

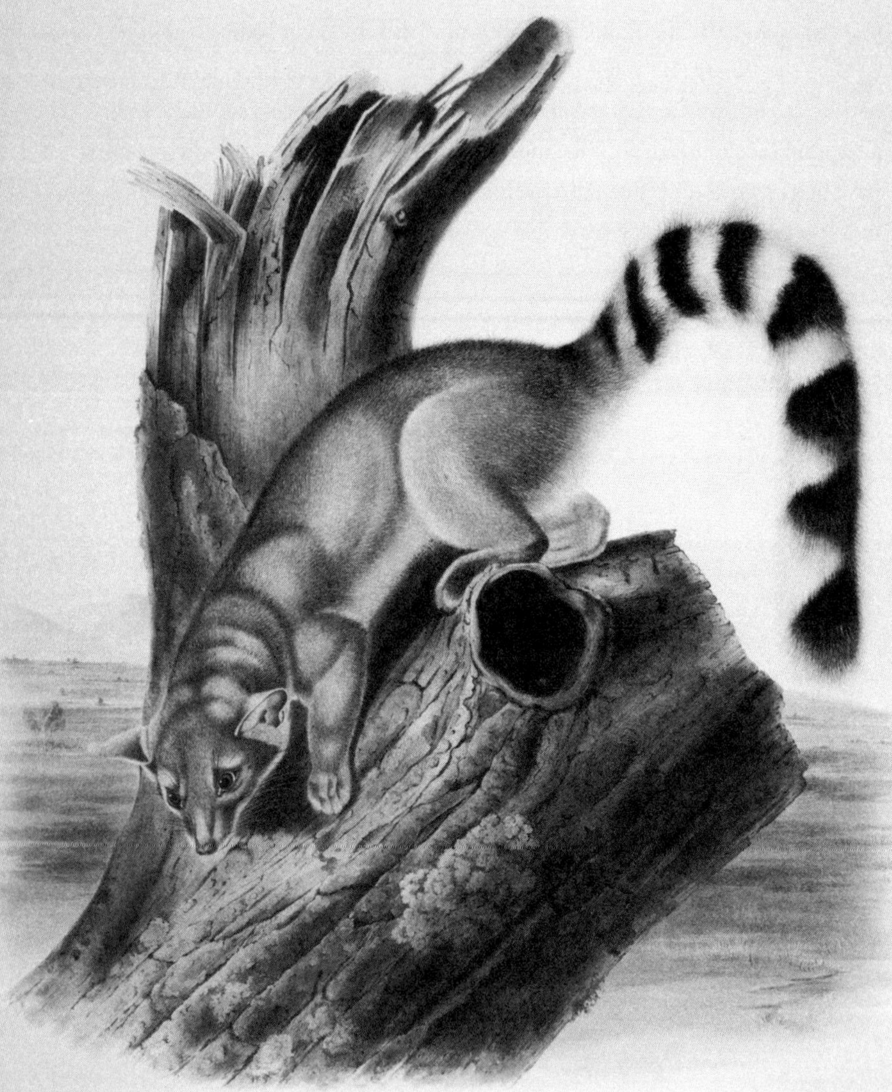

BASSARIS ASTUTA, LICHT.

RING-TAILED BASSARIS.

Ring-tailed Bassaris

THIS IS THE ONLY QUADRUPED ESSAY TO USE MATERIAL FROM John Woodhouse's western journals without explicitly stating that it is quoting from his journals. Here, then, the collective plural voice of the essays begins to include a third member. The opening of the essay emphasizes the causal relationship between a landscape and the mammal species that inhabit it. One does not expect "tree-climbing animals" in "prairie-land," but in the Texas prairies are some swampy areas that support a few species of trees, thereby making the arboreal bassaris possible. That John's eye for causation was good is also borne out by his observation that the trunks of the trees occupied by the bassaris are surrounded by dense growths of thorny plants that offer a natural protection from humans. This essay also raises our awareness of all the Audubon family journals that Maria Rebecca Audubon likely destroyed. We know she burned her grandfather's journals. We also know that she was in possession of those of her mother, her aunt Georgianna, and her father, John Woodhouse, none of which journals are known to have survived.

---

30. John Woodhouse Audubon, J. T. Bowen, lithographer, *Ring-tailed Bassaris*, 1846–48, hand-colored lithograph, plate 98 of *The Viviparous Quadrupeds of North America*. Courtesy of the Jule Collins Smith Museum of Fine Art, Auburn University; Louise Hauss and David Brent Miller Audubon Collection.

BASSARIS ASTUTA.—LICHT.

RING-TAILED BASSARIS.

PLATE XCVIII.—MALE.

HABITS.

The greater portion of Texas is prairie-land, and it is chiefly along the water courses, that trees are found growing together in numbers sufficient to constitute a "wood." On certain level and clayey portions of the prairie, however, the land is swampy, and is covered with several kinds of oaks and a few other trees. The well-known musquit tree or bush is found generally distributed in the western parts of the State. It resembles the acacia in leaf, and has a small white pea-shaped blossom; at a distance it looks something like an old peach tree. Its wood is similar to coarse mahogany in appearance, and burns well, in fact beautifully, as the coals keep in for a long time; and the wood gives out little or no smoke. The musquit bottoms are furnished with these trees, they are small, about the size of the alder, and grow much in the same way; the musquit has sharp thorns. The musquit grass, (*Holcus lanatus*), resembles what is called, *guinea* grass, it is broader, shorter, softer, and more curly.

The general features of the State of Texas, as it will be seen by the foregoing, do not indicate a country where many tree-climbing animals could be found, and the present beautiful species, which Professor LICHTENSTEIN most appropriately named *Bassaris astuta*, is by no means common. It is a lively, playful, and nimble creature, leaps about on the trees, and has very much the same actions as the squirrel, which it resembles in agility and grace, always having a hole in the tree upon which it resides, and betaking itself to that secure retreat at once if alarmed.

The *Bassaris Astuta* is shy and retired in its habits, and in the daytime often stays in its hole in some tree, so that we were only able to procure about half a-dozen of these animals during our stay in Texas; among which, to our regret, there was not a single female.

The food of this species is chiefly small animals, birds, and insects; they also eat nuts, as we were told, descending from their hiding place and travelling to the pecan and other trees, for the purpose of feeding on the nuts which, if true, is singular, as they are decidedly carnivorous in their dentition.

They are much attached to the tree on which they live, which is generally a post-oak, a live-oak, or other large tree, and they seldom quit the immediate vicinity of their hole, unless when driven out by thrusting a stick at them, when they ascend the trunk of the tree, and jump about among the higher branches so long as the pole is held close to their nest; as soon as this is withdrawn, they descend and at once re-enter their dwelling-place and hide themselves. These animals have a singular habit of eating or gnawing off the bark around the mouth of their holes, and where the bark does not appear freshly peeled off at their hole, you may be certain the animal is not at home, or has deserted the place. Their holes are generally the result of natural decay, and are situated on knobs, or at the ends of branches broken short off close to the main trunk.

They generally select a hole of this kind on the lower side of a leaning tree, probably for better protection from the rain; their holes vary in depth, but are seldom more than about a foot or eighteen inches to the bottom; they are usually furnished with moss or grass, for bedding. Sometimes pecan shells are found in these holes, which no doubt affords presumptive evidence that the Bassaris feeds upon this nut.

When scolding or barking at an intruder, the ring-tailed Raccoon, (as this animal is called by the Texans), holds the tail over its back, bending it squirrel fashion; this animal, however, does not stand upon his hind feet like a squirrel, and cannot jump or leap so far. We have not heard of their springing from one branch to another beyond the distance of about ten feet, and when frightened at the presence of a man, they will sometimes run along a branch even toward him, in order to get within jumping distance of another, evincing more timidity than a squirrel exhibits in springing among the boughs, although they run up the bark with ease, holding on with their claws.

Sometimes the Ring-tailed Bassaris may be seen squatted on the top of a branch, basking in the sun, and half rolled up, appearing almost asleep. On the slightest manifestation of danger, however, he darts into his hole, (which is always within a foot or two of his basking place), and he is seen no more. We have the impression that only one of these singular animals is to be found on a tree at a time—they, therefore, are not very social in their habits, and, as the live-oak and other trees are generally very much scattered, and many of them have no holes suitable for residences for the Bassaris, it is very difficult to procure one. At the foot of many of the trees whereon they dwell, the cactus, brush-wood, and chapperal generally are so thick and tangled, that a man would be pretty well scratched should he attempt to penetrate the thorny, prickly thicket which surrounds the dwelling-place of this solitary and singular animal.

Notwithstanding the shyness and retired habits of this species, it is easily tamed, and when it has been confined in a cage a sufficient length of time, is frequently let loose in the houses of the Mexicans, where it answers the purpose of a playful pet, and catches mice and rats. We have seen one that was thus domesticated, running about the streets of a little Mexican village, and we were informed that one was kept as a great pet in a Camanche camp, visited by the Indian who hunted for us during our explorations of the western part of Texas. As far as we could ascertain, the northern limit of the range of this species is somewhere in the neighbourhood of the southern branches of Red river. As you travel south they are more abundant, and probably are found throughout all Mexico; we were informed by our friend, the celebrated Col. HAYS,[41] the Ranger, that he saw them more abundant in the mountainous region near the head-waters of the San Saba river than at any other place.

The Bassaris produces three or four young at a birth, as has been ascertained from the animal kept in confinement.

N° 20. PLATE C.

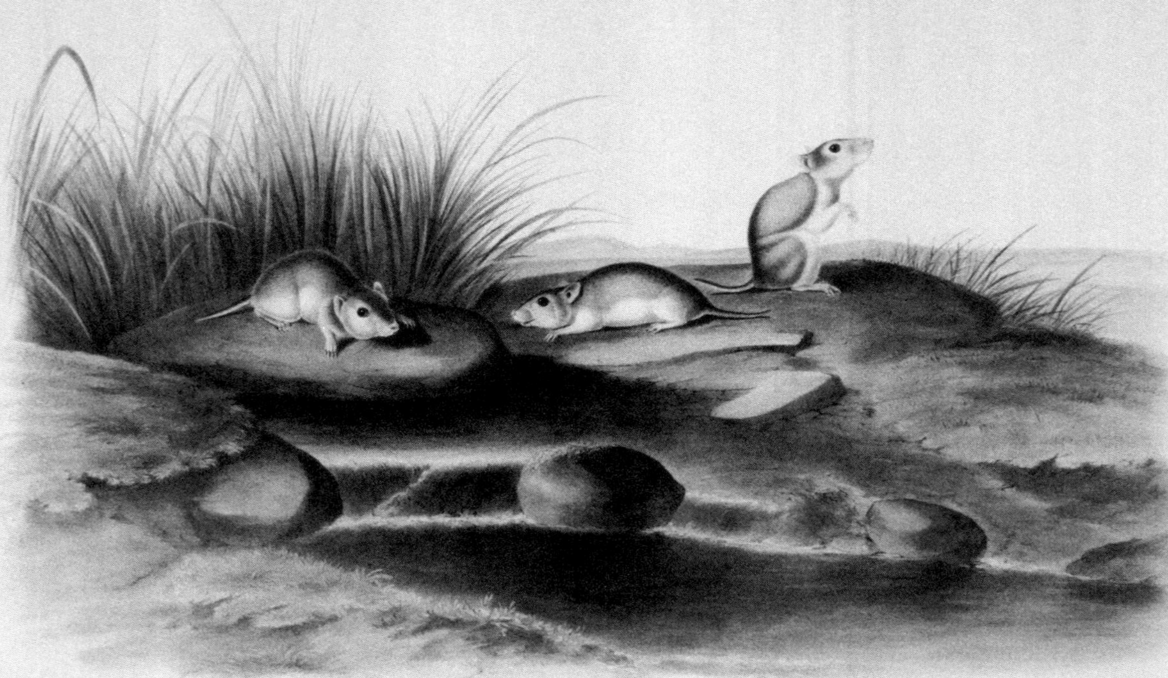

MUS MISSOURIENSIS, AUD. & BACH.

MISSOURI MOUSE.

Missouri Mouse

OUT OF SHEER WEARINESS AND FRUSTRATION, BACHMAN admonishes his old, declining friend for never having made any notes on this species' behavior when he was at Fort Union in 1843. To complete volume 2 of the letterpress, Bachman and his new wife and former sister-in-law, Maria Martin, have been working twelve-hour days, with Bachman dictating from any materials Victor has sent him. When they turn to the final species for the volume and are reminded that they have no information about it, Bachman stages a kind of public confession for Audubon: "At that time being in quest of antelopes and large animals, we did not give it that close attention, which we should have done." The cruel irony for Bachman would have been that the Missouri mouse was, he believed, the only new mammal species that Audubon brought back from that expedition. With the original Missouri River journals at hand, though, Bachman decided to include "a short description of our mode of travelling," which, while having no bearing on the species this essay should address, does preserve a day's entry from the journal that was later destroyed.

---

31. John Woodhouse Audubon, J. T. Bowen, lithographer, *Missouri Mouse*, 1846–48, hand-colored lithograph, plate 100 of *The Viviparous Quadrupeds of North America*. Courtesy of the Jule Collins Smith Museum of Fine Art, Auburn University; Louise Hauss and David Brent Miller Audubon Collection.

MUS MISSOURIENSIS.—AUD. AND BACH.

MISSOURI MOUSE

PLATE C.—FEMALES.

HABITS.

We close our second volume with this new species of mouse, of which we have given three figures. This pretty little animal was discovered for us by Mr. DENIG, during our sojourn at, and in the neighbourhood of Fort Union in 1843. It was in full summer pelage, having been killed on the 14th of July. At that time being in quest of antelopes and large animals, we did not give it that close attention, which we should have done. A glance at our plate, or an examination of our description, will suffice to convince any one of its being entirely new. This species is much larger, and has a thicker and shorter tail than *Mus leucopus*.

Expecting to get more of them we did not make any notes of the habits of those killed at that time, and which had doubtless been observed by the hunters, who procured them. The next day after they were brought in, we left the fort on an expedition to the Yellow-Stone river, from which we did not return for some time.

As a short description of our mode of travelling, &c., the first day's journal is here given. "July 15, Saturday, we were all up pretty early, making preparations for our trip to the Yellow-Stone river. After breakfast all the party who were going, announced themselves as ready, and with a wagon, a cart, and two extra men from the fort, we crossed the Missouri, and at 7 o'clock, were fairly under way; HARRIS, BELL, CULBERTSON, and ourself in the wagon, SQUIRES, PROVOST, and OWEN on horse-back, while the cart brought a skiff, to be launched on the Yellow-Stone when we should arrive at that river. We travelled rather slowly until we had crossed a point and headed the ponds on the prairie at the foot of the hills opposite the fort. We saw one sharp-tailed grouse, but although Mr. HARRIS searched for it diligently, it could not be started. Soon after this we got one of the wheels of our wagon fast in a crack or crevice in the ground, and wrenched it so badly that we were obliged to get out and walk, while the men set to work to repair the wheels which were all in a rickety condition; after the needful fixing-up had been done, the wagon overtook us, and we proceeded on. Saw some antelopes on the prairie, and many more on the tops of the hills bounding our view to the westward. We stopped to water the horses at a 'saline,' where we observed that buffaloes, antelopes, and other animals had been to drink, and had been lying down on the margin. The water was too hot for us to drink. After sitting for nearly an hour to allow the horses to get cool enough to take a bait, for it was very warm, we again proceeded on until we came to the bed of a stream, which during spring overflows its banks, but now exhibits only pools of water here and there. In one of these pools we soaked our dry wagon wheels, by way of tightening the 'tires,' and here we refreshed ourselves and quenched our thirst. SQUIRES, PROVOST, and OWEN, started on before us to reconnoitre, and we followed at a pretty good pace, as the prairie was hereabouts firm and tolerably smooth. Shot a red-winged black-bird. Heard the notes of NUTTALL's short-billed marsh-wren,—supposed by some of our party to be those of a new bird. Saw nothing else; reached our camping-place at about 6 o'clock. Unloaded the wagon and cart, hobbled the horses, and turned them out to grass. Two or three of the men went off to a point above our camp, in search of something for supper. We took the red-winged black-bird, and a fishing-line, and went to the bank of the famed Yellow-Stone river, (near the margin of which our tent was pitched,) and in this stream of the far west, running from the bases of the Rocky Mountains, we threw our line, and exercised our piscatory skill so successfully as to catch some cat fish. These fish we found would not bite at pieces of their own kind, with which we tried them; after expending our bird bait, we therefore gave up fishing. One of our men took a bath, while two others, having launched the skiff rowed across the river to seek for deer or other game on the opposite shore. Toward dark the hunting parties all returned to camp without success; and we found the cat-fish the principal portion of our supper, having no fresh meat at all.

"Our supper over, all parties shortly disposed themselves to sleep as they best could. About 10 o'clock, we were all disturbed by a violent thunder storm, accompanied by torrents of rain and vivid flashes of lightning; the wind arose and blew a gale; all of us were a-foot in a few moments; and amid some confusion, our guns, loaded with ball, and our ammunition, were placed under the best covering we could provide, our beds huddled together under the tent along with them, and some of us crawled in on top of all, while others sought shelter under the shelving bank of the river. This storm benefitted us, however, by driving before the gale the mosquitoes, to keep off which we had in vain made a large fire, before we laid ourselves down for the night."

As there is little grain of any kind grown in this part of the country, the Missouri Mouse no doubt exists on the seeds and roots of wild plants entirely, of which it is able to lay up a store for the winter in holes in the ground. It may, however, possibly resort to the patches of corn planted by the squaws of some of the Indian tribes, at the time that grain is ripe. We brought with us from this country, when we returned home, some ears of a very small corn, (maize,) which ripens early, and bears its fruit near the ground. Having planted it on our place, we found that it was advanced enough to be eaten at table as a vegetable, several weeks before the ordinary kinds of corn known about New-York. We, therefore, distributed some of the seed among our farming neighbours, and likewise sent some to England to Lord Derby and other friends, but this was unfortunately lost. We incline to believe that this corn would ripen well in the climate of England or Scotland. Unluckily, ours has become mixed by having been planted too near common corn, and is now depreciated or reduced to nearly the same thing as the latter.

Nº 22.

PLATE CVI.

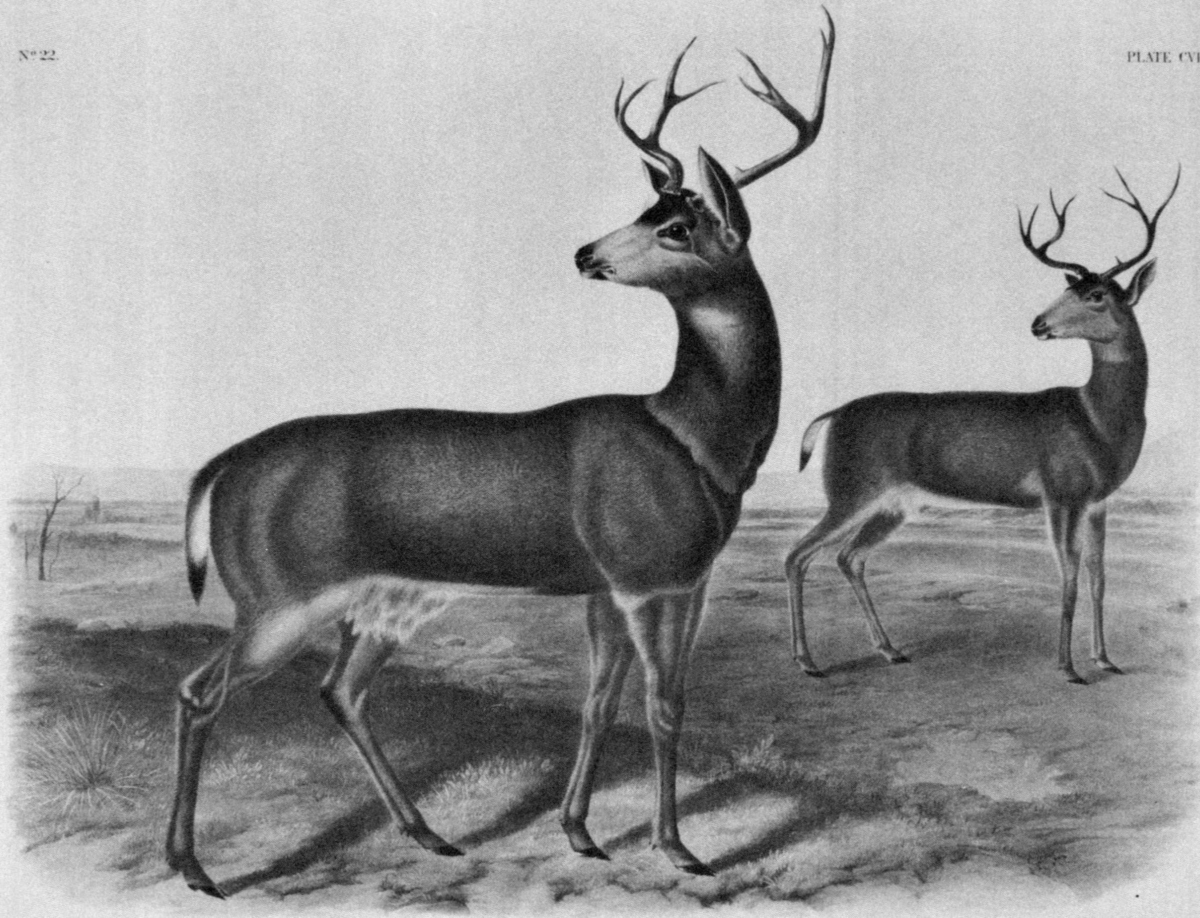

CERVUS RICHARDSONII, AUD & BACH.

COLUMBIAN BLACK-TAILED DEER.

Males.

Columbian Black-tailed Deer

THE CALIFORNIA GOLD RUSH, THE HISTORICAL BACKDROP OF this essay, precipitated this species being "generally known to white men." Indeed, the Columbian black-tailed deer provided venison for thousands of miners. John G. Bell, who accompanied Audubon up the Missouri in 1843, was swept west in the gold rush and provided Bachman with information concerning the deer. Bell's time as a skilled (and rather ruthless) hunter in Audubon's party on the prairies of the Upper Missouri seems to have set the taxidermist up for lucrative returns: he reported that through still-hunting one morning in California, he killed three of this species before breakfast "and sold them, after reserving some of the best parts for himself and companion, for eighty dollars apiece!" In 2019 that amounts to over $2,600 per deer. The fact that even a deer stripped of tenderloins could be so valuable points to an almost incomprehensible economic situation in California at the time.

---

32. John Woodhouse Audubon, J. T. Bowen, lithographer, *Columbian Black-tailed Deer*, 1846–48, hand-colored lithograph, plate 106 of *The Viviparous Quadrupeds of North America*. Courtesy of the Jule Collins Smith Museum of Fine Art, Auburn University; Louise Hauss and David Brent Miller Audubon Collection.

CERVUS RICHARDSONII.—AUD. AND BACH.

COLUMBIAN BLACK-TAILED DEER.

PLATE CVI.—MALES.

HABITS.

This beautiful Deer is found variously dispersed over the western portions of the North American continent, where it was first noticed by LEWIS and CLARK, near the mouth of the Columbia River; but not until the discovery of the golden treasures of California did it become generally known to white men. In that country, along the hill sides and in the woody dells and "gulches," the hardy miners have killed hundreds, nay thousands, of Black-tailed Deer; and it is from the accounts they have given that it is now known to replace, near the great Sierra Nevada, the common or Virginian Deer which is found east of the Rocky Mountains; all the hunters who have visited California, and whom we have seen, tell us that every Deer they shot there was the Black-tailed species.

J. W. AUDUBON killed a good many of these Deer, and describes them as tender and of good flavour; and during the time his party encamped on the Tuolome River, and in the "dry diggings" near Stockton, when he kept two of his men busy shooting for the support of the others, they generally had one or two Deer brought into camp every day. The mode of hunting them was more similar to what is called Deer-stalking in Scotland than to the methods used for killing Deer in the eastern part of the Union. Sometimes the hunters (who had no dogs) would start before day, and, gaining the hills, anxiously search for fresh tracks in the muddy soil (for it was then the rainy season, and the ground everywhere wet and soft), and, having found a trail, cautiously follow; always trying to keep the wind in such a direction as not to carry the scent to the animals. After discovering a fresh track, a search of a most tedious and toilsome nature awaited them, as the unsuspecting Deer might be very near, or miles off, they knew not which; at every hill-top they approached, they were obliged to lie down and crawl on the earth, pausing when they could command the view to the bottom of the valley which lay beyond the one they had just quitted; and after assuring themselves none were in sight, carefully following the zigzag trail, proceed to the bottom. Again another summit has been almost reached; now the hunters hope for a shot: eye and ear are strained to the utmost, and they move slowly forward; the ridge of the next hill breaks first upon their sight beyond a wide valley. The slope nearest them is still hidden from their view. On one side the mountains rise in steeper and more irregular shapes; pine-trees and oaks are thickly grown in the deepest and most grassy spot far below them. The track trends that way, and silently they proceed, looking around at almost every step, and yet uncertain where their game has wandered. Once the trail has been almost lost in the stony, broken ground they pass, but again they have it; now they approach and search in different directions the most likely places to find the Deer, but in vain; at last they gain the next summit: the object of their chase is at hand; suddenly they see him—a fine buck—he is yet on the declivity of the hill, and they cautiously observe his motions. Now they see some broken ground and rocky fragments scattering towards the left; they redouble their caution; locks are ready cocked; and, breathing rapidly, they gain the desired spot. One instant—the deadly rifle has sent its leaden messenger and the buck lies struggling in his gore.

Short work is made of the return to camp if no more Deer signs are about; and a straight cut may bring the hunters home in less than an hour, even should they have been two or three in following their prize.

Sometimes the Deer start up suddenly, quite near, and are shot down on the instant; occasionally, after a long pursuit, the crack of a rifle from an unknown hunter deprives the others of their chance; and—must we admit it—sometimes they miss; and not unfrequently they see no game at all.

Mr. J. G. Bell informed us that while he was digging gold in a sequestered and wild cañon, in company with a young man with whom he was associated in the business, they used to lie down to rest during the heat of the day, and occasionally he shot a Black-tailed Deer, which unsuspectingly came within shooting distance down the little brook that flowed in the bottom of the ravine. He also used to rise very early in the mornings occasionally, and seek for the animals in the manner of still-hunting, as practised in the United States. One morning he killed three in this manner, before his breakfast-time, and sold them, after reserving some of the best parts for himself and companion, for eighty dollars apiece! He frequently sold Deer subsequently, as well as hares and squirrels, birds, &c., which he shot at different times, for enormous prices. Many of the miners, indeed, turned their attention to killing Deer, elk, bear, antelopes, geese, ducks, and all sorts of game and wild fowl, by which they realized considerable sums from selling them at San Francisco and other places. We have heard of one person who, after a luckless search for gold, went to killing Deer and other game, and in the course of about eighteen months had made five thousand dollars by selling to the miners at the diggings.

The gait of this species is not so graceful as that of the Virginian Deer; it bounds rather more like the roebuck of Europe than any other of our Deer except the Long-tailed Deer, and is reported to be very swift. The season of its breeding is earlier than that of the common Deer, and it no doubt brings forth the same number of young at a time.

N°24.

PLATE CXVI.

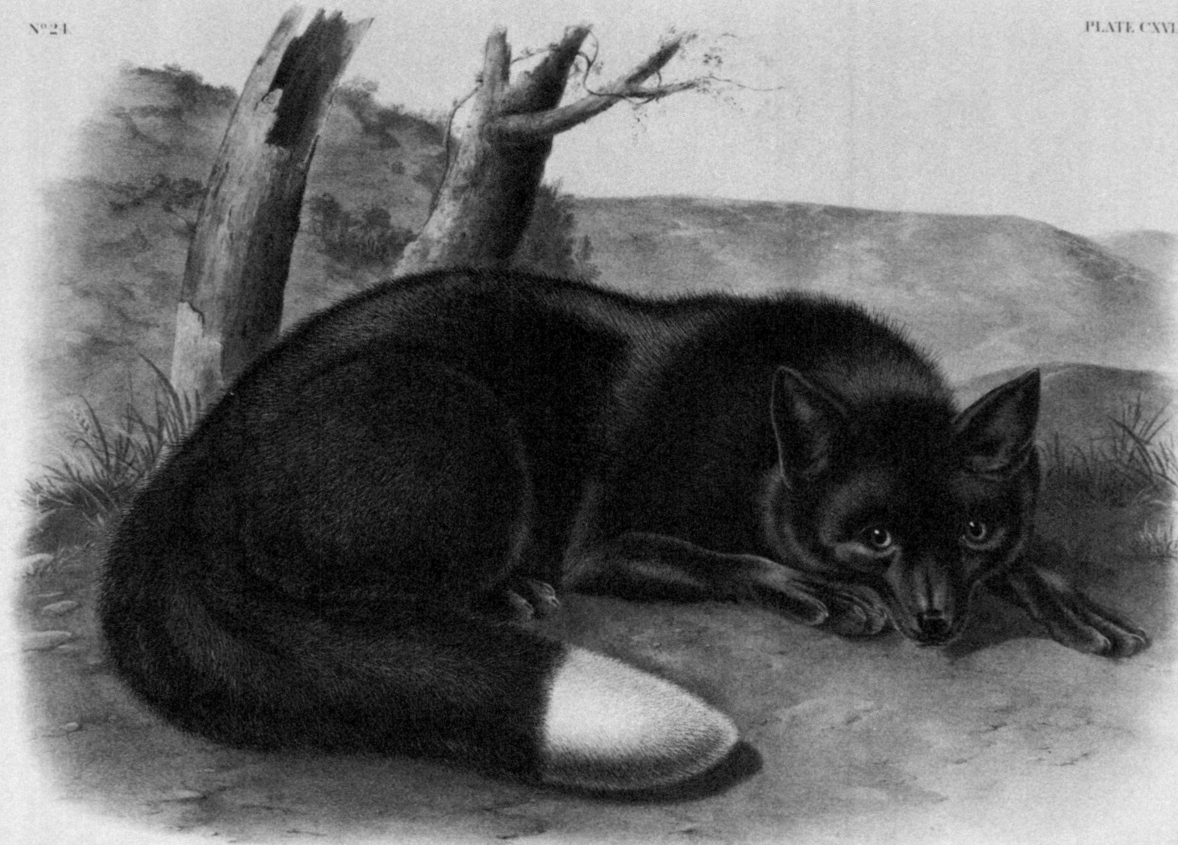

VULPES FULVUS, DESM. VAR. ARGENTATUS, RICH.

AMERICAN BLACK OR SILVER FOX.

American Black or Silver Fox

IN HIS LABRADOR JOURNAL OF 1833, AUDUBON HAD NOTHING positive to say about fur and fishing companies, primarily because their agents kept trade secrets and followed policies that prevented him from accessing the country as freely as he wished.[42] Nearly two decades later, Bachman was more benevolent as he wrote the third and final volume of *Quadrupeds*. He reports that when John Woodhouse Audubon was in London painting specimens for *Quadrupeds* at the Zoological Gardens in 1846–47, it gave him "pleasure to render our thanks to the Hon. Hudson's Bay Company for a superb female Black or Silver-gray Fox which was procured for us, and sent to the Zoological Gardens in London alive." The implication of the difference in Audubon's and Bachman's perspectives may be that fur companies are uncomfortable with outsiders' boots on the ground in their territory, but sharing specimens with naturalists back in London is good public relations. The generosity of the HBC and the Zoological Society prompted Bachman to close the anecdote, "When shall we have a Zoological Garden in the United States?" Bachman was interested in producing natural history that competed with or exceeded the quality of European works, so he never resisted pointing out the U.S. government's lack of serious support for American science.

---

33. John Woodhouse Audubon, J. T. Bowen, lithographer, *American Black or Silver Fox*, 1846–48, hand-colored lithograph, plate 116 of *The Viviparous Quadrupeds of North America*. Courtesy of the Jule Collins Smith Museum of Fine Art, Auburn University; Louise Hauss and David Brent Miller Audubon Collection.

VULPES FULVUS.—DESM. (VAR. ARGENTATUS.—RICH.)
AMERICAN BLACK OR SILVER FOX
PLATE CXVI.—FEMALE.

HABITS.

Our account of the habits of this beautiful Fox will be perhaps less interesting to many than our description of its skin; for, as is well known, the Silver-gray Fox supplies one of the most valuable furs in the world, not only for the luxurious nobles of Russia and other parts of Europe, but for the old-fashioned, never-go-ahead Chinese, and other Eastern nations.

In the richness and beauty of its splendid fur the Silver-gray Fox surpasses the beaver or the sea-otter, and the skins are indeed so highly esteemed that the finest command extraordinary prices, and are always in demand.

The Silver-gray Fox is by no means abundant, and presents considerable variations both in colour and size. Some skins are brilliant black (with the exception of the end of the tail, which is invariably white); other specimens are bluish-gray, and many are tinged with a cinereous colour on the sides: it perhaps is most commonly obtained with parts of its fur hoary, the shiny black coat being thickly interspersed with white or silvery-blue tipped hairs.

According to Sir JOHN RICHARDSON, a greater number than four or five of these Foxes is seldom taken in a season at any one post in the fur countries, though the hunters no sooner find out the haunts of one than they use every art to catch it. From what he observed, Sir JOHN does not think this Fox displays more cunning in avoiding a snare than the red one, but the rarity of the animal, and the eagerness of the hunters to take it, make them think it peculiarly shy.

This animal appears to be as scarce in northern Europe as in America; but we do not mean by this to be understood as considering the European Black Fox identical with ours.

The Black or Silver Fox is sometimes killed in Labrador, and on the Magdeleine Islands, and occasionally—very rarely—in the mountainous parts of Pennsylvania and the wilder portions of the northern counties of New York, where, however, PENNANT's marten is generally called the "Black Fox," by the hunters and farmers.

It gives us pleasure to render our thanks to the Hon. Hudson's Bay Company for a superb female Black or Silver-gray Fox which was procured for us, and sent to the Zoological Gardens in London alive, where J. W. AUDUBON was then making figures of some of the quadrupeds brought from the Arctic regions of our continent for this work. Having drawn this beautiful animal, which was at the time generously tendered us, but thinking it should remain in the Zoological Gardens, as we have no such establishment in America, J. W. AUDUBON declined the gift in favor of the Zoological Society, in whose interesting collection we hope it still exists. When shall we have a Zoological Garden in the United States?

This variety of the Fox does not differ in its propensities from the red Fox or the cross Fox, and its extraordinary cunning is often equalled by the tricks of these sly fellows.

The white tip at the end of the tail appears to be a characteristic of the Silver-gray Fox, and occurs in every specimen we have seen.

It is stated in MORTON's New England Canaan (p. 79), that the skin of the Black Fox was considered by the Indians, natives of that part of the colonies, as equivalent to forty beaver skins; and when offered and accepted by their kings, it was looked upon as a sacred pledge of reconciliation.[43]

The present species has been seen "mousing" in the meadows, near Ipswich, Massachusetts, as we were informed by the late WILLIAM OAKES, who also wrote to us that "the common and cross Foxes were abundant about the White Mountains, and that they were most easily shot whilst scenting and following game, when their whole attention appears to be concentrated on that one object."

This Fox is occasionally seen in Nova Scotia, and a friend there informs us that some have been shot in his vicinity.

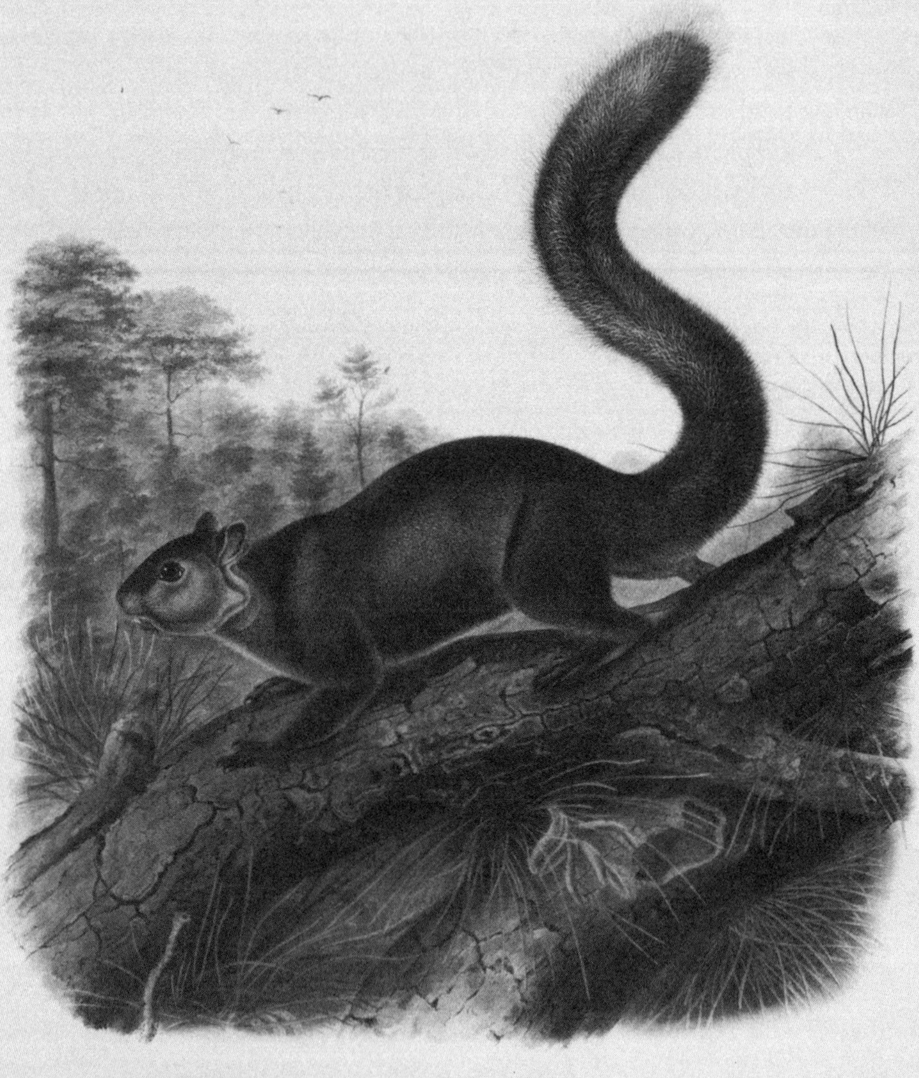

BENNETT

DUSKY SQUIRREL

Dusky Squirrel

ANYONE FAMILIAR WITH AUDUBON'S ORNITHOLOGICAL WRITings and journals is aware of his disdain for "closet naturalists" who pursue natural history exclusively in studies and libraries. Bachman shares the sentiment in this short essay. Perhaps more intriguing is his observation that the profusion of nut-bearing woods in North America has made the continent suitable habitat for "an unusual number of species of squirrels." More intriguing still is his envisioning of deeper zoological time in which there may have been more species of squirrel in Europe than the single known species. Although the essay says nothing of the actual habits of the dusky squirrel, it is a concise representation of Bachman's basic scientific philosophy: direct observation and logical inferences based on a knowledge of natural laws, as well as a little imagination. Ironically, as he informs us in the essay's "Geographical Distribution" section and "General Remarks," Bachman never saw a living specimen of this squirrel. Instead, he saw the only stuffed specimen available at the time; it was supposedly obtained in California and was housed at the Zoological Society of London.

---

34. John Woodhouse Audubon, J. T. Bowen, lithographer, *Dusky Squirrel*, 1846–48, hand-colored lithograph, plate 117 of *The Viviparous Quadrupeds of North America*. Courtesy of the Jule Collins Smith Museum of Fine Art, Auburn University; Louise Hauss and David Brent Miller Audubon Collection.

SCIURUS NIGRESCENS.—BENNETT.

DUSKY SQUIRREL

PLATE CXVII.—MALE.

HABITS.

The existence in North America of an unusual number of species of squirrels has been made known to our subscribers in the course of this publication. There are many closely allied, and many very beautiful species among them; all are graceful and agile, and possess very similar habits.

The great number of these nut-eating animals in North America would be a proof (were any such wanting) that nature has been more bountiful to our country in distributing nut-bearing trees over the whole extent of our continent than to other parts of the globe, and this in connexion with the fact that so great a proportion of wood-land cannot be found in any other part of the world of similar extent, marks America as intended for a very dense population hereafter. In Europe there is only one well determined species of squirrel known, at present at least, although at some remote period there may have been more.

In regard to the peculiar habits of the Dusky Squirrel, we have nothing to say. It is one of the species which, being shot or procured by collectors of objects of natural history, and sent to Europe, have there been described by naturalists who, having the advantages of museums which contain specimens from every part of the globe, and the largest libraries in the world also to which they can refer, may sometimes discover new species with much less difficulty, but also less certainty, than the student of nature must encounter while seeking for knowledge in the woods.

But the naturalist who learns from books only, and describes from dried skins, is at best liable to mistakes. We have in fact always found that where young animals, or accidental varieties, have been described as new species, this has been the result of study in the museum or cabinet, not in the fields.

N° 26.

PLATE CXXVII.

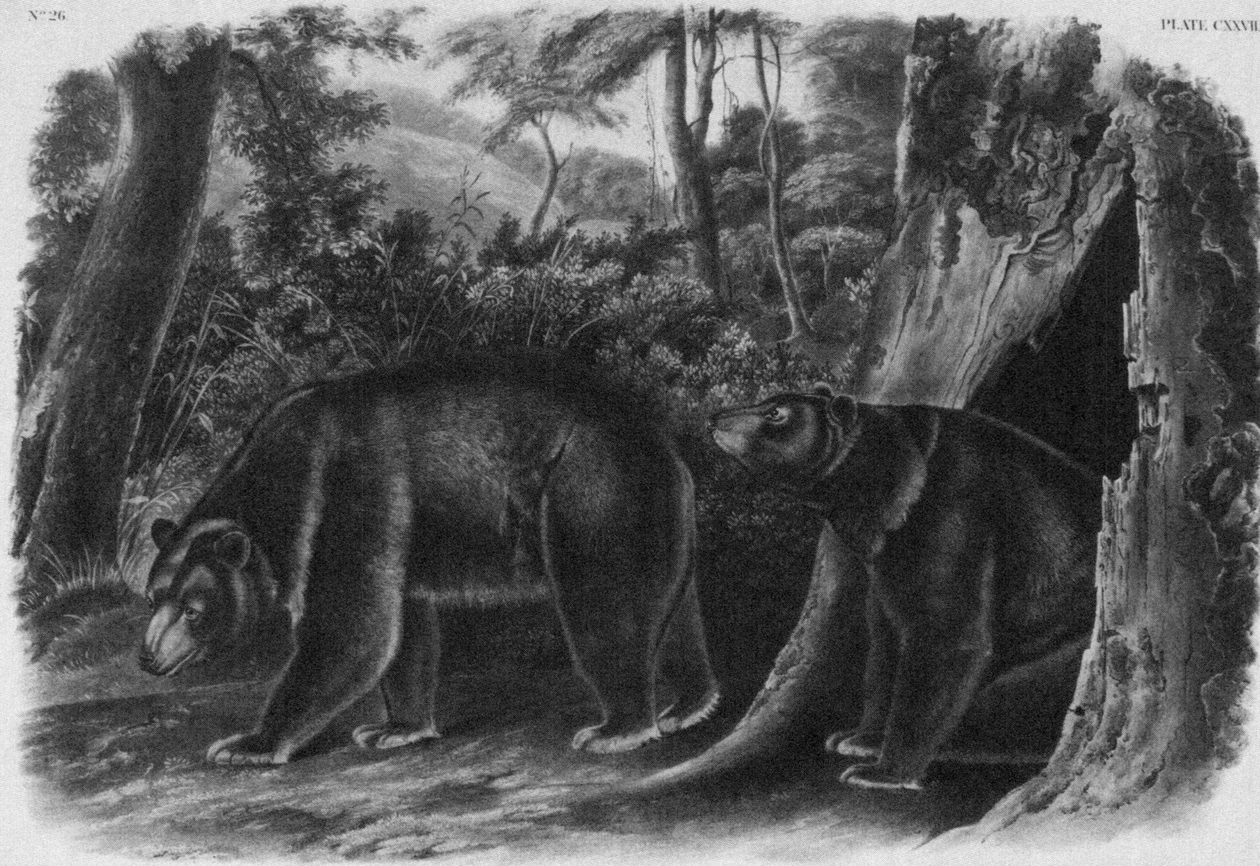

URSUS AMERICANUS, PALLAS.

VAR. CINNAMOMUM, AUD & BACH.

CINNAMON BEAR.

MALE & FEMALE.

Cinnamon Bear

BACHMAN IS NOT COMPLETELY SURE IF *URSUS AMERICANUS cinnamonum* is a subspecies of black bear or a new species altogether. He highly, and correctly, suspects the former, designating the cinnamon bear as a "permanent variety" of the black bear. Bachman is dismissive of the notion that species could morph into new species, so his permanent variety allowed for variance within a single species while also permitting him to maintain that new species could not emerge beyond God's initial creation. As Bachman eloquently puts it in the "General Remarks," "We conceive that figures of the permanent variety may be useful to future observers in order to awaken inquiry and enable them to decide whether they are true species or mere varieties." Interestingly, when Lewis and Clark consulted the Nez Perce ("Choppunish") on this issue half a century prior, the Nez Perce asserted that the cinnamon bear was indeed a variety of the black bear.

---

35. John Woodhouse Audubon, J. T. Bowen, lithographer, *Cinnamon Bear*, 1846–48, hand-colored lithograph, plate 127 of *The Viviparous Quadrupeds of North America*. Courtesy of the Jule Collins Smith Museum of Fine Art, Auburn University; Louise Hauss and David Brent Miller Audubon Collection.

URSUS AMERICANUS.—PALLAS.

(VAR. CINNAMOMUM.—AUD. AND BACH.)

CINNAMON BEAR

PLATE CXXVII.—MALE AND FEMALE.

HABITS.

LEWIS and CLARK (Expedition, vol. ii. p. 303) mention that one of their men purchased a Bear-skin of a uniform pale reddish-brown colour, which the Indians (Chopunnish) distinguished from every variety of the Grizzly Bear: this induced those travellers to inquire more particularly into the opinions held by the Indians as to the several species of Bears, and they exhibited all the Bear-skins they had killed in that neighbourhood, which the Indians immediately classed into two species—the Grizzly Bear, including all those with the extremities of the hair of a white or frosty colour, under the name of *hohhost*, and the black skins, those which were black with a number of entire white hairs intermixed, or white with a white breast, uniform bay, brown, and light reddish-brown, were ranged under the name of *Yackkah*. These we refer to the Cinnamon and other varieties of the Black Bear. LEWIS and CLARK, however, appear not to have considered these Bear-skins as belonging to the Black Bear, owing merely to the differences in colour, for they say the common Black Bear is "indeed unknown in that country." Their account of the fur of the brown Bears above mentioned corresponds, however, with the description of the Cinnamon Bear, they remarking that the skins of the Bears in that region differ from those of the Black Bears "in having much finer, thicker, and longer hair, with a greater proportion of fur mixed with it." LEWIS and CLARK considered that the Black Bear was always black, whereas it varies very considerably: they say nothing in regard to the sizes of the various coloured Bears above alluded to.

The Cinnamon Bear has long been known to trappers and fur traders, and its skin is much more valuable than that of the Black Bear. We have seen in the warehouse of Messrs. P. CHOUTEAU, JR., and Co., in New

York, some beautiful skins of this animal, and find that those gentlemen receive some every year from their posts near the Rocky Mountains. Being a permanent variety, and having longer and finer hair than the common Black Bear, we might possibly have elevated it into a distinct species but that in every other particular it closely resembles the latter animal. By the Indians (according to Sir JOHN RICHARDSON) it is considered to be an accidental variety of the Black Bear.

The Cinnamon Bear, so far as we have been able to ascertain, is never found near the sea coast, nor even west of the Ohio valley until you approach the Rocky Mountain chain, and it is apparently quite a northern animal.

Of the habits of this variety we have no accounts, but we may suppose that they do not differ in any essential particulars from those of the Black Bear, which we shall shortly describe.

Our figures were made from living specimens in the gardens of the Zoological Society of London, which manifested all the restlessness usually exhibited by this genus when in a state of captivity.

We are inclined to consider Sir JOHN RICHARDSON's "Barren-ground Bear" a variety of the common black Bear,—perhaps our present animal; but not having seen any specimen of his *Ursus Arctos ? Americanus*, we do not feel justified in expressing more than an opinion on this subject, which indeed is founded on the description of the colour of the Barren-ground Bear as given by RICHARDSON himself (see Fauna Boreali Americana, pp. 21, 22).

GENERAL REMARKS.

We have given a figure of this permanent variety of Bear, not because we felt disposed to elevate it into a species, but because it is a variety so frequently found in the collections of skins made by our fur companies, and which is so often noticed by travellers in the northwest, that errors might be made by future naturalists were we to omit mentioning it and placing it where it should be. Whilst we are not disposed to figure an occasional variety in any species, and have throughout our work rather declined doing this, yet we conceive that figures of the permanent varieties may be useful to future observers in order to awaken inquiry and enable them to decide whether they are true species or mere varieties. We have done this in the case of some species of squirrel, the otter, and the wolves, as well as this variety of Bear. The yellow Bear of Carolina no doubt belongs to this variety, and probably the brown Barren-ground Bear of RICHARDSON may be referred to the same species, as all Bears vary very greatly in size.

N° 26

PLATE CXXVIII.

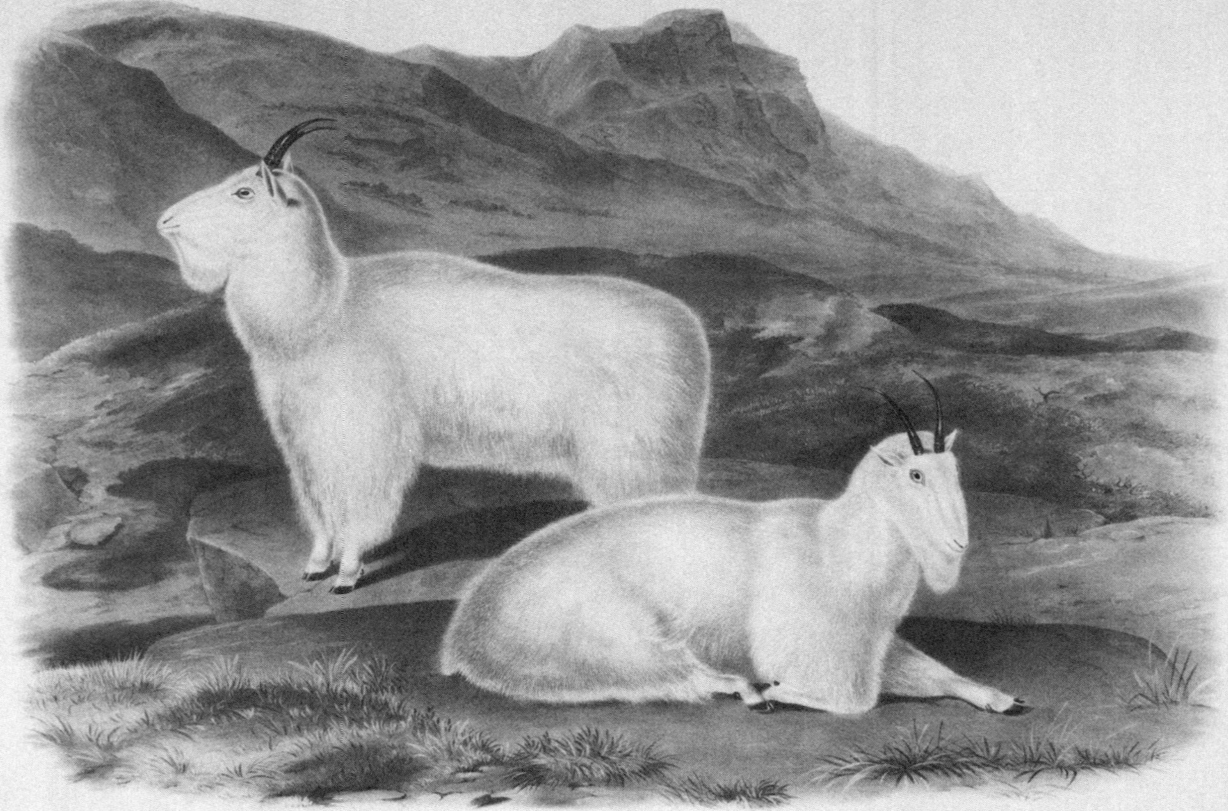

CAPRA AMERICANA, BLAINVILLE.

ROCKY MOUNTAIN GOAT.

Rocky Mountain Goat

THE WESTERN SUBLIME CONSTRUCTED IN THIS ESSAY DRAWS an implicit connection between God's catastrophic but generative might and this elusive animal. When standing far atop one of the Rocky Mountains, we are told that the imagination "may go back without a check through ages of time to the period when an Almighty power upheaved the gigantic masses" surrounding the observer. It is in this landscape that the Rocky Mountain goat thrives and evades the smartest and toughest of hunters, even in death. Despite the fictional hunter's well-placed rifle shot, the goat tumbles into an inaccessible ravine, highlighting the alienation of humans in a landscape so reflective of God's power that only the mind can wrestle with it, the physical body proving insufficient. Also of note is Bachman's conception of Native peoples, common for Americans at the time: in this sublime place, the observer can "find no spot upon which to arrest the eye as a place where once dwelt man! No—we only know the Indian as a wanderer, and we cannot say here stood the strong fortress, the busy city, or even the humble cot." Despite Bachman's ability to envision geological beginnings, Native people remain nomadic caricatures with no claim to the land, land that is completely ahistorical aside from its biblical origins.

---

36. John Woodhouse Audubon, J. T. Bowen, lithographer, *Rocky Mountain Goat*, 1846–48, hand-colored lithograph, plate 128 of *The Viviparous Quadrupeds of North America*. Courtesy of the Jule Collins Smith Museum of Fine Art, Auburn University; Louise Hauss and David Brent Miller Audubon Collection.

CAPRA AMERICANA.—BLAINVILLE.

ROCKY MOUNTAIN GOAT

PLATE CXXVIII.—MALE AND FEMALE.

HABITS.

Standing "at gaze," on a table-rock projecting high above the valley beyond, and with a lofty ridge of stony and precipitous mountains in the background, we have placed one of our figures of the Rocky Mountain Goat; and lying down, a little removed from the edge of the cliff, we have represented another.

In the vast ranges of wild and desolate heights, alternating with deep valleys and tremendous gorges, well named the Rocky mountains, over and through which the adventurous trapper makes his way in pursuit of the rich fur of the beaver or the hide of the bison, there are scenes which the soul must be dull indeed not to admire. In these majestic solitudes all is on a scale to awaken the sublimest emotions and fill the heart with a consciousness of the infinite Being "whose temple is all space, whose altar earth, sea, skies."[44]

Nothing indeed can compare with the sensations induced by a view from some lofty peak of these great mountains, for there the imagination may wander unfettered, may go back without a check through ages of time to the period when an Almighty power upheaved the gigantic masses which lie on all sides far beneath and around the beholder, and find no spot upon which to arrest the eye as a place where once dwelt man! No—we only know the Indian as a wanderer, and we cannot say here stood the strong fortress, the busy city, or even the humble cot. Nature has here been undisturbed and unsubdued, and our eyes may wander all over the scene to the most distant faint blue line on the horizon which encircles us, and forget alike the noisy clamour of toiling cities and the sweet and smiling quiet of the well cultivated fields, where man has made a "home" and dwelleth in peace. But in these regions we may find the savage grizzly bear, the huge bison, the elegant and fleet antelope, the large-horned sheep of the mountains, and the agile fearless climber of the steeps—the Rocky Mountain Goat.

This snow-white and beautiful animal appears to have been first described, from skins shown to LEWIS and CLARK, as "the Sheep," in their general description of the beasts, birds, and plants found by the party in their expedition. They say, "The Sheep is found in many places, but mostly in the timbered parts of the Rocky Mountains. They live in greater numbers on the chain of mountains forming the commencement of the woody country on the coast, and passing the Columbia between the falls and the rapids. We have only seen the skins of these animals, which the natives dress with the wool, and the blankets which they manufacture from the wool. The animal from this evidence appears to be of the size of our common sheep, of a white colour. The wool is fine on many parts of the body, but in length not equal to that of our domestic sheep. On the back, and particularly on the top of the head, this is intermixed with a considerable portion of long straight hairs. From the Indian account these animals have erect pointed horns."

The Rocky Mountain Goat wanders over the most precipitous rocks, and springs with great activity from crag to crag, feeding on the plants, grasses, and mosses of the mountain sides, and seldom or never descends to the luxuriant valleys, as the Big-Horn does. This Goat indeed resembles the wild Goat of Europe, or the chamois, in its habits, and is very difficult to procure. Now and then the hunter may observe one browsing on the extreme verge of some perpendicular rock almost directly above him, far beyond gun-shot, and entirely out of harm's way. At another time, after fatiguing and hazardous efforts, the hungry marksman may reach a spot from whence his rifle will send a ball into the unsuspecting Goat; then slowly he rises from his hands and knees, on which he has been creeping, and the muzzle of his heavy gun is "rested" on a loose stone, behind which he has kept his movements from being observed, and now he pulls the fatal trigger with deadly aim. The loud sharp crack of the rifle has hardly rung back in his ear from the surrounding cliffs when he sees the Goat in its expiring struggles reach the verge of the dizzy height: a moment of suspense and it rolls over, and swiftly falls, striking perchance here and there a projecting point, and with the clatter of thousands of small stones set in motion by its rapid passage down the steep slopes which incline outward near the base of the cliff, disappears, enveloped in a cloud of dust in the deep ravine beneath;

where a day's journey would hardly bring an active man to it, for far around must he go to accomplish a safe descent, and toilsome and dangerous must be his progress up the gorge within whose dark recesses his game is likely to become the food of the ever prowling wolf or the solitary raven. Indeed cases have been mentioned to us in which these Goats, when shot, fell on to a jutting ledge, and there lay fifty or a hundred feet below the hunter, in full view, but inaccessible from any point whatever.

Notwithstanding these difficulties, as portions of the mountains are not so precipitous, the Rocky Mountain Goat is shot and procured tolerably easily, it is said, by some of the Indian tribes, who make various articles of clothing out of its skin, and use its soft woolly hair for their rude fabrics.

According to Sir John Richardson, this animal has been known to the members of the Northwest and Hudson's Bay Companies from the first establishment of their trading posts on the banks of the Columbia River and in New Caledonia, and they have sent several specimens to Europe. The wool being examined by a competent judge, under the instructions of the Wernerian Society of Edinburgh, was reported to be of great fineness and fully an inch and a half long. "It is unlike the fleece of the common sheep, which contains a variety of different kinds of wool suitable to the fabrication of articles very dissimilar in their nature, and requires much care to distribute them in their proper order. The fleece under consideration is wholly fine. That on the fore part of the skin has all the apparent qualities of wool. On the back part it very much resembles cotton. The whole fleece is much mixed with hairs, and on those parts where the hairs are long and pendant, there is almost no wool.

"Mr. Drummond saw no Goats on the eastern declivity of the mountains, near the sources of the Elk river, where the sheep are numerous, but he learned from the Indians that they frequent the steepest precipices, and are much more difficult to procure than the sheep. Their manners are said to greatly resemble those of the domestic Goat. The exact limits of the range of this animal have not been ascertained, but it probably extends from the fortieth to the sixty-fourth or sixty-fifth degree of latitude. It is common on the elevated part of the Rocky Mountain range that gives origin to four great tributaries to as many different seas, viz. the Mackenzie, the Columbia, the Nelson, and the Missouri rivers."—*F. B. A.*, p. 269.

The flesh of this species is hard and dry, and is not so much relished as that of the Big-Horn, the Elk, &c., by the hunters or travellers who have journeyed towards the Pacific across the wild ranges of mountains inhabited by these animals.

N° 27

PLATE CXXXI

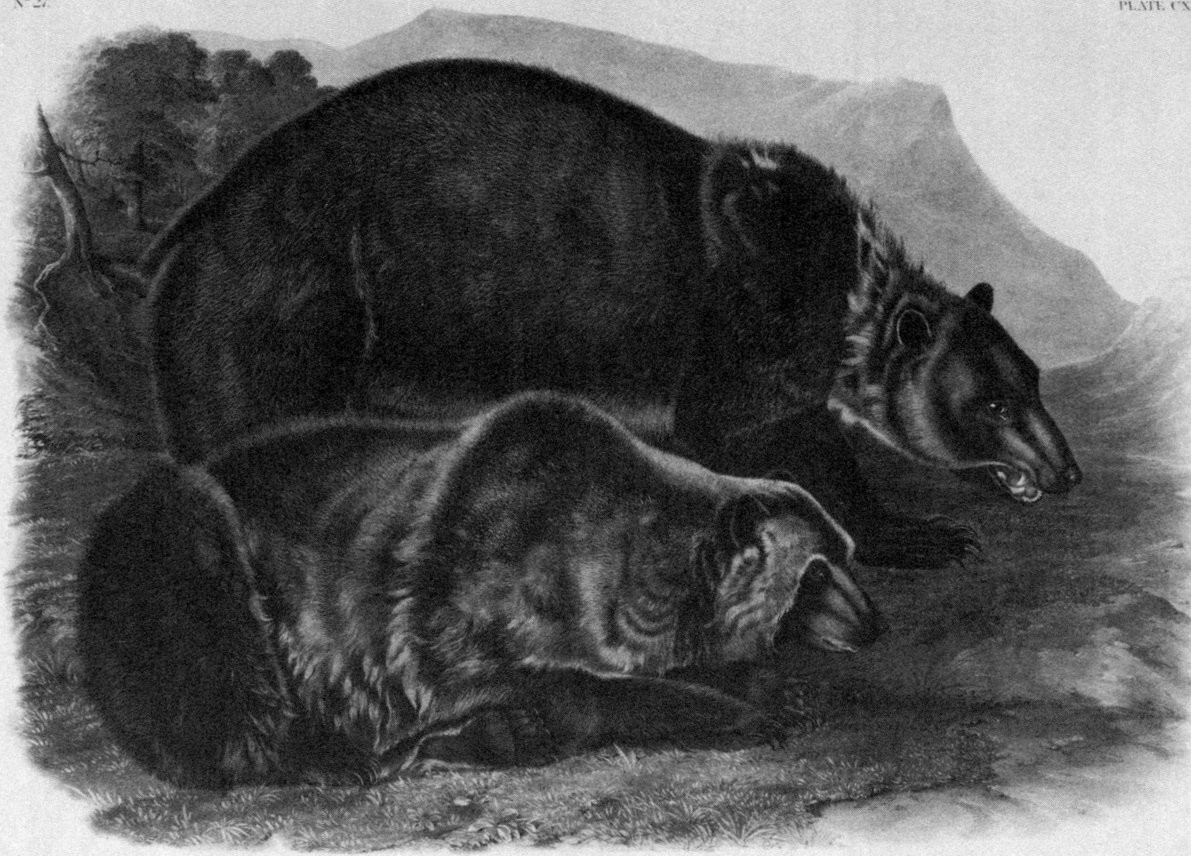

URSUS FEROX, LEWIS & CLARK.

GRIZZLY BEAR.

MALES.

Grizzly Bear

THIS ESSAY CONTAINS A THIRD VERSION OF A STORY THAT occurred during Audubon's expedition up the Missouri in 1843. On August 22, Audubon describes in his field notebook how Isaac Sprague alerts the party to the presence of a grizzly bear, which is shot and killed by Alexander Culbertson, John G. Bell, and Audubon, which infuriates Sprague and causes a noteworthy argument: Sprague promises not to mention any other bear he sees and "was quite abusive in other words"; Audubon tells him he can leave the party on the spot; and Sprague threatens to do so the following morning, a threat Audubon was right to doubt. In the Newberry Partial Copy of the journal, the argument is considerably reduced, as is Audubon's level of annoyance: "Sprague abused me greatly, but let it pass!"[45] In the *Quadrupeds* version, the argument completely disappears, even though we are told that the authors' "journals are before us" as the story commences. Audubon likely died before Bachman wrote the essay, so perhaps Bachman was honoring the memory of his friend when he omitted the argument with Sprague. It is just as likely, however, that Bachman wanted to avoid having juvenile passions distract readers from a good hunting story's embellishment of science. A third possibility is that Victor Audubon made sure this blemish on his father's expedition never made it into the essay. Whatever the case, Audubon's experiences are another contribution to the wealth of stories

---

37. John Woodhouse Audubon, J. T. Bowen, lithographer, *Grizzly Bear*, 1846–48, hand-colored lithograph, plate 131 of *The Viviparous Quadrupeds of North America*. Courtesy of the Jule Collins Smith Museum of Fine Art, Auburn University; Louise Hauss and David Brent Miller Audubon Collection.

that had already given "the huge shaggy monster disputing possession of the wilderness against all comers" elevated status in American lore, many of which are anthologized in this essay. Faithful to his principles, however, Bachman opts to include only stories that "we believe are true."

URSUS FEROX.—LEWIS AND CLARK.

GRIZZLY BEAR.

PLATE CXXXI.—MALES.

HABITS.

We have passed many hours of excitement, and some, perchance, of danger, in the wilder portions of our country; and at times memory recalls adventures we can now hardly attempt to describe; nor can we ever again feel the enthusiasm such scenes produced in us. Our readers must therefore imagine, the startling sensations experienced on a sudden and quite unexpected face-to-face meeting with the savage Grizzly Bear—the huge shaggy monster disputing possession of the wilderness against all comers, and threatening immediate attack!

Whilst in a neighbourhood where the Grizzly Bear may possibly be hidden, the excited nerves will cause the heart's pulsations to quicken if but a startled ground-squirrel run past; the sharp click of the lock is heard, and the rifle hastily thrown to the shoulder, before a second of time has assured the hunter of the trifling cause of his emotion.

But although dreaded alike by white hunter and by red man, this animal is fortunately not very abundant to the eastward of the Rocky Mountains, and the chance of encountering him does not often occur. We saw only a few of these formidable beasts during our expedition up the Missouri river and in the country over which we hunted during our last journey to the west.

The Indians, as is well known, consider the slaughter of a Grizzly Bear a feat second only to scalping an enemy, and necklaces made of the claws of this beast are worn as trophies by even the bravest among them.

On the 22d of August, 1843, we killed one of these Bears, and as our journals are before us, and thinking it may be of interest, we will extract the account of the day's proceedings, although part of it has no connection with our present subject.[46] We were descending the Upper Missouri river.

"The weather being fine we left our camp of the previous night early, but had made only about twelve miles when the wind arose and prevented our men from making any headway with the oars; we therefore landed under a high bank amongst a number of fallen trees and some drifted timber. All hands went in search of elks. Mr. CULBERTSON killed a deer, and with the help of Mr. SQUIRES brought the meat to the boat. We saw nothing during a long walk we took, but hearing three or four gun-shots which we thought were fired by some of our party, we hastened in the direction from whence the reports came, running and hallooing, but could find no one. We then made the best of our way back to the boat and despatched three men, who discovered that the firing had been at an elk, which was however not obtained. Mr. BELL killed a female elk and brought a portion of its flesh to the boat. After resting ourselves a while and eating dinner, Mr. CULBERTSON, SQUIRES, and ourselves walked to the banks of the Little Missouri, distant about one mile, where we saw a buffalo bull drinking at the edge of a sand-bar. We shot him, and fording the stream, which was quite shallow, took away the 'nerf;' the animal was quite dead. We saw many ducks in this river. In the course of the afternoon we started in our boat, and rowed about half a mile below the Little Missouri. Mr. CULBERTSON and ourselves walked to the body of the bull again and knocked off his horns, after which Mr. CULBERTSON endeavoured to penetrate a large thicket in hopes of starting a Grizzly Bear, but found it so entangled with briars and vines that he was obliged to desist, and returned very soon. Mr. HARRIS, who had gone in the same direction and for the same purpose, did not return with him. As we were approaching the boat we met Mr. SPRAGUE, who informed us that he thought he had seen a Grizzly Bear walking along the upper bank of the river, and we went towards the spot as fast as possible. Meantime the Bear had gone down to the water, and was clumsily and slowly proceeding on its way. It was only a few paces from and below us, and was seen by our whole party at the same instant. We all fired, and the animal dropped dead without even the power of uttering a groan. Mr. CULBERTSON put a rifle ball through its neck, BELL placed two large balls in its side, and our bullet entered its belly. After shooting the Bear we proceeded to a village of 'prairie dogs' (*Spermophilus Ludovicianus*), and set traps in hopes of catching some of them. We were inclined to think they had all left, but Mr. BELL seeing two, shot them. There were thousands of their burrows in sight. Our 'patroon,' assisted by one of the men, skinned the Bear, which weighed, as we thought, about four hundred pounds. It appeared to be between four and five years old, and was a male. Its lard was rendered, and filled sundry bottles with 'real Bear's grease,' whilst we had the skin preserved by our accomplished taxidermist, Mr. BELL.

"The following afternoon, as we were descending the stream, we saw another Grizzly Bear, somewhat smaller than the one mentioned above.[47] It was swimming towards the carcass of a dead buffalo lodged in the prongs of a 'sawyer' or 'snag,' but on seeing us it raised on its hind feet until quite erect, uttered a loud grunt or snort, made a leap from the water, gained the upper bank of the river, and disappeared in an instant amid the tangled briars and bushes thereabouts. Many wolves of different colours—black, white, red, or brindle—were also intent on going to the buffalo to gorge themselves on the carrion, but took fright at our approach, and we saw them sneaking away with their tails pretty close to their hind-legs."

The Grizzly Bear generally inhabits the swampy, well covered portions of the districts where it is found, keeping a good deal among the trees and bushes, and in these retreats it has its "beds" or lairs. Some of these we passed by, and our sensations were the reverse of pleasant whilst in such thick, tangled, and dangerous neighbourhoods; the Bear in his concealment having decidedly the advantage in case one should come upon him unawares. These animals ramble abroad both by day and night. In many places we found their great tracks along the banks of the rivers where they had been prowling in search of food. There are seasons during the latter part of summer, when the wild fruits that are eagerly sought after by the Bears are very abundant. These beasts then feed upon them, tearing down the branches as far as they can reach whilst standing in an upright posture. They in this manner get at wild plums, service berries, buffalo berries, and the seeds of a species of *cornus* or dog-wood which grows in the alluvial bottoms of the northwest. The Grizzly Bear is also in the habit of scratching the gravelly earth on the sides of hills where the vegetable called "pomme blanche" is known to grow,[48] but the favourite food of these animals is the more savoury flesh of such beasts as are less powerful, fleet, or cunning than themselves. They have been known to seize a wounded buffalo, kill it, and partially bury it in the earth for future use, after having gorged themselves on the best parts of its flesh and lapped up the warm blood.

We have heard many adventures related, which occurred to hunters either when surprised by these Bears, or when approaching them with the intention of shooting them. A few of these accounts, which we believe are true, we will introduce: During a voyage (on board one of the steamers belonging to the American Fur Company) up the Missouri river, a large she-Bear with two young was observed from the deck, and several gentlemen proposed to go ashore, kill the dam, and secure her cubs. A small boat was lowered for their accommodation, and with guns and ammunition they pushed off to the bank and landed in the mud. The old Bear had observed them and removed her position to some distance, where she stood near the bank, which was there several feet above the bed of the river. One of the hunters having neared the animal, fired at her, inflicting a severe wound. Enraged with pain the Bear rushed with open jaws towards the sportsmen at a rapid rate, and with looks that assured them she was in a desperate fury. There was but a moment's time; the party, too much frightened to stand the charge, "ingloriously turned and fled," without even pulling another trigger, and darting to the margin of the river jumped into the stream, losing their guns, and floundering and bobbing under, while their hats floated away with the muddy current. After swimming a while they were picked up by the steamer, as terrified as if the Bear was even then among them, though the animal on seeing them all afloat had made off, followed by her young.

The following was related to us by one of the "engagés" at Fort Union:[49] A fellow having killed an Indian woman, was forced to run away, and fearing he would be captured, started so suddenly that he took neither gun nor other weapon with him; he made his way to the Crow Indians, some three hundred miles up the Yellow Stone river, where he arrived in a miserable plight, having suffered from hunger and exposure. He escaped the men who were first sent after him, by keeping in ravines and hiding closely; but others were despatched, who finally caught him. He said that one day he saw a dead buffalo lying near the river bank, and going towards it to get some of the meat, to his utter astonishment and horror a young Grizzly Bear which was feeding on the carcass, raised up from behind it and so suddenly attacked him that his face and hands were lacerated by its claws before he had time to think of defending himself. Not daunted, however, he gave the cub a tremendous jerk, which threw it down, and took to his heels, leaving the young savage in possession of the prize.

The audacity of these Bears in approaching the neighbourhood of Fort Union at times was remarkable. The waiter, "Jean Battiste," who had been in the employ of the company for upwards of twenty years, told us that while one day picking peas in the garden, as he advanced towards the end of one of the rows, he saw a large Grizzly Bear gathering that excellent vegetable also. At this unexpected and startling discovery, he dropped his bucket, peas and all, and fled at his fastest pace to the Fort. Immediately the hunters turned out on their best horses, and by riding in a circle, formed a line which enabled them to approach the Bear on all sides. They found the animal greedily feasting on the peas, and shot him without his apparently caring for their approach. We need hardly say the bucket was empty.[50]

In GODMAN's Natural History there are several anecdotes connected with the Grizzly Bear. The first is as follows: A Mr. JOHN DOUGHERTY, a very experienced and respectable hunter belonging to Major LONG's expedition, relates that once, while hunting with another person on one of the upper tributaries of the Missouri, he heard the report of his companion's rifle, and when he looked round, beheld him at a short distance endeavouring to escape from one of these beasts, which he had wounded as it was coming towards him. DOUGHERTY, forgetful of every thing but the preservation of his friend, hastened to call off the attention of the Bear, and arrived in rifle-shot distance just in time to effect his generous purpose. He discharged his ball at the animal, and was obliged in his turn to fly; his friend, relieved from immediate danger, prepared for another attack by charging his rifle, with which he again wounded the Bear, and saved Mr. DOUGHERTY from peril. Neither received any injury from this encounter, in which the Bear was at length killed.

On another occasion, several hunters were chased by a Grizzly Bear, which rapidly gained upon them. A boy of the party, who could not run so fast as his companions, perceiving the Bear very near him, fell with his face towards the ground. The animal reared up on his hind feet, stood for a moment, and then bounded over him, impatient to catch the more distant fugitives.

Mr. DOUGHERTY, the hunter before mentioned, relates the following instance of the great muscular strength of the Grizzly Bear: Having killed a bison, and left the carcass for the purpose of procuring assistance to skin and cut it up, he was very much surprised on his return to find that it had been dragged off whole, to a considerable distance by a Grizzly Bear, and was then placed in a pit which the animal had dug with his claws for its reception.

The following is taken from Sir JOHN RICHARDSON's Fauna Boreali Americana: "A party of voyagers, who had been employed all day in tracking a canoe up the Saskatchewan, had seated themselves in the bright light by a fire, and were busy in preparing their supper, when a large Grizzly Bear sprung over their canoe, that was placed behind them, and seizing one of the party by the shoulder, carried him off. The rest fled in terror, with the exception of a Metis, named BOURAPO, who, grasping his gun, followed the Bear as it was retreating leisurely with its prey. He called to his unfortunate comrade that he was afraid of hitting him if he fired at the Bear, but the latter entreated him to fire immediately, without hesitation, as the Bear was squeezing him to death. On this he took a deliberate aim and discharged the contents of his piece into the body of the Bear, which instantly dropped its prey to pursue BOURAPO. He escaped with difficulty, and the Bear ultimately retired to a thicket, where it was supposed to have died; but the curiosity of the party not being a match for their fears, the fact of its decease was not ascertained. The man who was rescued had his arm fractured, and was otherwise severely bitten by the Bear, but finally recovered. I have seen BOURAPO, and can add that the account which he gives is fully credited by the traders resident in that part of the country, who are best qualified to judge of its truth from the knowledge of the parties. I have been told that there is a man now living in the neighbourhood of Edmonton-house who was attacked by a Grizzly Bear, which sprang out of a thicket, and with one stroke of its paw completely scalped him, laying bare the skull and bringing the skin of the forehead down over the eyes. Assistance coming up, the Bear made off without doing him further injury, but the scalp not being replaced, the poor man has lost his sight, although he thinks that his eyes are uninjured."

Mr. DRUMMOND, in his excursions over the Rocky Mountains, had frequent opportunities of observing the manners of the Grizzly Bear, and it often happened that in turning the point of a rock or sharp angle of a valley, he came suddenly upon one or more of them. On such occasions they reared on their hind legs and made a loud noise like a person breathing quick, but much harsher. He kept his ground without attempting to molest them, and they, on their part, after attentively regarding him for some time, generally wheeled round and galloped off, though, from their disposition, there is little doubt but he would have been torn in pieces had he lost his presence of mind and attempted to fly. When he discovered them from a distance, he generally frightened them away by beating on a large tin box, in which he carried his specimens of plants. He never saw more than four together, and two of these he supposes to have been cubs; he more often met them singly or in pairs. He was only once attacked, and then by a female, for the purpose of allowing her cubs time to escape. His gun on this occasion missed fire, but he kept her at bay with the stock of it, until some gentlemen of the Hudson's Bay Company, with whom he was travelling at the time, came up and drove her off. In the latter end of June, 1826, he observed a male caressing a female, and soon afterwards they both came towards him, but whether accidentally, or for the purpose of attacking him, he was uncertain. He ascended a tree, and as the female drew near, fired at and mortally wounded her. She uttered a few loud screams, which threw the male into a furious rage, and he reared up against the trunk of the tree in which Mr. DRUMMOND was seated, but never attempted to ascend it. The female, in the meantime, retired to a short distance, lay down, and as the male was proceeding to join her, Mr. DRUMMOND shot him also.

The young Grizzly Bears and gravid females hibernate, but the older males often come abroad in the winter in quest of food. MACKENZIE mentions the den or winter retreat of a Grizzly Bear, which was ten feet wide, five feet high, and six feet long.

This species varies very much in colour; we have skins in our possession collected on the Upper Missouri, some of which are nearly white, whilst others are as nearly of a rufous tint. The one that was killed by our party (of which we have also the skin) was a dark brown one.

The following is from notes of J. W. AUDUBON, made in California in 1849 and 1850: "High up on the waters of the San Joaquin, in California, many of these animals have been killed by the miners now overrunning all the country west of the Sierra Nevada. Greatly as the Grizzly Bear is dreaded, it is hunted with all the more enthusiasm by these fearless pioneers in the romantic hills, valleys, and wild mountains of the land of gold, as its flesh is highly prized by men who have been living for months on salt pork or dry and tasteless deer-meat. I have seen two dollars a pound paid for the leaf-fat around the kidneys. If there is time, and the animal is not in a starving condition, the Grizzly Bear always runs at the sight of man; but should the hunter come too suddenly on him, the fierce beast always commences the engagement—And the first shot of the hunter is a matter of much importance, as, if unsuccessful, his next move must be to look for a sapling to climb for safety. It is rare to find a man who would willingly come into immediate contact with one of these powerful and vindictive brutes. Some were killed near 'Green Springs,' on the Stanislaus, in the winter of 1849–50, that were nearly eight hundred pounds weight. I saw many cubs at San Francisco, Sacramento city, and Stockton, and even those not larger than an ordinary sized dog, showed evidence of their future fierceness, as it required great patience to render them gentle enough to be handled with impunity as pets. In camping at night, my friend ROBERT LAYTON, and I too, often thought what sort of defence we could make should an old fellow come smelling round our solitary tent for supper; but as 'Old Riley,' our pack-mule, was always tied near, we used to quiet ourselves with the idea that while Riley was snorting and kicking, we might place a couple of well aimed balls from our old friend Miss Betsey (as the boys had christened my large gun), so that our revolvers, COLT's dragoon pistols, would give us the victory; but really a startling effect would be produced by the snout of a Grizzly Bear being thrust into your tent, and your awaking at the noise of the sniff he might take to induce his appetite.

"I was anxious to purchase a few of the beautiful skins of this species, but those who had killed 'an old Grizzly,' said they would take his skin *home*. It makes a first rate bed under the thin and worn blanket of the digger.

"The different colours of the pelage of this animal, but for the uniformity of its extraordinary claws, would puzzle any one not acquainted with its form, for it varies from jet black in the young of the first and second winter to the hoary gray of age, or of summer."

In TOWNSEND's "Narrative of a Journey across the Rocky Mountains to the Columbia River, &c." (Philadelphia, 1839), we find two adventures with the Grizzly Bear. The first is as follows: The party were on Black Foot river, a small stagnant stream which runs in a northwesterly direction down a valley covered with quagmires through which they had great difficulty in making their way. "As we approached our encampment, near a small grove of willows on the margin of the river, a tremendous Grizzly Bear rushed out upon us. Our horses ran wildly in every direction, snorting with terror, and became nearly unmanageable. Several balls were instantly fired into him, but they only seemed to increase his fury. After spending a moment in rending each wound (their invariable practice), he selected the person who happened to be nearest, and darted after him, but before he proceeded far he was sure to be stopped again by a ball from another quarter. In this way he was driven about amongst us for perhaps fifteen minutes, at times so near some of the horses that he received several severe kicks from them. One of the pack-horses was fastened upon by the brute, and in the terrified animal's efforts to escape the dreaded gripe, the pack and saddle were broken to pieces and disengaged. One of our mules also lent him a kick in the head, while pursuing it up an adjacent hill, which sent him rolling to the bottom. Here he was finally brought to a stand. The poor animal was so completely surrounded by enemies that he became bewildered. He raised himself upon his hind feet, standing almost erect, his mouth partly open, and from his protruding tongue the blood fell fast in drops. While in this position he received about six more balls, each of which made him reel. At last, as in complete desperation, he dashed into the water, and swam several yards with astonishing strength and agility, the guns cracking at him constantly. But he was not to proceed far. Just then, RICHARDSON, who had been absent, rode up, and fixing his deadly aim upon him, fired a ball into the back of his head, which killed him instantly. The strength of four men was required to drag the ferocious brute from the water, and upon examining his body he was found completely riddled; there did not appear to be four inches of his shaggy person, from the hips upward, that had not received a ball. There must have been at least thirty shots made at him, and probably few missed him, yet such was his tenacity of life that I have no doubt he would have succeeded in crossing the river, but for the last shot in the brain. He would probably weigh, at the least, six hundred pounds, and was about the height of an ordinary steer. The spread of the foot, laterally, was ten inches, and the claws measured seven inches in length. This animal was remarkably lean; when in good condition he would doubtless much exceed in weight the estimate I have given."

At p. 68, TOWNSEND says: "In the afternoon one of our men had a somewhat perilous adventure with a Grizzly Bear. He saw the animal crouching his huge frame in some willows which skirted the river, and approaching him on horseback to within twenty yards, fired upon him. The Bear was only slightly wounded by the shot, and with a fierce growl of angry malignity, rushed from his cover, and gave chase. The horse happened to be a slow one, and for the distance of half a mile the race was hard contested, the Bear frequently approaching so near the terrified animal as to snap at his heels, whilst the equally terrified rider, who had lost his hat at the start, used whip and spur with the most frantic diligence, frequently looking behind, from an influence which he could not resist, at his rugged and determined foe, and shrieking in an agony of fear, 'shoot him! shoot him!' The man, who was one of the greenhorns, happened to be about a mile behind the main body, either from the indolence of his horse or his own carelessness; but as he approached the party in his desperate flight, and his lugubrious cries reached the ears of the men in front, about a dozen of them rode to his assistance, and soon succeeded in diverting the attention of his pertinacious foe. After he had received the contents of all the guns, he fell, and was soon despatched. The man rode in among his fellows, pale and haggard from overwrought feelings, and was probably effectually cured of a propensity for meddling with Grizzly Bears."

N°27

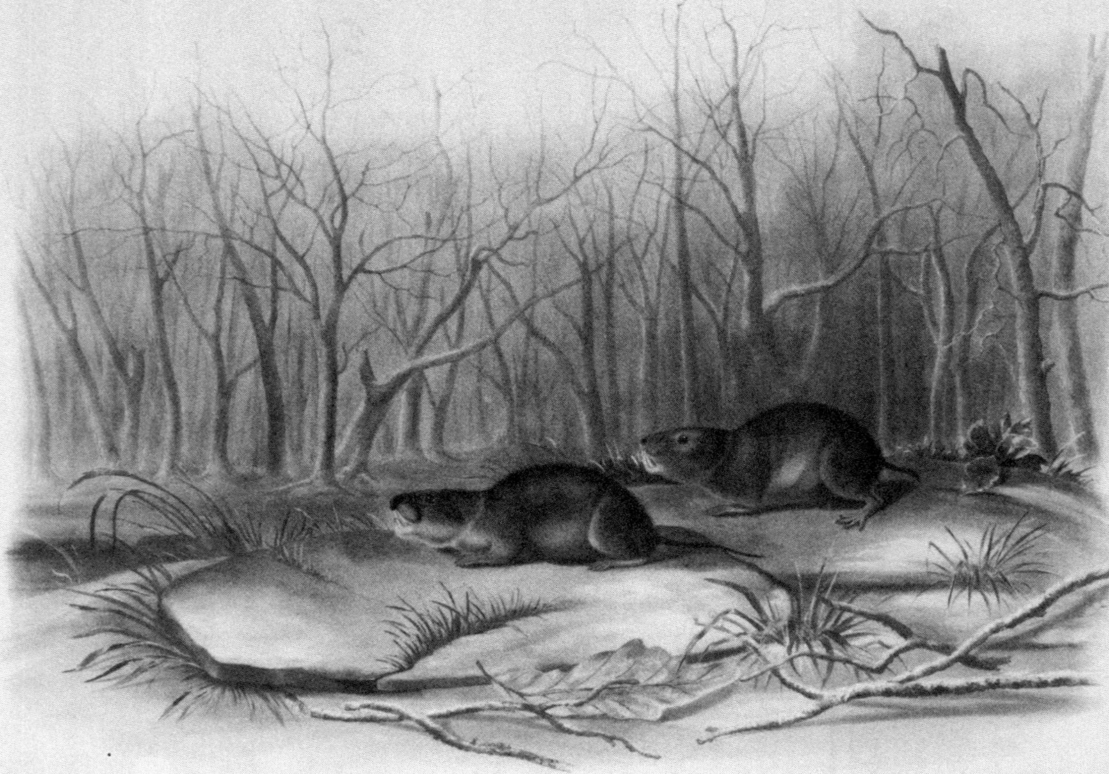

ARVICOLA RICHARDSONII, AUD & BACH.

RICHARDSON'S MEADOW MOUSE.

Richardson's Meadow-Mouse

A MEADOW MOUSE, A CISTERN, AND A WOODEN PLANK ALL come together for a rare occasion in *Quadrupeds*: that of Bachman resolutely, patiently attempting to save the life of a wild animal.

---

38. John Woodhouse Audubon, J. T. Bowen, lithographer, *Richardson's Meadow Mouse*, 1846–48, hand-colored lithograph, plate 135 of *The Viviparous Quadrupeds of North America*. Courtesy of the Jule Collins Smith Museum of Fine Art, Auburn University; Louise Hauss and David Brent Miller Audubon Collection.

ARVICOLA RICHARDSONII.—AUD. AND BACH.

RICHARDSON'S MEADOW-MOUSE

PLATE CXXXV.—FIG. I.

HABITS.

DRUMMOND, who procured this Meadow-Mouse, states that its habits are analogous to those of the common water-rat of Europe (*Arvicola amphibius*), with which it may be easily confounded, although the shortness of its tail may serve as a mark by which to distinguish it.

It frequents moist meadows amongst the Rocky Mountains, and swims and dives well, taking to the water at once when pursued. All Meadow-Mice indeed are capital swimmers. We some time since amused ourselves watching one that had fallen into a circular cistern partly built up with stone and partly excavated out of the solid rock by blasting, and which was plastered with cement on the inside to make it water-tight. This cistern had about four feet of water in it. On one side there was a projecting rounded knob of stone some five or six inches long and about two wide, which slanted out of the water so that the upper edge of it was dry. Upon this little resting-place there was a large *Arvicola pennsylvanica* (Wilson's Meadow-Mouse) seated very quietly, having probably tumbled in the preceding night. When we approached the edge and looked down into the clear element we at first did not observe the Rat, but as soon as we espied him he saw us, immediately dived, and swam around underneath the surface quite rapidly; he soon arose, however, and regained his position on the ledge, and we determined to save him from what had been his impending fate—drowning or starving, or both. We procured a plank, and gently lowering one end of it towards the ledge, thought he would take advantage of the inclined plane thus afforded him, to come out; but in our awkwardness we suffered the plank to slip, and at the plash in the water the little fellow dived and swam around several times before he again returned to his resting place, where we now had the end of the board fixed, so that he could get upon it. As soon as he was on it, we began to raise the plank, but when we had him about three feet above the surface he dashed off into the water, making as pretty a dive as need be. He always looked quite dry, and not a hair of his coat was soiled or turned during these frequent immersions, and it was quite interesting to see the inquisitive looks he cast towards us, turning his head and appearing to have strong doubts whether we meant to help, or to make an end of him. We put down the plank again, and after two attempts, in both of which his timidity induced him to jump off it when he was nearly at the edge of the cistern, he at last reached the top, and in a moment disappeared amid the weeds and grasses around.

N° 28

PLATE CXXXVIII

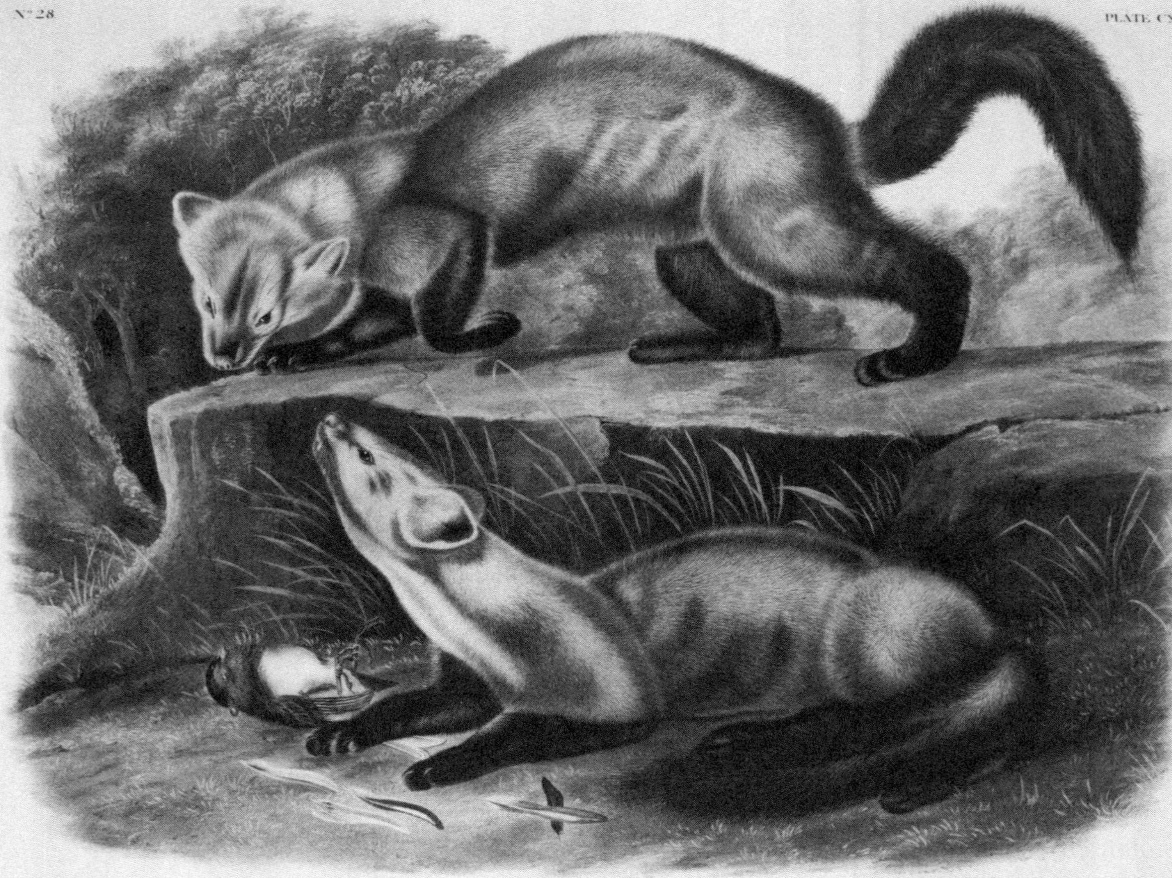

MUSTELA MARTES, LINN.

PINE MARTEN.

Pine Marten

IN BACHMAN'S ESTIMATION, THE PINE MARTEN IS SOMETHING of a composite of other familiar animals. Its character is a mixture of traits it shares with the fox, the weasel, and the mink; physically, it is a good climber like the raccoon and has "a little of the fetid odour" of the mink. The "cunning" of the fox and pine marten seems to have rubbed off on furriers: even when it was fashionable, the pine marten's fur was frequently marketed "as fur of a more costly kind" and dyed to fool buyers. If the pine marten serves as a convergence of animal personalities, it also marks a convergence of the nonhuman world, natural history, and retail.

---

39. John Woodhouse Audubon, J. T. Bowen, lithographer, *Pine Marten*, 1846–48, hand-colored lithograph, plate 138 of *The Viviparous Quadrupeds of North America*. Courtesy of the Jule Collins Smith Museum of Fine Art, Auburn University; Louise Hauss and David Brent Miller Audubon Collection.

MUSTELA MARTES.—LINN.—GMEL.

PINE MARTEN.

PLATE CXXXVIII.—MALE AND FEMALE. WINTER PELAGE.

HABITS.

Let us take a share of the cunning and sneaking character of the fox, as much of the wide-awake and cautious habits of the weasel, a similar proportion of the voracity (and a little of the fetid odour) of the mink, and add thereto some of the climbing propensities of the raccoon, and we have a tolerable idea of the attributes of the little prowler of which we have just given the description and dimensions. The Pine Marten, as may be inferred from this compound, is shy, cruel, cunning, and active, and partakes of the habits of the predacious animals above mentioned, with the exception that it is not known to approach the residences of man like the fox, weasel, or mink, but rather keeps in dense woods where it can prey upon birds, their eggs and young, squirrels, the white-footed and other mice, shrews, wood-rats, &c., together with beetles and other insects, larvae of different species, toads, frogs, lizards, water reptiles, and fish. It is also an eater of some kinds of berries and nuts (as we are informed), and is said to be fond of honey like the bear.

It has been supposed that the name Pine Marten was given to this animal because it inhabits the pine forests of the northern parts of this continent, and shows a preference for those trees, in the lofty tops of which it frequently resides. The Pine Marten, however, is often called the American Sable or the Sable, and in fact is more generally known to the country people of our northern States, and also to the furriers, by the latter name than by any other.

Sprightly and agile in its movements, the Pine Marten commonly procures abundance of food. It is prolific, bringing forth from six to eight young at a time, so that notwithstanding the value of its fur and the consequent pursuit of it during the proper season, it is still by no means a scarce animal. We have had several specimens sent to us by friends residing in the State of New York and in the wilder portions of our Canada frontier, which were procured among the woody hills of these districts.

According to Dr. DEKAY (New York Fauna, p. 33), this species is so active as to destroy great quantities of squirrels, the red squirrel (*Sciurus Hudsonius*) only escaping by its superior agility. Dr. GODMAN remarks that the "Pine Marten frequently has its den in the hollows of trees, but very commonly takes possession of the nest of some industrious squirrel, which it enlarges to suit its own convenience, after putting the builder to death."

Sir JOHN RICHARDSON says that "particular races of Martens, distinguished by the fineness and dark colour of their fur, appear to inhabit certain, rocky districts." "A partridge's head, with the feathers, is the best bait for the log traps in which this animal is taken. It does not reject carrion, and often destroys the hoards of meat and fish laid up by the natives, when they have accidentally left a crevice by which it can enter. The Marten, when its retreat is cut off, shows its teeth, sets up its hair, arches its back, and makes a hissing noise like a cat. It will seize a dog by the nose and bite so hard, that unless the latter is accustomed to the combat, it suffers the little animal to escape."

The Indians sometimes eat the Pine Marten, but its flesh is rank and coarse. We have seen this species in confinement, when it appeared tolerably gentle, and had lost much of its snappish character.

The Pine Marten burrows in the ground at times, and the female brings forth her young in a fallen hollow log, a hole under rocks, or in a burrow, generally in April or May. These animals are chiefly caught with dead-falls baited with meat of any kind, birds, rabbits, squirrels, &c., and generally a hunter has many traps set, each of which he visits as often as once or twice a week. The Martens are sometimes devoured by larger animals after they have been caught. They are only trapped in the autumn and winter.

The fur of this species has been considered valuable, and when in fashion the skins were worth good prices. It is often palmed off on purchasers as fur of a more costly kind, and for this purpose is dyed any desired colour.

Nº 29

PLATE CXLI

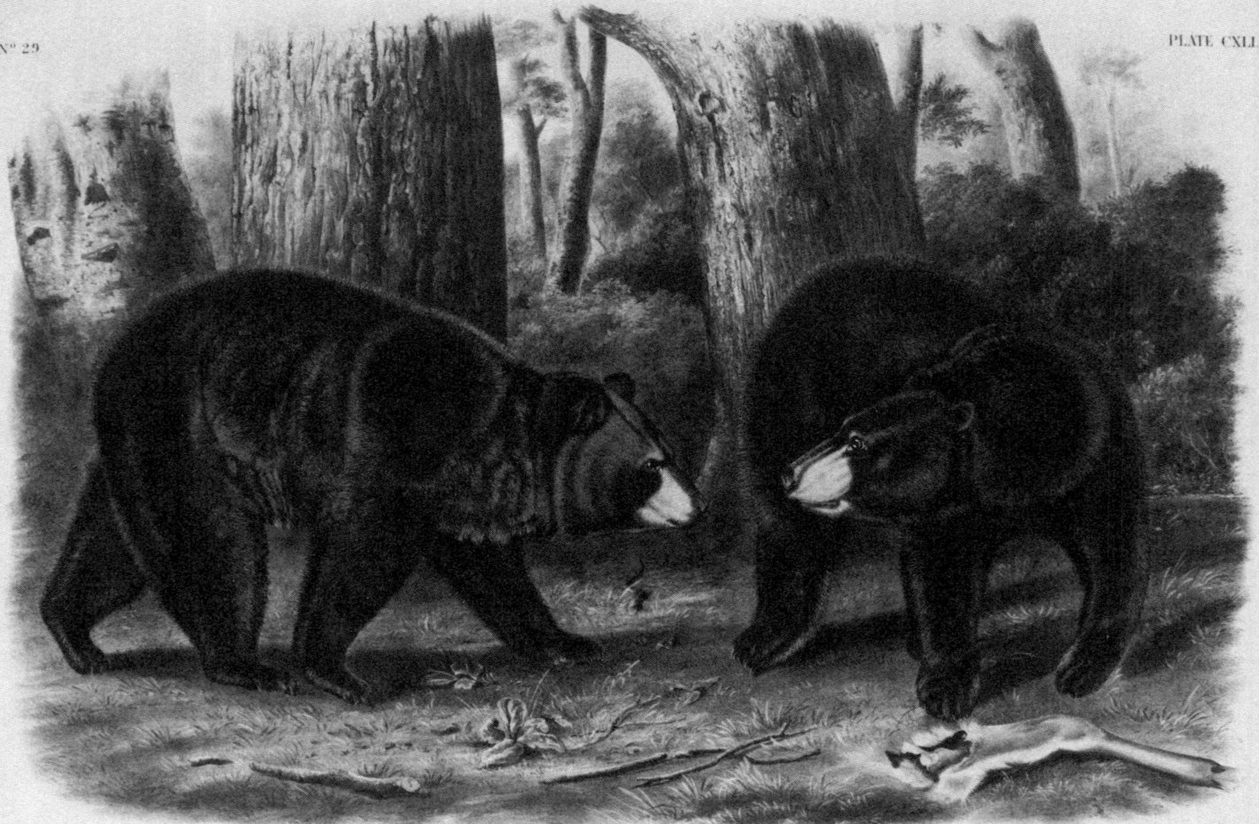

URSUS AMERICANUS, PALLAS.

AMERICAN BLACK BEAR.

American Black Bear

THIS ESSAY IS A TOUR DE FORCE THAT SHOWS THE PERVASIVE presence of the black bear in the American imagination. Borrowing the first five paragraphs of natural history and the story of killing bears on a plantation in Louisiana from Audubon's earlier published episode "Scipio and the Bear," Bachman relates perspectives on the bear from Labrador to Texas and from Georgia to Canada.[51] No other animal in America, he suggests, "has been more frequently the theme of adventure or anecdote than the Bear." One might add that in no other quadruped essay is the pervasive anthropocentric ethic more fully expressed than in the narrative of the Louisiana night hunt, when the hunters deliberately wound a mother bear to prevent her possible escape in order to "procure as much sport as possible"—and this with no apparent concern that readers may be revolted by the human behavior. The crowning achievement of this essay may be one of dramatic irony: while the essay claims that bear stories are really about people, offering readers insights into "the characteristics of the bold and hardy huntsman," they can also reveal the human cruelty possible when mammals are seen as undeserving of ethical consideration.

---

40. John Woodhouse Audubon, J. T. Bowen, lithographer, *American Black Bear*, 1846–48, hand-colored lithograph, plate 141 of *The Viviparous Quadrupeds of North America*. Courtesy of the Jule Collins Smith Museum of Fine Art, Auburn University; Louise Hauss and David Brent Miller Audubon Collection.

URSUS AMERICANUS.—PALLAS.

AMERICAN BLACK BEAR.

PLATE CXLI.—MALE AND FEMALE.

HABITS.

The Black Bear, however clumsy in appearance, is active, vigilant, and persevering, possesses great strength, courage, and address, and undergoes with little injury the greatest fatigues and hardships in avoiding the pursuit of the hunter. Like the deer it changes its haunts with the seasons, and for the same reason, viz. the desire of obtaining suitable food, or of retiring to the more inaccessible parts, where it can pass the time in security, unobserved by man, the most dangerous of its enemies.

During the spring months it searches for food in the low rich alluvial lands that border the rivers, or by the margins of such inland lakes as, on account of their small size, are called by us ponds. There it procures abundance of succulent roots and tender juicy plants, upon which it chiefly feeds at that season. During the summer heat, it enters the gloomy swamps, passes much of its time in wallowing in the mud like a hog, and contents itself with crayfish, roots, and nettles, now and then seizing on a pig, or perhaps a sow, a calf, or even a full-grown cow. As soon as the different kinds of berries which grow on the mountains begin to ripen, the Bears betake themselves to the high grounds, followed by their cubs.

In retired parts of the country, where the plantations are large and the population sparse, it pays visits to the corn-fields, which it ravages for a while. After this, the various species of nuts, acorns, grapes; and other forest fruits, that form what in the western States is called *mast*, attract its attention. The Bear is then seen rambling singly through the woods to gather this harvest, not forgetting, meanwhile, to rob every *bee-tree* it meets with, Bears being expert at this operation.

The Black Bear is a capital climber, and now and then *houses* itself in the hollow trunk of some large tree for weeks together during the winter, when it is said to live by sucking its paws.

At one season, the Bear may be seen examining the lower part of the trunk of a tree for several minutes with much attention, at the same time looking around, and snuffing the air. It then rises on its hind-legs, approaches the trunk, embraces it with the fore-legs, and scratches the bark with its teeth and claws for several minutes in continuance. Its jaws clash against each other until a mass of foam runs down on both sides of the mouth. After this it continues its rambles.

The female Black Bear generally brings forth two cubs at a time, although, as we have heard, the number is sometimes three or four. The period of gestation is stated to be from six to seven weeks, but is mentioned as one hundred days by some authors. When born the young are exceedingly small, and if we may credit the accounts of hunters with whom we have conversed on the subject, are not larger than kittens. They are almost invariably brought forth in some well concealed den, or great hollow tree, and so cautious is the dam in selecting her place of accouchment, that it is extremely difficult to discover it, and consequently very rarely that either the female or her cubs are seen until the latter have obtained a much larger size than when born, are able to follow their dam, and can climb trees with facility.

Most writers on the habits of this animal have stated that the Black Bear does not eat animal food from choice, and never unless pressed by hunger. This we consider a great mistake, for in our experience we have found the reverse to be the case, and it is well known to our frontier farmers that this animal is a great destroyer of pigs, hogs, calves, and sheep, for the sake of which we have even known it to desert the pecan groves in Texas. At the same time, as will have been seen by our previous remarks, its principal food generally consists of berries, roots, and other vegetable substances. It is very fond also of fish, and during one of our expeditions to Maine and New Brunswick, we found the inhabitants residing near the coast unwilling to eat the flesh of the animal on account of its fishy taste. In our western forests, however, the Bear feeds on so many nuts and well tasted roots and berries, that its meat is considered a great delicacy, and in the city of New York we have generally found its market price three or four times more than the best beef per pound. The fore-paw of the Bear when cooked presents a striking resemblance to the hand of a child or young person, and we have known some individuals to be hoaxed by its being represented as such.

Perhaps the most acrid vegetable eaten by the Bear is the Indian turnip (*Arum triphyllum*), which is so pungent that we have seen people almost distracted by it, when they had inadvertently put a piece in their mouth.

The Black Bear is a remarkably swift runner when first alarmed, although it is generally "treed," that is, forced to ascend a tree, when pursued by dogs and hunters on horseback. We were, not very long since, when on an expedition in the mountains of Virginia, leisurely making our way along a road through the forest after a long hunt for deer and turkeys, with our gun thrown behind our shoulders and our arms resting on each end of it, when, although we had been assured there were no Bears in that neighbourhood, we suddenly perceived one above us on a little acclivity at one side of the road, where it was feeding, and nearly concealed by the bushes. The bank was only about fifteen feet high, and the Bear not more than twenty paces from us, so we instantly disengaged our gun, and cocking both barrels, expected to "fill our bag" at one shot, but at the instant and before we could fire, the Bear, with a celerity that astonished us, disappeared. We rushed up the bank and found the land on the top nearly level for a long distance before us, and neither very thickly wooded nor very bushy; but no Bear was to be seen, although our eye could penetrate the woods for at least two hundred yards. After the first disappointing glance around, we thought Bruin might have mounted a tree, but such was not the case, as on looking everywhere nothing could be seen of his black body, and we were obliged to conclude that he had run out of sight in the brief space of time we occupied in ascending the little bank.

As we were once standing at the foot of a large sycamore tree on the borders of a long and deep pond, on the edge of which, in our rear, there was a thick and extensive "cane-brake," we heard a rushing roaring noise, as if some heavy animal was bearing down and passing rapidly through the canes, directly towards us. We were not kept long in suspense, for in an instant or two, a large Bear dashed out of the dense cane, and plunging into the pond without having even seen us, made off with considerable speed through the water towards the other shore. Having only bird-shot in our gun we did not think it worth while to call his attention to us by firing at him, but turned to the cane-brake, expecting to hear either dogs or men approaching shortly. No further noise could be heard, however, and the surrounding woods were as still as before this adventure. We supposed the Bear had been started at some distance, and that his pursuers, not being able to follow him through the almost impenetrable canes, had given up the hunt.

Being one night sleeping in the house of a friend who was a Planter in the State of Louisiana, we were awakened by a servant bearing a light, who gave us a note, which he said his master had just received. We found it to be a communication from a neighbour, requesting our host and ourself to join him as soon as possible, and assist in killing some Bears at that moment engaged in destroying his corn. We were not long in dressing, and on entering the parlour, found our friend equipped. The overseer's horn was heard calling up the negroes. Some were already saddling our horses, whilst others were gathering all the cur-dogs of the plantation. All was bustle. Before half an hour had elapsed, four stout negro men, armed with axes and knives, and mounted on strong nags, were following us at a round gallop through the woods, as we made directly for the neighbour's plantation.

The night was none of the most favourable, a drizzling rain rendering the atmosphere thick and rather sultry; but as we were well acquainted with the course, we soon reached the house, where the owner was waiting our arrival. There were now three of us armed with guns, half a dozen servants, and a good pack of dogs of all kinds. We jogged on towards the detached field in which the Bears were at work. The owner told us that for some days several of these animals had visited his corn, and that a negro who was sent every afternoon to see at what part of the enclosure they entered, had assured him there were at least five in the field that night. A plan of attack was formed: the bars at the usual entrance of the field were to be put down without noise; the men and dogs were to divide, and afterwards proceed so as to surround the Bears, when, at the sounding of our horns, every one was to charge towards the centre of the field, and shout as loudly as possible, which it was judged would so intimidate the animals as to induce them to seek refuge upon the dead trees with which the field was still partially covered.

The plan succeeded: the horns sounded, the horses galloped forward, the men shouted, the dogs barked and howled. The shrieks of the negroes were enough to frighten a legion of bears, and by the time we reached the middle of the field we found that several had mounted the trees, and having lighted fires, we now saw them crouched at the junction of the larger branches with the trunks. Two were immediately shot down. They were cubs of no great size, and being already half dead, were quickly dispatched by the dogs.

We were anxious to procure as much sport as possible, and having observed one of the Bears, which from its size we conjectured to be the mother of the two cubs just killed, we ordered the negroes to cut down the tree on which it was perched, when it was intended the dogs should have a tug with it, while we should support them, and assist in preventing the Bear from escaping, by wounding it in one of the hind-legs. The surrounding woods now echoed to the blows of the axemen. The tree was large and tough, having been girded more than two years, and the operation of felling it seemed extremely tedious. However, at length it began to vibrate at each stroke; a few inches alone now supported it, and in a short time it came crashing to the ground.

The dogs rushed to the charge, and harassed the Bear on all sides, whilst we surrounded the poor animal. As its life depended upon its courage and strength, it exercised both in the most energetic manner. Now and then it seized a dog and killed him by a single stroke. At another time, a well administered blow of one of its fore-legs sent an assailant off, yelping so piteously that he might be looked upon as *hors de combat*. A cur had daringly ventured to seize the Bear by the snout, and was seen hanging to it, covered with blood, whilst several others scrambled over its back. Now and then the infuriated animal was seen to cast a revengeful glance at some of the party, and we had already determined to dispatch it, when, to our astonishment, it suddenly shook off all the dogs, and before we could fire, charged upon one of the negroes, who was mounted on a pied horse. The Bear seized the steed with teeth and claws, and clung to its breast. The terrified horse snorted and plunged. The rider, an athletic young man and a capital horseman, kept his seat, although only saddled on a sheep-skin tightly girthed, and requested his master not to fire at the Bear. Notwithstanding his coolness and courage, our anxiety for his safety was raised to the highest pitch, especially when in a moment we saw rider and horse come to the ground together; but we were instantly relieved on witnessing the masterly manner in which Scipio dispatched his adversary, by laying open his skull with a single well directed blow of his axe, when a deep growl announced the death of the Bear.

In our country no animal, perhaps, has been more frequently the theme of adventure or anecdote than the Bear, and in some of our southwestern States it is not uncommon to while away the winter evenings with Bear stories that are not only interesting on account of the traits of the habits of the animal with which they are interspersed, but from the insight they afford the listener into the characteristics of the bold and hardy huntsmen of those parts.

In the State of Maine the lumbermen (wood-cutters) and the farmers set guns to kill this animal, which are arranged in this way: A funnel-shaped space about five feet long is formed by driving strong sticks into the ground in two converging lines, leaving both the ends open, the narrow end being wide enough to admit the muzzle of an old musket, and the other extremity so broad as to allow the head and shoulders of the Bear to enter. The gun is then loaded and fastened securely so as to deliver its charge facing the wide end of the enclosure. A round and smooth stick is now placed behind the stock of the gun, and a cord leading from the trigger passed around it, the other end of which, with a piece of meat or a bird tied to it (an owl is a favourite bait), is stretched in front of the gun, so far that the Bear can reach the bait with his paw. Upon his pulling the meat towards him, the string draws the trigger and the animal is instantly killed.

On the coast of Labrador we observed the Black Bear catching fish with great dexterity, and the food of these animals in that region consisted altogether of the fishes they seized in the edge of the water inside the surf. Like the Polar Bear, the present species swims with ease and rapidity, and it is a difficult matter to catch a full grown Bear with a skiff, and a dangerous adventure to attempt its capture in a canoe, which it could easily upset.

We were once enjoying a fine autumnal afternoon on the shores of the beautiful Ohio, with two acquaintances who had accompanied us in quest of some swallows that had built in a high sandy bank, when we observed three hunters about the middle of the river in a skiff, vigorously rowing, the steersman paddling too, with all his strength, in pursuit of a Bear which, about one hundred and fifty yards ahead of them, was cleaving the water and leaving a widening wake behind him on its unrippled surface as he made for the shore, directly opposite to us. We all rushed down to the water at this sight, and launching a skiff we then kept for fishing, hastily put off to intercept the animal, which we hoped to assist in capturing. Both boats were soon nearing the Bear, and we, standing in the bow of our skiff, commenced the attack by discharging a pistol at his head. At this he raised one paw, brushed it across his forehead, and then seemed to redouble his efforts. Repeated shots from both boats were now fired at him, and we ran alongside, thinking to haul his carcass triumphantly on board; but suddenly, to our dismay, he laid both paws on the gunwale of the skiff, and his great weight brought the side for an instant under water, so that we expected the boat would fill and sink. There was no time to be lost: we all threw our weight on to the other side, to counterpoise that of the animal, and commenced a pell-mell battery on him with the oars and a boat-hook; the men in the other boat also attacked him, and driving the bow of their skiff close to his head, one of them laid his skull open with an axe, which killed him instanter. We jointly hurraed, and tying a rope round his neck, towed him ashore behind our boats.

The Black Bear is very tenacious of life, and like its relative, the Grizzly Bear, is dangerous when irritated or wounded. It makes large beds of leaves and weeds or grasses, in the fissures of rocks, or sleeps in hollow logs, when no convenient den can be found in its neighbourhood; it also makes lairs in the thick cane-brakes and deep swamps, and covers itself with a heap of leaves and twigs, like a wild sow when about to litter.

The skin of the Black Bear is an excellent material for sleigh-robes, hammer-cloths, caps, &c., and makes a comfortable bed for the backwoodsman or Indian; and the grease procured from this species is invaluable to the hair-dresser, being equal if not superior to "Thine incomparable oil Macassar!"[52] which we (albeit unacquainted with the mode of preparing it) presume to be a compound much less expensive to the manufacturer than would be the "genuine real Bear's grease"—not of the shops, but of the prairies and western woods.

The Black Bear is rather docile when in confinement, and a "pet" Bear is occasionally seen in various parts of the country.[53] In our large cities, however, where civilization (?) is thought to have made the greatest advances, this animal is used to amuse the gentlemen of the fancy, by putting its strength and "pluck" to the test, in combat with bull-dogs or mastiffs. When the Bear has not been so closely imprisoned as to partially destroy his activity, these encounters generally end with the killing of one or more dogs; but occasionally the dogs overpower him, and he is rescued for the time by his friends, to "fight (again) some other day."

We are happy to say, however, that Bear-baiting and bull-baiting have not been as yet fully naturalized amongst us, and are only popular with those who, perhaps, in addition to the natural desire for excitement, have the hope and intention of winning money, to draw them to such cruel and useless exhibitions.

Among the many Bear stories that have been published in the newspapers, and which, whether true or invented, are generally interesting, the following is one of the latest, the substance of which we will give, as nearly as we can recollect it:

A young man in the State of Maine, whilst at work in a field, accompanied only by a small boy, was attacked by a Bear which suddenly approached from the edge of the forest, and quite unexpectedly fell upon him with great fury. Almost at the first onset the brute overthrew the young farmer, who fell to the ground on his back, with the Bear clutching him, and biting his arm severely. Nothing but the utmost presence of mind could have saved the young man, as he was unarmed with the exception of a knife, which he could not get out of his pocket owing to the position in which he had fallen. Perceiving that his chance of escape was desperate, he rammed his hand and arm so far down the throat of the Bear as to produce the effect of partial strangulation, and whilst the beast became faint from consequent loss of breath, called to the boy to come and hand him the knife. The latter bravely came to the rescue, got the knife, opened it, and gave it to him, when he succeeded in cutting the Bear's throat, and with the exception of a few severe bites, and some lacerations from the claws of the animal, was not very much injured. The Bear was carried next day in triumph to a neighbouring village, and weighed over four hundred pounds.

Such assaults are, however, exceedingly rare, and it is seldom that even a wounded Bear attacks man.

Captain J. P. McCown has furnished us with the following remarks: "In the mountains of Tennessee the Bear lives principally upon mast and fruits. It is also fond of a bee-tree, and is often found seeking even a wasp's or yellow-jacket's nest. In the autumn the Bear is hunted when 'lopping' for chestnuts. Lopping consists in breaking off the branches by the Bear to procure the mast before it falls. When pursued by the dogs the Bear sometimes backs up against a tree, when it exhibits decided skill as a boxer, all the time looking exceedingly good-natured; but woe to the poor dog that ventures within its reach!

"The dogs generally employed for pursuing the Bear are curs and fice, as dogs of courage are usually killed or badly injured, while the cur will attack the Bear behind, and run when he turns upon him. No number of dogs can kill a Bear unless assisted by man.

"In 1841, the soldiers of my regiment had a pet he-Bear (castrated) that was exceedingly gentle and playful with the men. It becoming necessary to sell or kill it, one of the soldiers led it down the streets of Buffalo and exposed it for sale. Of course it attracted a large crowd, and was bid for on all sides on account of its gentleness. But unfortunately Bruin was carried near a hogshead of sugar, and not disposed to lose so tempting a repast, quietly upset it, knocking out the head, and commenced helping himself in spite of the soldier's efforts to prevent the depredation. The owner of the sugar rushed out and kicked the Bear, which, not liking such treatment, gave in return for the assault made upon him, a blow that sent his assailant far into the street, to the terror of the crowd, which scattered, leaving him to satisfy his appetite for sugar unmolested."

The number of Black Bears is gradually decreasing in the more settled parts of the "back woods," but in some portions of Carolina and Georgia, where the vast swamps prevent any attempt to settle or cultivate the land, they have within a few years been on the increase, and have become destroyers of the young stock of the Planter (which generally range through the woods) to a considerable extent.

Sir JOHN RICHARDSON says that when resident in the fur countries this Bear almost invariably hibernates, and that about one thousand skins are annually procured by the Hudson's Bay Company from those that are destroyed in their winter retreats. "It generally selects a spot for its den under a fallen tree, and having scratched away a portion of the soil, retires to it at the commencement of a snow-storm, when the snow soon furnishes it with a close, warm covering. Its breath makes a small opening in the den, and the quantity of hoar-frost which occasionally gathers round the aperture serves to betray its retreat to the hunter."

The Black Bear is somewhat migratory, and in hard winters is found to move southwardly in considerable numbers, although not in company. They couple in September or October, after which the females retire to their dens before the setting in of very cold weather.

It is said that the males do not so soon resort to winter quarters as the females, and require some time after the love season to recover their lost fat. The females bring forth about the beginning of January.

The Indian tribes have many superstitions concerning the Bear, and it is with some of them necessary to go through divers ceremonies before proceeding to hunt the animal.

Textual Notes

1. John Bachman, *The Doctrine of the Unity of the Human Race Examined on the Principles of Science* (Charleston SC: C. Canning, 1850), 27.
2. "begins to give tongue": To bay while following a scent.
3. "the trail becomes a drag": The line of scent left by a pursued animal; the trail.
4. "like the specimen figured in our work": The original edition of 1846 reads: "like the specimen figured in Dr. GODMAN's work." Having obtained a better specimen of the cross fox, Bachman emended the text for the Octavo edition and deleted the following short paragraph: "In describing the habits of the Red Fox, (*V. fulvus*,) which we trust to be able to do hereafter, we conceive that we shall have a further and better opportunity of giving the characteristics of the species of which this is a variety."
5. Lester D. Stephens, *Science, Race, and Religion in the American South: John Bachman and the Charleston Circle of Naturalists, 1815–1895* (Chapel Hill: University of North Carolina Press, 2000), 47.
6. Bachman, *The Doctrine of the Unity*, 7.
7. Stephens, *Science, Race, and Religion*, 48.
8. "'shining morning faces'": From Shakespeare's *As You Like It*: "Then the whining schoolboy, with his satchel / And shining morning face, creeping like snail / Unwillingly to school" (2.7.145–47).
9. "a national point of view": Bachman wrote the first essays for volume 1 in 1844, the year of the highly contentious presidential election between Polk and Clay. The nation was deeply divided over major issues like slavery, nullification, the annexation of Texas, and a looming war against Mexico. The degree of division in the country at the time is reflected in the closeness of the vote: Polk received 49.5 percent to Clay's 48.1. Some historians argue that if Clay had won, there would have been no Mexican War and no Civil War in the 1860s (see Daniel Walker Howe, *What Hath God Wrought: The Transformation of America, 1815–1848* [New York: Oxford University Press,

2007], 688–90). Bachman here likens the North and the South to battling shrew moles "falling victims to their own passions."

10. "'the Scalops'": Brewer's Shrew Mole (*Scalops breweri*; *Quadrupeds*, plate 74).
11. "FABRICIUS, SPENCE, or KIRBY": See "Works of Natural History Cited in the Essays" in this volume. The works of all authors cited in the quadruped essays can be found there.
12. Charles Wilkins Webber, "The Viviparous Quadrupeds of North America," *Southern Quarterly Review* 12, no. 24 (October 1847): 273–306, 273–75.
13. "which we have laid before our readers, in pages 49 and 50": Pages 20–22 in the present volume.
14. On this subject, see Eric Russell, "Modern Violence: Animal Studies, Speciesism, and the Writings of John James Audubon," *ESQ: A Journal of Nineteenth-Century American Literature and Culture* 65, no. 3 (2019): 470–510.
15. "some rats destroyed many of our valued drawings": John James Audubon, *Ornithological Biography*, 5 vols. (Edinburgh: Adam Black, 1831–39), 1:xiii–xiv.
16. See John James Audubon, *The Missouri River Journals of John James Audubon*, ed. Daniel Patterson (Lincoln: University of Nebraska Press, 2016), 302, 93–96.
17. "secundum artem": According to standard practice.
18. "Mr. SQUIRE": Bachman consistently misspells the name of Lewis M. Squires, whom Audubon took on as his assistant on the 1843 expedition.
19. "the Mamelles": For Audubon's field sketches and subsequent drawing of this favored camp site, see Audubon, *The Missouri River Journals*, figures 19–21.
20. "his wish to kill a cow": In his journal entry for July 27, 1843, Harris vociferously complains when a Fort Union hunter violates an expressed rule by firing into more than one of four bison cows being pursued by four hunters: "There was a cow for each of us" (Edward Harris, *Up the Missouri with Audubon: The Journal of Edward Harris*, ed. John Francis McDermott [Norman: University of Oklahoma Press, 1951], 159).
21. "towards the coupee": Audubon also sketched this landscape feature in his field notebook. See Audubon, *The Missouri River Journals*, figure 31.
22. "taken to the press post": Bison hides were folded and pressed into bundles and bound for transport. The hide press stands just outside the fort's bankside gate. A long wooden pole to which a weight can be attached acts as a lever to exert pressure on the folded hides in the box of the press.
23. "the above letter": Here we omit the lengthy letter "respecting the breeding of the Bison in a domesticated state" by a correspondent in Kentucky.
24. "a singular variety of the Bison": The only known other "variety" of bison is known as the wood bison (*B. bison athabascae*). It is associated with northern sedge meadows, and its range is in the northern bison range. Whether it is a

distinct subspecies is debated (see Dale F. Lott, *American Bison: A Natural History* [Berkeley: University of California Press, 2002], 67–68).

25. "figure of 4 trigger": This is a deadfall trap in which three sticks are fitted into one another in the form of the numeral 4; when an animal triggers it, a large flat rock is allowed to fall upon the animal.
26. "article on the raccoon, pp. 80, 81": Page 143 in the present volume.
27. "the humble poor": From Thomas Gray's "Elegy Written in a Country Churchyard" (1750). The fourth line of this quote, however, differs from Gray's "The short and simple annals of the poor." If Bachman wrote the line, and we cannot find it in any other source, he is substituting his slaves for Gray's poor country farm families, all of whom were white.
28. Audubon, *The Missouri River Journals*, 151–55.
29. Audubon, *The Missouri River Journals*, 44–45.
30. Audubon, *The Missouri River Journals*, 1–23.
31. "taken from WESSON's patent": Daniel Baird Wesson (1825–1906), cofounder of Smith and Wesson firearms manufacturer.
32. The three spaced asterisks seem to indicate that some portion of the original journal is not reproduced here.
33. "'Phantom Bark,' for it 'seemed never the nigher'": John Woodhouse Audubon quotes from a popular ballad, "The Indian Boat," or "The Phantom Ship," which was set to music and sung. See, for example, *The Phantom Ship from Legendary Ballads by Thomas More by Henry R. Bishop* (New York: William Hall and Son, n.d.).
34. On bringing home the young deer, see Audubon, *The Missouri River Journals*, 115, 129.
35. "the horns": In "General Remarks," Bachman quotes Meriwether Lewis's claim that the "hams" of a supposed separate deer species resemble "in form and colour those of the Mule Deer." Lewis wrote "horns," not "hams." Nicholas Biddle, from whose *History of the Expedition under the Command of Captains Lewis and Clark* Bachman quotes (2:166), misread Lewis's hand. See Lewis's entry for February 19, 1806, in Gary Moulton's edition of the journals (*The Definitive Journals of Lewis and Clark*, 8 vols. [Lincoln: University of Nebraska Press, 1983–2001]).
36. These opening lines of "General Remarks" appeared originally in John James Audubon and John Bachman, "Descriptions of New Species of Quadrupeds Inhabiting North America," *Journal of the Academy of Natural Sciences* 8, no. 2 (1842): 321–22.
37. "cat's-paw": One used by another to serve his own purposes.
38. "with 'a cat' over his head": Probable pun on cat-o'-nine-tails.

39. "the reading of the 'Mysteries of Udolpho' or the 'Castle of Otranto'": Two popular gothic novels: Ann Radcliffe's *The Mysteries of Udolpho* (London, 1794) and Horace Walpole's *The Castle of Otranto* (London, 1764).
40. On Audubon's protest against regarding bird behavior as stupid, see *The Missouri River Journals*, 292–93.
41. "the celebrated Col. HAYS": John Coffee Hays (1817–83), celebrated for his leading role in the Texas Rangers' campaign against the Comanches in Texas and during the war against Mexico.
42. The Labrador journal of 1833 is in John James Audubon, *Audubon and His Journals*, ed. Maria R. Audubon, 2 vols. (New York: Charles Scribner's Sons, 1897), 1:343–445.
43. The prized skin that Morton refers to is that of the black wolf, not the black fox. Morton refers only to red and gray foxes (see Thomas Morton, *New English Canaan* [Amsterdam, 1637], 79). Why Bachman misreports this (as well as Morton's title) is unclear.
44. "whose temple is all space": From Alexander Pope, "The Universal Prayer" (1738), set to music in the hymn "The Children of the Kingdom," by Jeremiah Clark.
45. Audubon, *The Missouri River Journals*, 166.
46. "our journals are before us": For the original entries from Audubon's original field notebook and a partial copy of the full journal, see Audubon, *The Missouri River Journals*, 165–67.
47. "The following afternoon": In the Newberry Partial Copy, this event is included in the entry for August 22 rather than "the following afternoon" (Audubon, *The Missouri River Journals*, 167).
48. "pomme blanche": The prairie turnip (*Psoralea esculents*).
49. "engagés": Engagés were common laborers engaged in St. Louis and transported up the Missouri to the various forts of the fur trade.
50. Audubon originally recorded this story in his Missouri River journal (see *The Missouri River Journals*, 140).
51. "Scipio and the Bear": See Audubon, *Ornithological Biography*, 1:479–83.
52. "Thine incomparable oil Macassar!": The London barber Alexander Rowland (1747–1823) began producing Macassar oil as a hair dressing for men in 1783; it became very popular. The quoted line is from the first canto of Lord Byron's epic satire *Don Juan* (London, 1819).
53. "'pet' Bear is occasionally seen": For Audubon's story of having encountered one such "pet" bear, see "The Pet Bear" in *The Missouri River Journals*, 427–31.

Works of Natural History Cited in the Essays

Audubon, John James. *Birds of America*. 4 vols. London: J. J. Audubon, 1827–38.

———. *Ornithological Biography*. 5 vols. Edinburgh: Adam Black, 1831–39.

Audubon, John James, and John Bachman. "Descriptions of New Species of Quadrupeds Inhabiting North America." *Journal of the Academy of Natural Sciences of Philadelphia* 8 (1842): 280–323.

Bachman, John. "Additional Remarks on the Genus *Lepus* . . ." *Journal of the Academy of Natural Sciences of Philadelphia* 8 (1839): 75–105.

———. "Notes on the Generation of the Virginia Opossum (*Didelphis Virginiana*)." *Proceedings of the Academy of Natural Sciences of Philadelphia* 4 (1848): 40–47.

———. "Observations on the Different Species of Hares . . ." *Journal of the Academy of Natural Sciences of Philadelphia* 7 (1837): 282–361.

Barton, Benjamin Smith. *Facts, Observations, and Conjectures Relative to the Generation of the Opossum*. Philadelphia: Thomas and George Palmer, 1806.

Beverley, Robert. *The History and Present State of Virginia*. London: R. Parker, 1705.

Bewick, Thomas. *A General History of Quadrupeds*. Newcastle: S. Hodgson, 1790.

Blainville, Henri Marie Ducrotay de. *De l'organisation des animaux*. Paris: F. G. Levrault, 1822.

Blumenbach, Johann Friedrich. *Abbildungen naturhistorischer Gegenstände*. Göttingen: Dieterich, 1810.

Boddaert, Pieter. *Elenchus Animalium*. Rotterdam, 1784.

Cartwright, George. *A Journal of Transactions and Events . . . on the Coast of Labrador*. 3 vols. Newark: Allin & Ridge, 1792.

Catesby, Mark. *The Natural History of Carolina, Florida, and the Bahama Islands*. 2 vols. London: Mark Catesby, 1731, 1743.

Chastellux, Marquis de. *Voyages . . . dans l'Amérique septentrionale*. 2 vols. Paris: Prault, 1788.

DeKay, James. *Natural History of New York*. Albany: D. Appleton & Co. and Wiley & Putnam, 1842.

Desmarest, Anselme Gaëtan. *Mammalogie, ou, Description des espèces de mammifères.* 2 vols. Paris: Chez Mme Veuve Agasse, 1820–22.

Emmons, Ebenezer. *Report on the Quadrupeds of Massachusetts.* Cambridge: Zoological and Botanical Survey of Massachusetts, 1840.

Erxleben, Johann. *Systema Regni Animalis.* Leipzig: Weygands, 1777.

Fabricius, Johan Christian. *Systema Glossatorum.* Brunovici: C. Reichard, 1807.

Forster, Johann Reinhold. "Account of Several Quadrupeds from Hudson's Bay." *Philosophical Transactions of the Royal Society* 62 (1772): 370–81.

Gmelin, Johann Friedrich. *Caroli a Linné: Systema Naturae.* Leipzig: Georg Emanuel Beer, 1788.

Godman, John. *American Natural History.* 3 vols. Philadelphia: H. C. Carey and I. Lea, 1826–28.

Harlan, Richard. *Fauna Americana.* Philadelphia: A. Finley, 1825.

Harriot, Thomas. *Third Voyage of the English to Virginia, 1586.* N.p.

Hearne, Samuel. *A Journey from Prince of Wales's Fort* . . . London: A. Strahan and T. Cadell, 1795.

Hernández de Toledo, Francisco. *Plantas y animales de la Nueva España, y sus virtudes por Francisco Hernandez, y de Latin en Romance por Fr. Francisco Ximenez.* Nardo Antonio Recchi, 1615.

Humboldt, Alexander von. *Ansichten der Natur.* Tübingen: J. G. Cotta, 1808.

Kalm, Peter. *Travels into North America.* 2 vols. London: T. Lowndes, 1772.

Kirby, William, and William Spence. *An Introduction to Entomology.* 4 vols. London: Longman, Hurst, Rees, Orme & Brown, 1816.

Klein, Jacob Theodore. *Quadrupedum Dispositio Brevisque Historia Naturalis.* Leipzig: Schmidt, 1751.

Lawson, John. *A New Voyage to Carolina.* London, 1709.

Lewis, Meriwether, and William Clark. *History of the Expedition under the Command of Captains Lewis and Clark.* Edited by Nicholas Biddle [without attribution]. 2 vols. Philadelphia: Bradford and Inskeep, 1814.

Lichtenstein, Hinrich. *Darstellung neuer oder wenig bekannter Säugethiere* . . . Berlin: C. G. Lüderitz, 1827–34.

Lyon, George Francis. *The Private Journal of Captain G. F. Lyon, of H.M.S.* Hecla, *during the Recent Voyage of Discovery under Captain Parry.* London: John Murray, 1824.

Marcgrave, Georg, and Willem Piso. *Historia Naturalis Brasiliae.* 8 vols. Leiden: Johannes de Laet, 1648.

Michaux, André. *Flora Boreali-Americana.* 2 vols. Paris and Strasbourg: Levrault, 1803.

Michel, Middleton. Letter in *Proceedings of the Academy of Natural Sciences of Philadelphia* 4 (1848): 42–46.
Morton, Thomas. *New English Canaan*. Amsterdam, 1637.
Pallas, Peter Simon. *Zoographia Rosso-Asiatica*. 3 vols. Petersburg: Academy of Sciences, 1811.
Pennant, Thomas. *Arctic Zoology*. 2 vols. London: Henry Hughes, 1784.
Ray, John. *Synopsis Methodica Animalium Quadrupedum et Serpentini Generis*. London: Robert Southwell, 1693.
Richardson, John. *Fauna Boreali-Americana*. London: John Murray, 1829.
Sabine, Joseph. *Zoological Appendix to the Narrative of Captain Franklin's Journey to the Shores of the Polar Sea, in the Years 1819, 20, 21, and 22. Being Notices of the Quadrupeds and Birds of Which Specimens Were Collected during the Journey*. London, 1823.
Sagard, Gabriel. *Histoire du Canada et voyages que les Frères Mineurs Recollets ont faits pour la conversion des infidèles*. Paris, 1636.
———. *Le grand voyage au pays des Hurons*. Paris: Moreau, 1632.
Saint-Hilaire, Geoffroy. "Si les animaux à bourse naissent aux tétines de leur mère?" *Journal complémentaire du dictionnaire des sciences médicales* 3 (1819): 193.
Say, Thomas. In *Journals of 1817 and 1823* by Stephen H. Long. N.p.
Schoepff, Johan David. *Reise durch einige der mittlern und südlichen Vereinigten Staaten*. Erlangen, 1788.
Schreber, Johann von. *Die Säugthiere in Abbildungen nach der Natur mit Beschreibungen*. 64 vols. Erlangen: Walther, 1774–1804.
Smith, John. *A Map of Virginia* . . . Oxford: Joseph Barnes, 1612.
Townsend, John K. *Narrative of a Journey across the Rocky Mountains* . . . Philadelphia: Henry Perkins, 1839.

Index